IET ENGINEERING SERIES 144

Photovoltaic Technology for Hot and Arid Environments

Other volumes in this series:

Volume 1   **Power Circuit Breaker Theory and Design** C.H. Flurscheim (Editor)
Volume 4   **Industrial Microwave Heating** A.C. Metaxas and R.J. Meredith
Volume 7   **Insulators for High Voltages** J.S.T. Looms
Volume 8   **Variable Frequency AC Motor Drive Systems** D. Finney
Volume 10   **SF$_6$ Switchgear** H.M. Ryan and G.R. Jones
Volume 11   **Conduction and Induction Heating** E.J. Davies
Volume 13   **Statistical Techniques for High Voltage Engineering** W. Hauschild and W. Mosch
Volume 14   **Uninterruptible Power Supplies** J. Platts and J.D. St. Aubyn (Editors)
Volume 15   **Digital Protection for Power Systems** A.T. Johns and S.K. Salman
Volume 16   **Electricity Economics and Planning** T.W. Berrie
Volume 18   **Vacuum Switchgear** A. Greenwood
Volume 19   **Electrical Safety: A guide to causes and prevention of hazards** J. Maxwell Adams
Volume 21   **Electricity Distribution Network Design, 2nd Edition** E. Lakervi and E.J. Holmes
Volume 22   **Artificial Intelligence Techniques in Power Systems** K. Warwick, A.O. Ekwue and R. Aggarwal (Editors)
Volume 24   **Power System Commissioning and Maintenance Practice** K. Harker
Volume 25   **Engineers' Handbook of Industrial Microwave Heating** R.J. Meredith
Volume 26   **Small Electric Motors** H. Moczala *et al.*
Volume 27   **AC–DC Power System Analysis** J. Arrillaga and B.C. Smith
Volume 29   **High Voltage Direct Current Transmission, 2nd Edition** J. Arrillaga
Volume 30   **Flexible AC Transmission Systems (FACTS)** Y.-H. Song (Editor)
Volume 31   **Embedded Generation** N. Jenkins *et al.*
Volume 32   **High Voltage Engineering and Testing, 2nd Edition** H.M. Ryan (Editor)
Volume 33   **Overvoltage Protection of Low-Voltage Systems, Revised Edition** P. Hasse
Volume 36   **Voltage Quality in Electrical Power Systems** J. Schlabbach *et al.*
Volume 37   **Electrical Steels for Rotating Machines** P. Beckley
Volume 38   **The Electric Car: Development and future of battery, hybrid and fuel-cell cars** M. Westbrook
Volume 39   **Power Systems Electromagnetic Transients Simulation** J. Arrillaga and N. Watson
Volume 40   **Advances in High Voltage Engineering** M. Haddad and D. Warne
Volume 41   **Electrical Operation of Electrostatic Precipitators** K. Parker
Volume 43   **Thermal Power Plant Simulation and Control** D. Flynn
Volume 44   **Economic Evaluation of Projects in the Electricity Supply Industry** H. Khatib
Volume 45   **Propulsion Systems for Hybrid Vehicles** J. Miller
Volume 46   **Distribution Switchgear** S. Stewart
Volume 47   **Protection of Electricity Distribution Networks, 2nd Edition** J. Gers and E. Holmes
Volume 48   **Wood Pole Overhead Lines** B. Wareing
Volume 49   **Electric Fuses, 3rd Edition** A. Wright and G. Newbery
Volume 50   **Wind Power Integration: Connection and system operational aspects** B. Fox *et al.*
Volume 51   **Short Circuit Currents** J. Schlabbach
Volume 52   **Nuclear Power** J. Wood
Volume 53   **Condition Assessment of High Voltage Insulation in Power System Equipment** R.E. James and Q. Su
Volume 55   **Local Energy: Distributed generation of heat and power** J. Wood
Volume 56   **Condition Monitoring of Rotating Electrical Machines** P. Tavner, L. Ran, J. Penman and H. Sedding
Volume 57   **The Control Techniques Drives and Controls Handbook, 2nd Edition** B. Drury
Volume 58   **Lightning Protection** V. Cooray (Editor)
Volume 59   **Ultracapacitor Applications** J.M. Miller
Volume 62   **Lightning Electromagnetics** V. Cooray

Volume 63   **Energy Storage for Power Systems, 2nd Edition** A. Ter-Gazarian
Volume 65   **Protection of Electricity Distribution Networks, 3rd Edition** J. Gers
Volume 66   **High Voltage Engineering Testing, 3rd Edition** H. Ryan (Editor)
Volume 67   **Multicore Simulation of Power System Transients** F.M. Uriate
Volume 68   **Distribution System Analysis and Automation** J. Gers
Volume 69   **The Lightening Flash, 2nd Edition** V. Cooray (Editor)
Volume 70   **Economic Evaluation of Projects in the Electricity Supply Industry, 3rd Edition** H. Khatib
Volume 72   **Control Circuits in Power Electronics: Practical issues in design and implementation** M. Castilla (Editor)
Volume 73   **Wide Area Monitoring, Protection and Control Systems: The enabler for smarter grids** A. Vaccaro and A. Zobaa (Editors)
Volume 74   **Power Electronic Converters and Systems: Frontiers and applications** A.M. Trzynadlowski (Editor)
Volume 75   **Power Distribution Automation** B. Das (Editor)
Volume 76   **Power System Stability: Modelling, analysis and control** A.A. Sallam and B. Om P. Malik
Volume 78   **Numerical Analysis of Power System Transients and Dynamics** A. Ametani (Editor)
Volume 79   **Vehicle-to-Grid: Linking electric vehicles to the smart grid** J. Lu and J. Hossain (Editors)
Volume 81   **Cyber-Physical-Social Systems and Constructs in Electric Power Engineering** S. Suryanarayanan, R. Roche and T.M. Hansen (Editors)
Volume 82   **Periodic Control of Power Electronic Converters** F. Blaabjerg, K. Zhou, D. Wang and Y. Yang
Volume 86   **Advances in Power System Modelling, Control and Stability Analysis** F. Milano (Editor)
Volume 87   **Cogeneration: Technologies, optimisation and implantation** C.A. Frangopoulos (Editor)
Volume 88   **Smarter Energy: From smart metering to the smart grid** H. Sun, N. Hatziargyriou, H.V. Poor, L. Carpanini and M.A. Sánchez Fornié (Editors)
Volume 89   **Hydrogen Production, Separation and Purification for Energy** A. Basile, F. Dalena, J. Tong and T.N.Veziroğlu (Editors)
Volume 90   **Clean Energy Microgrids** S. Obara and J. Morel (Editors)
Volume 91   **Fuzzy Logic Control in Energy Systems with Design Applications in MATLAB®/Simulink®** İ. H. Altaş
Volume 92   **Power Quality in Future Electrical Power Systems** A.F. Zobaa and S.H.E.A. Aleem (Editors)
Volume 93   **Cogeneration and District Energy Systems: Modelling, analysis and optimization** M. A. Rosen and S. Koohi-Fayegh
Volume 94   **Introduction to the Smart Grid: Concepts, technologies and evolution** S.K. Salman
Volume 95   **Communication, Control and Security Challenges for the Smart Grid** S.M. Muyeen and S. Rahman (Editors)
Volume 96   **Industrial Power Systems with Distributed and Embedded Generation** R. Belu
Volume 97   **Synchronized Phasor Measurements for Smart Grids** M.J.B. Reddy and D.K. Mohanta (Editors)
Volume 98   **Large Scale Grid Integration of Renewable Energy Sources** A. Moreno-Munoz (Editor)
Volume 100   **Modeling and Dynamic Behaviour of Hydropower Plants** N. Kishor and J. Fraile-Ardanuy (Editors)
Volume 101   **Methane and Hydrogen for Energy Storage** R. Carriveau and D.S.-K. Ting
Volume 104   **Power Transformer Condition Monitoring and Diagnosis** A. Abu-Siada (Editor)
Volume 106   **Surface Passivation of Industrial Crystalline Silicon Solar Cells** J. John (Editor)
Volume 107   **Bifacial Photovoltaics: Technology, applications and economics** J. Libal and R. Kopecek (Editors)

Volume 108   **Fault Diagnosis of Induction Motors** J. Faiz, V. Ghorbanian and G. Joksimović
Volume 109   **Cooling of Rotating Electrical Machines: Fundamentals, modelling, testing and design** D. Staton, E. Chong, S. Pickering and A. Boglietti
Volume 110   **High Voltage Power Network Construction** K. Harker
Volume 111   **Energy Storage at Different Voltage Levels: Technology, integration, and market aspects** A.F. Zobaa, P.F. Ribeiro, S.H.A. Aleem and S.N. Afifi (Editors)
Volume 112   **Wireless Power Transfer: Theory, technology and application** N. Shinohara
Volume 114   **Lightning-Induced Effects in Electrical and Telecommunication Systems** Y. Baba and V.A. Rakov
Volume 115   **DC Distribution Systems and Microgrids** T. Dragičević, F.Blaabjerg and P. Wheeler
Volume 116   **Modelling and Simulation of HVDC Transmission** M. Han (Editor)
Volume 117   **Structural Control and Fault Detection of Wind Turbine Systems** H.R. Karimi
Volume 118   **Modelling and Simulation of Complex Power Systems** A. Monti and A. Benigni
Volume 119   **Thermal Power Plant Control and Instrumentation: The control of boilers and HRSGs, 2nd Edition** D. Lindsley, J. Grist and D. Parker
Volume 120   **Fault Diagnosis for Robust Inverter Power Drives** A. Ginart (Editor)
Volume 121   **Monitoring and Control using Synchrophasors in Power Systems with Renewables** I. Kamwa and C. Lu (Editors)
Volume 123   **Power Systems Electromagnetic Transients Simulation, 2nd Edition** N. Watson and J. Arrillaga
Volume 124   **Power Market Transformation** B. Murray
Volume 125   **Wind Energy Modeling and Simulation Volume 1: Atmosphere and plant** P.Veers (Editor)
Volume 126   **Diagnosis and Fault Tolerance of Electrical Machines, Power Electronics and Drives** A.J. M. Cardoso
Volume 127   **Lightning Electromagnetics Volume 1: Return stroke modelling and electromagnetic radiation 2nd Edition** V. Cooray, F. Rachidi, M. Rubinstein (Editors)
Volume 127   **Lightning Electromagnetics Volume 2: Electrical processes and effects 2nd Edition** V. Cooray, F. Rachidi and M. Rubinstein (Editors)
Volume 128   **Characterization of Wide Bandgap Power Semiconductor Devices** F. Wang, Z. Zhang and E.A. Jones
Volume 129   **Renewable Energy from the Oceans: From wave, tidal and gradient systems to offshore wind and solar** D. Coiro and T. Sant (Editors)
Volume 130   **Wind and Solar Based Energy Systems for Communities** R. Carriveau and D. S-K. Ting (Editors)
Volume 131   **Metaheuristic Optimization in Power Engineering** J. Radosavljević
Volume 132   **Power Line Communication Systems for Smart Grids** I.R.S. Casella and A. Anpalagan
Volume 134   **Hydrogen Passivation and Laser Doping for Silicon Solar Cells** B. Hallam and C. Chan (Editors)
Volume 139   **Variability, Scalability and Stability of Microgrids** S.M. Muyeen, S.M. Islam and F. Blaabjerg (Editors)
Volume 142   **Wind Turbine System Design: Volume 1: Nacelles, drive trains and verification** J. Wenske (Editor)
Volume 143   **Medium Voltage DC System Architectures** B. Grainger and R.D. Doncker (Editors)
Volume 145   **Condition Monitoring of Rotating Electrical Machines** P. Tavner, L. Ran, and C. Crabtree
Volume 146   **Energy Storage for Power Systems, 3rd Edition** A.G. Ter-Gazarian
Volume 147   **Distribution Systems Analysis and Automation 2nd Edition** J. Gers
Volume 151   **SiC Power Module Design: Performance, robustness and reliability** A. Castellazzi and A. Irace (Editors)
Volume 152   **Power Electronic Devices: Applications, failure mechanisms and reliability** F. Iannuzzo (Editor)

Volume 153   **Signal Processing for Fault Detection and Diagnosis in Electric Machines and Systems** M. Benbouzid (Editor)

Volume 155   **Energy Generation and Efficiency Technologies for Green Residential Buildings** D. Ting and R. Carriveau (Editors)

Volume 156   **Lithium-Ion Batteries Enabled by Silicon Anodes** C. Ban and K. Xu (Editors)

Volume 157   **Electrical Steels, 2 Volumes** A. Moses, K. Jenkins, Philip Anderson and H. Stanbury

Volume 158   **Advanced Dielectric Materials for Electrostatic Capacitors** Q Li (Editor)

Volume 159   **Transforming the Grid Towards Fully Renewable Energy** O. Probst, S. Castellanos and R. Palacios (Editors)

Volume 160   **Microgrids for Rural Areas: Research and case studies** R.K. Chauhan, K. Chauhan and S.N. Singh (Editors)

Volume 161   **Artificial Intelligence for Smarter Power Systems: Fuzzy logic and neural networks** M. G. Simoes

Volume 165   **Digital Protection for Power Systems 2nd Edition** Salman K Salman

Volume 166   **Advanced Characterization of Thin Film Solar Cells** N. Haegel and M. Al-Jassim (Editors)

Volume 167   **Power Grids with Renewable Energy**: Storage, integration and digitalization A.A. Sallam and B. OM P. Malik

Volume 169   **Small Wind and Hydrokinetic Turbines** P. Clausen, J. Whale and D. Wood (Editors)

Volume 170   **Reliability of Power Electronics Converters for Solar Photovoltaic Applications** F. Blaabjerg, A.I. Haque, H. Wang, Z. Abdin Jaffery and Y. Yang (Editors)

Volume 171   **Utility-Scale Wind Turbines and Wind Farms** A. Vasel-Be-Hagh and D.S.-K. Ting

Volume 172   **Lighting Interaction with Power Systems, 2 volumes** A. Piantini (Editor)

Volume 174   **Silicon Solar Cell Metallization and Module Technology** T. Dullweber (Editor)

Volume 175   **n-Type Crystalline Silicon Photovoltaics: Technology, applications and economics** D. Munoz and R. Kopecek (Editors)

Volume 178   **Integrated Motor Drives** Xu Deng and B. Mecrow (Editors)

Volume 180   **Protection of Electricity Distribution Networks, 4th Edition** J. Gers and E. Holmes

Volume 182   **Surge Protection for Low Voltage Systems** A. Rousseau (Editor)

Volume 184   **Compressed Air Energy Storage: Types, systems and applications** D. Ting and J. Stagner

Volume 186   **Synchronous Reluctance Machines: Analysis, optimization and applications** N. Bianchi, C. Babetto and G. Bacco

Volume 190   **Synchrophasor Technology: Real-time operation of power networks** N. Kishor and S.R. Mohanty (Editors)

Volume 191   **Electric Fuses: Fundamentals and new applications 4th Edition** N. Nurse, A. Wright and P. G. Newbery

Volume 193   **Overhead Electric Power Lines: Theory and practice** S. Chattopadhyay and A. Das

Volume 194   **Offshore Wind Power Reliability, Availability and Maintenance, 2nd edition** P. Tavner

Volume 196   **Cyber Security for Microgrids** S. Sahoo, F. Blaajberg and T. Dragicevic

Volume 198   **Battery Management Systems and Inductive Balancing** A. Van den Bossche and A. Farzan Moghaddam

Volume 199   **Model Predictive Control for Microgrids: From power electronic converters to energy management** J. Hu, J. M. Guerrero and S. Islam

Volume 201   **Design, Control and Monitoring of Tidal Stream Turbine Systems** M. Benbouzid (Editor)

Volume 204   **Electromagnetic Transients in Large HV Cable Networks: Modeling and calculations** Ametani, Xue, Ohno and Khalilnezhad

Volume 208   **Nanogrids and Picogrids and their Integration with Electric Vehicles** S. Chattopadhyay

Volume 210   **Superconducting Magnetic Energy Storage in Power Grids** M. H. Ali

Volume 211   **Blockchain Technology for Smart Grids: Implementation, management and security** H.L. Gururaj, K.V. Ravi, F. Flammini, H. Lin, B. Goutham, K.B.R. Sunil and C. Sivapragash

Volume 212   **Battery State Estimation: Methods and models** S. Wang

Volume 215   **Industrial Demand Response: Methods, best practices, case studies, and applications** H. H. Alhelou, A. Moreno-Muñoz and P. Siano (Editors)

Volume 213   **Wide Area Monitoring of Interconnected Power Systems 2nd Edition** A. R. Messina

Volume 214   **Distributed Energy Storage in Urban Smart Grids** P.F. Ribeiro and R.S. Salles (Editors)

Volume 217   **Advances in Power System Modelling, Control and Stability Analysis 2nd Edition** F. Milano (Editor)

Volume 225   **Fusion-Fission Hybrid Nuclear Reactors: For enhanced nuclear fuel utilization and radioactive waste reduction** W.M. Stacey

Volume 228   **Digital Technologies for Solar Photovoltaic Systems: From general to rural and remote installations** S. Motahhir (Editor)

Volume 238   **AI for Status Monitoring of Utility Scale Batteries** S. Wang, K. Liu, Y. Wang, D. Stroe, C. Fernandez and J.M. Guerrero

Volume 239   **Intelligent Control of Medium and High Power Converters** M. Bendaoud, Y. Maleh and S. Padmanaban (Editors)

Volume 905   **Power System Protection, 4 volumes**

Photovoltaic Technology for Hot and Arid Environments

Edited by
Brahim Aïssa and Nouar Tabet

The Institution of Engineering and Technology

Published by The Institution of Engineering and Technology, London, United Kingdom

The Institution of Engineering and Technology is registered as a Charity in England & Wales (no. 211014) and Scotland (no. SC038698).

The Institution of Engineering and Technology
Futures Place
Kings Way, Stevenage
Hertfordshire SG1 2UA, United Kingdom

www.theiet.org

British Library Cataloguing in Publication Data
A catalogue record for this product is available from the British Library

ISBN 978-1-78561-911-3 (hardback)
ISBN 978-1-78561-912-0 (PDF)

Typeset in India by MPS Limited

Cover Image: alexsl/E+ via Getty Images

Contents

About the editors   xv
List of figures   xvii
List of tables   xxix
Preface   xxxi
Acknowledgment   xxxiii

1   Solar energy resources and harvesting technologies   1
    Daniel Perez Astudillo and Nouar Tabet
    1.1   Introduction   1
    1.2   Solar resources   1
          1.2.1   Sun–earth system   1
          1.2.2   Effects of earth's atmosphere on sunlight   3
          1.2.3   Standard solar spectrum   4
          1.2.4   Solar light components: GHI, DHI, DNI   4
          1.2.5   Estimating insolation   6
          1.2.6   Ground measurements, instruments   6
          1.2.7   Satellite-based data   10
          1.2.8   Insolation maps and data sets   11
    1.3   Harvesting technologies   13
          1.3.1   Solar energy conversion   13
          1.3.2   Photovoltaics   14
          1.3.3   CSP   17
    References   20

2   Solar cell fundamentals   23
    Brahim Aïssa, Fahhad Alharbi and Nouar Tabet
    2.1   Introduction   23
    2.2   Cell structure and light conversion   23
    2.3   Current–voltage (I–V) characteristics   25
    2.4   Factors limiting the cell performance   27
          2.4.1   Effect of the energy gap   29
          2.4.2   Current losses through defects   30
          2.4.3   Auger recombination   31
          2.4.4   Surface recombination   32
          2.4.5   Effect of temperature   33

2.5 Theoretical limit of silicon cell efficiency: Shockley–Queisser
limit (SQ limit)   35
References   36

3 Thermodynamics of solar energy conversion   **39**
Fahhad Alharbi, Fedwa EL Mellouhi, Brahim Aïssa and Nouar Tabet
3.1 Introduction   39
3.2 The thermodynamics of solar energy conversion   39
   3.2.1 Thermodynamics limit of solar energy conversion:
      AM1.5g   40
   3.2.2 Single junction solar cells   43
   3.2.3 Thermodynamics limit of solar energy conversion:
      the effect of increased intensity in hot areas 43
   3.2.4 Hot climate implications on common solar
      cell technologies   46
3.3 Materials for solar cells   50
   3.3.1 Conventional materials   50
   3.3.2 Emerging materials   51
   3.3.3 Computational materials design for solar cells   53
References   56

4 Solar cell technologies   **59**
Brahim Aïssa, Marie Buffiere and Mohammad I. Hossain
4.1 Introduction   59
4.2 Discussion about the main parameters affecting the silicon
solar cell device performance under desert conditions
(V_{oc}, temperature coefficient TC_η)   60
4.3 Silicon-based photovoltaic technologies   61
   4.3.1 Aluminum back surface field (Al-BSF) cells   63
   4.3.2 PERC   64
   4.3.3 Silicon heterojunction solar cells: concept and status   64
   4.3.4 Configuration and fabrication   66
4.4 Device operation under various temperatures—experimental results   68
   4.4.1 Comparison of different technologies   68
4.5 CIGS and CdTe thin film solar cells   72
   4.5.1 Device configuration   72
   4.5.2 Device fabrication   75
   4.5.3 Device performances   78
   4.5.4 Device reliability   83
4.6 Organo-metal halide perovskites based solar cells   85
   4.6.1 Solar cell architectures and designs   86
   4.6.2 Device fabrication   87
      4.6.3 Stability and toxicity issues   89
4.7 Multijunction solar cells   89
   4.7.1 Basic principles of multi-junction solar cells   89

4.7.2   Fabrication of multijunction solar cells   92
4.7.3   Multijunctions solar cell under high
        temperature operation   93
References   96

5   **PV module technology and energy yield under desert
    environment conditions**   **111**
    Ahmer A.B. Baloch, Brahim Aïssa, Amir A. Abdallah and Nouar Tabet
    5.1   Introduction   111
    5.2   PV module materials   113
    5.3   PV module design   113
    5.4   Cell-to-module (CTM) performance   115
          5.4.1   Geometry-related losses   118
          5.4.2   Electrical losses   120
          5.4.3   Optical losses and gains   124
    5.5   Module energy yield and reliability under desert environment 125
          5.5.1   Module installation parameters for mono-facial and
                  bifacial modules   125
          5.5.2   Effect of mounting height   127
          5.5.3   Effect of ground albedo   128
          5.5.4   Effect of tilt angle   128
          5.5.5   Effect of module temperature   129
          5.5.6   Effect of ground coverage ratio   129
          5.5.7   Effect of azimuth orientation   130
    5.6   Reliability issue for hot arid desert   132
          5.6.1   Infrared thermography of PV modules to identify defects 132
    5.7   Performance and reliability of crystalline-silicon photovoltaics
          in desert climate   139
          5.7.1   PV performance monitoring and degradation rate   139
          5.7.2   Temperature coefficients (TC) measurement   141
          5.7.3   Visual inspection   142
          5.7.4   Electroluminescence imaging   145
    5.8   Conclusions   146
    References   147

6   **Bifacial solar technology and module installation**   **153**
    Ahmer A.B. Baloch, Brahim Aïssa and Nouar Tabet
    6.1   Introduction   153
    6.2   Technological progress   156
    6.3   Bifacial performance parameters   158
    6.4   Potential applications   160
    6.5   Solar cell device optimization   161
    6.6   Theoretical and practical efficiency limits of solar cells 163
    6.7   Bifacial perovskite silicon tandem   165
    6.8   Material characterization model for device assessment   167

6.9   Cell to module to field performance   169
6.10   Module installation   173
       6.10.1 Outdoor module performance monitoring   174
       6.10.2 Bifacial versus monofacial energy yield gain and
               standalone versus in-array mounted losses   175
       6.10.3 Effect of mounting height   176
       6.10.4 Effect of natural and synthetic ground albedo   178
       6.10.5 Effect of module temperature   179
       6.10.6 Optimization of height, tilt, and albedo   179
       6.10.7 Effect of azimuth orientation   181
       6.10.8 Effect of device parameters: bifaciality and
               temperature coefficient   182
6.11   Conclusions   186
References   186

7   **Photostatic soiling in desert environment**   **197**
    Brahim Aïssa, Benjamin W. Figgis and Klemens Isle
    7.1   Introduction   197
    7.2   Description of the OTF   201
    7.3   Impact of the soiling on the attenuation of solar radiation
           in a desert environment   203
    7.4   Soiling rates in different countries   205
    7.5   Dust characteristics in different countries   207
           7.5.1   Particle size distribution   207
           7.5.2   Particle-surface adhesion forces   209
    7.6   Dew, cementation, particle caking, capillary aging   212
    7.7   Mechanics of dust accumulation   215
           7.7.1   Particle deposition   215
           7.7.2   Particle rebound and resuspension   217
           7.7.3   Net soiling rate   220
           7.7.4   Collector geometry   221
    7.8   Field measurement of soiling   224
           7.8.1   Soil mass   224
           7.8.2   Light transmission   225
           7.8.3   PV module output   226
    7.9   Analysis of PV field data   227
           7.9.1   Angle-of-incidence effects   227
           7.9.2   Surface imaging   229
    7.10   Cleaning and soiling mitigation   229
           7.10.1   Cleaning economics   232
           7.10.2   Abrasion effects   232
           7.10.3   Development of anti-soiling coating   233
           7.10.4   Development of a TiO_2-based self-cleaning coating   235
           7.10.5   Electrodynamic shield/screen   236
    7.11   Impact on global solar power production and energy costs   239

7.12 Renewable energy and soiling in the Gulf Cooperation
Council (GCC) context 241
7.13 Conclusions 243
References 244

8 Desert PV applications **255**
Ali Elrayyah and Mohd Zamri Che Wanik
8.1 Introduction 255
8.2 Desert climate and solar resource 256
8.2.1 General weather profile of a desert 256
8.2.2 Arabian and Qatar desert climate 256
8.2.3 Solar resource in the Qatar desert 257
8.2.4 Daily PV profile in the desert environment 258
8.3 PV strings and arrays 264
8.3.1 Partial shading impact on PV system performance 264
8.3.2 Performance of PV inverters in desert environments 266
8.4 Loads and energy consumption in Qatar 267
8.4.1 Animal barns 268
8.4.2 Water pumps 269
8.4.3 Water desalination 269
8.4.4 HVAC systems 270
8.4.5 Additional loads 271
8.4.6 Urban energy consumption of the Qatar Desert 271
8.5 PV energy system design, control and operation 272
8.5.1 PV-powered water pumping system 273
8.5.2 PV-powered lighting systems 273
8.5.3 Remote residential PV system 273
8.5.4 PV-hybrid system 274
8.5.5 A system with energy storage 274
8.5.6 Off-grid vs. grid connected system 274
8.5.7 Energy management and control 275
8.6 Techno-economic benefits and case study 276
8.7 Conclusions 281
8.8 Term definitions 281
References 283

**9 PV systems in Australia: market evolution and performance
in desert applications** **285**
Jose Bilbao, Sharon Young and Bram Hoex
9.1 Introduction 285
9.2 Evolution of the PV market in Australia 285
9.2.1 International context 286
9.2.2 Off-grid uses (pre-2000) 287
9.2.3 Development of a grid-connected industry 287
9.2.4 The dominance of grid-connected PV 287

9.2.5 Grid parity	290
9.2.6 The new investors	291
9.3 Policies and their effects on the market: lessons learned	291
9.3.1 Small-scale programs	291
9.3.2 The Renewable Energy Target: a national initiative	292
9.3.3 Certificate multipliers and adaptability	293
9.3.4 Feed in tariffs: boom and bust	293
9.3.5 The cost of finance	295
9.3.6 The importance of the political environment	295
9.4 Australian climate: challenges and opportunities	296
9.5 PV systems across the Australian network	299
9.6 Case study: Desert Knowledge Australia Solar Centre	303
9.7 Conclusions	309
References	310
10 Conclusions, learned lessons and outlook into the future	**313**
Brahim Aïssa	
10.1 Summary of the book	314
10.2 Looking into the future: PV in harsh environments at large	316
References	318
Index	**319**

About the editors

Brahim Aïssa is a senior scientist at the Qatar Environment and Energy Research Institute (QEERI). He earned his PhD with highest honors from INRS-EMT in Canada, then joined QEERI in 2014. In collaboration with the École Polytechnique Fédérale de Lausanne, he participated to hit, twice, the new world record for power conversion efficiencies in quasi-mono silicon solar cells. He published over 180 refereed papers, three books, and holds many patents, and has numerous international awards and fellowships including the NPI award from European Space Agency.

Nouar Tabet is a professor in the Physics Department and Dean of the college of Science at the University of Sharjah, United Arabic Emirates. His research interests include solar cells, defects in semiconductors and oxides, thin films, nanostructured materials, X-ray, electron and ion spectroscopies. He published more than 120 journals papers, edited and authored books, and won numerous awards. Previous positions include chief scientist at Qatar Environment and Energy Research Institute (QEERI), visiting scientist at Berkeley National Laboratory and MIT, USA, and rector of the University of Constantine in Algeria.

List of figures

Figure 1.1   The zenith and azimuth angles commonly used to describe the
               sun's position in the sky dome of an observer; in this case,
               azimuth is measured clockwise from north, from 0° to 360°.
               The finest dotted line illustrates the sun's path through a
               random day, from sunrise to sunset, in the northern hemisphere.   3
Figure 1.2   Spectral distributions of the solar radiation before
               (extraterrestrial) and after (direct normal irradiance at ground
               level) crossing the earth's atmosphere at an air mass of 1.5;
               both were generated with SMARTS 2.9.5 using a US Standard
               Atmosphere in the "test for July" example in SMARTS   5
Figure 1.3   (a) and (b) Samples of calculated vs. measured GHI at
               one-minute resolution for two sites [7]. Large deviations from
               the one-to-one line are an indication of measurement issues.   7
Figure 1.4   Main components of standard thermoelectric (left)
               pyrheliometers to measure DNI and (right) pyranometers
               for GHI and DHI   9
Figure 1.5   Average profiles of measured and predicted GHI for Doha,
               Qatar, for August 2015, in local time (UTC+3) [10]   11
Figure 1.6   World map of long-term GHI, obtained from the Global Solar
               Atlas, owned by the World Bank Group and provided by
               Solargis   12
Figure 1.7   Conversion of sun light to heat for water desalination   13
Figure 1.8   Electricity production by sources: data adapted from Ref. [13] 14
Figure 1.9   Fabrication process of conventional solar cells and
               heterojunction solar cells: (1) Wafer preparation.
               (2) Cell fabrication. (3) Electric contacts and testing.   15
Figure 1.10   Configuration of PV module layers for monofacial and bifacial
               module technologies   16
Figure 1.11   Sumimoto CPV installation of 1 MWp in Morocco, using
               two axis tracking system and Fresnel lenses, 2016
               https://global-sei.com/company/press/2016/05/prs034.html   18
Figure 1.12   Parabolic trough solar concentrator   18
Figure 1.13   CSP with molten salts for energy storage during the day and
               electricity supply overnight   19
Figure 2.1   Schematic illustrating the cross section of a p–n junction
               solar cell   24

Figure 2.2   The basic operational concept of solar cells   25
Figure 2.3   (a) p–n junction at equilibrium under dark, (b) photocurrent
             and carrier storage in both sides of the junction under
             illumination. Electrons and holes are injected from the n-side
             and p-side, respectively, due to the forward bias which
             weakens the built in electric field (i.e. reduces the
             barrier height).   26
Figure 2.4   I–V characteristics and power curve under illumination   30
Figure 2.5   Electrons in the conduction bands and holes in the valence
             band, trap energy level (E_t) associated with a defect,
             (b) electron trapped from the conduction band,
             (c) recombination of the trapped electron with a hole from
             the valence band. The complete process leads to the
             annihilation of one electron hole pair.   32
Figure 2.6   (a) Electrons in the conduction band and holes in the valence
             band, (b) electron hole pair recombination with excitation
             of an electron (Auger electron) to upper level in the
             conduction band, and (c) Auger recombination with
             excitation of a hole to upper level in the valence band   34
Figure 2.7   Schematic effect of temperature on the $I-V$ characteristics
             of a solar cell   34
Figure 2.8   Effect of light intensity on the I–V characteristics of a
             solar cell   35
Figure 2.9   SQ efficiency limit in function of the energy band gap and
             the correspondence for different solar cells materials   36
Figure 3.1   The theoretical limit of η for single junction solar cells with
             the associated losses vs. E_g at temperatures 300 K, 200 K,
             100 K, and at 1 K. (η is in yellow and the losses in gray.
             The darkest is for L_{Ph}, the middle is for L_{Th}, and the
             lightest is for L_{RR}).   42
Figure 3.2   The contours of η, L_{Ph}, L_{Th}, and L_{RR} vs. E_g and temperature   42
Figure 3.3 η_{max} vs. X for single junction solar cells with 1.0, 1.2, and
             1.4 eV energy gaps   44
Figure 3.4 η_{max} and maximum power yield vs. X and E_g for single
             junction solar cells   44
Figure 3.5   Si, GaAs, and CdTe solar cells performances vs. temperature
             in the range of 300–350 K: (a) η, (b) J_{sc}, (c) V_{oc}, and (d) FF 49
Figure 3.6   Si, GaAs, and CdTe solar cells performances vs. E_g at
             different temperatures: (a) η, (b) J_{sc}, (c) V_{oc}, and (d) FF 50
Figure 3.7   Relative efficiency for Si, GaAs, and CdTe solar cells
             temperature in the range of 300–350 (red dots are for
             experimental data and the blue ones are by simulations)   51
Figure 3.8   Market share of solar cell technologies, adapted from [24].   52

Figure 3.9   State-of-the-art of single-junction hybrid halide perovskite
               solar cells. (A) Scheme of the standard architecture of a
               perovskite solar cell. (B and C) Calculated ideal short-circuit
               current. (B) Open-circuit voltage (C) as a function of the
               bandgap energy of the absorber material according to
               Shockley–Queisser theory. A selection of the best experimental
               parameters found in the literature is also plotted [27].   53
Figure 3.10   Summary of different solar cell absorbers and their PCEs
               achieved in solar cells for halide and non-halide perovskites.
               Major institutions holding the record efficiencies are listed.
               Adapted from [28].   54
Figure 3.11   An example of a workflow of materials screening for
               solar cells   55
Figure 4.1   Record efficiencies for different photovoltaic technologies
               over the last 50 years. This plot is courtesy of the National
               Renewable Energy Laboratory, Golden, CO, USA [9].   62
Figure 4.2   Sketch of a cross section of an (a) aluminum back surface
               field solar cell in comparison with (b) a PERC cell.
               Reproduced from Ref. [20] with permission from
               The Royal Society of Chemistry.   63
Figure 4.3   Comparison of the structure of (a) n-PERC and (b) SHJ
               solar cell. Adapted from Ref. [20] with permission from
               The Royal Society of Chemistry.   65
Figure 4.4   (a) Sketch of SHJ as first developed by Sanyo, Japan.
               The wafer is n-type. (b) Basic process steps for SHJ devices.   67
Figure 4.5   Band diagram of amorphous/crystalline silicon heterojunction
               solar cell   68
Figure 4.6   Schematic sketches of the different device architectures
               investigated in this study. Reproduced from Ref. [20] with
               permission from The Royal Society of Chemistry.   69
Figure 4.7   Injection-dependent effective minority carrier lifetime, t_{min},
               of symmetrically passivated samples as obtained from
               temperature-dependent minority carrier lifetime measurements
               on a Sinton WCT-120TS setup. The lifetime curves indicate
               the passivation schemes in the investigated cells. (a) On p-type
               c-Si, homojunction passivation, (b) on n-type c-Si
               (homojunction), and (c) on n-type c-Si (a-Si:H(i/p) and
               a-Si:H(i/n) passivation). Reproduced from Ref. [20] with
               permission from The Royal Society of Chemistry.   69
Figure 4.8   Temperature-dependent J (V) parameters of the investigated
               solar cell architectures. Reproduced from Ref. [20] with
               permission from The Royal Society of Chemistry.   70
Figure 4.9   Typical structures of CIGS and CdTe heterojunction thin film
               solar cells   73

Figure 4.10   Typical CIGS and CdTe solar cells processing baselines
at the laboratory scale   78
Figure 4.11   Evolution of the performances of CIGS and CdTe solar
cells over the last decades (adapted from [135])   79
Figure 4.12   EQE and IV curves of the certified record CIGS and CdTe
solar cells (adapted from Refs [138,139])   80
Figure 4.13   Possible recombination paths in typical CdS/CIGS and
CdS/CdTe solar cells; note that these band diagram have
been simplified. With E_g, the band gap of the absorber,
E_v the maximum energy level of the valence band,
E_c the minimum energy level of the conduction band, E_{fn}
and E_{fp} the quasi fermi level for the electrons and the holes,
respectively.   81
Figure 4.14   Absolute temperature coefficients of the efficiency, open
circuit voltage, short circuit current density, and fill factor
as a function of these parameters at room temperature for
CIGS solar cells processed using different conditions.
Reproduced with permission from Ref. [156].   82
Figure 4.15   Degradation mechanisms identified on CIGS solar cells
under ATL tests   85
Figure 4.16   Different configurations of perovskite solar cells. n–i–p (left),
p–i–n (middle), and dye (right).   86
Figure 4.17   The AM1.5 solar spectrum and the parts of the spectrum
that can be used theoretically by (a) Si solar cells, and
(b) $Ga_{0.35}In_{0.65}P/Ga_{0.83}In_{0.17}As/Ge$ solar cells, adapted
from Ref. [204].   90
Figure 4.18   Dependency of the conversion efficiency on the semiconductor
bandgap. Adapted from Ref. [205]   91
Figure 4.19   Absorption coefficient versus wavelength for various
semiconductor materials. Adapted from Ref. [213].   92
Figure 4.20   Quantum efficiency of each layer of a GaInP/GaAs/Ge
triple-junction solar cell. Adapted from Ref. [202].   94
Figure 4.21   (a) Temperature dependence of the open circuit voltage (V_{oc})
for the GaInP/GaAs/GaInNAsSb multijunction cell and single
junction cells in the temperature range from 25°C to 90°C.
All the cells were measured at AM1.5G except the triple
junction cell which was measured at AM1.5D. (b) Measured
I–V characteristics of GaInP/GaAs/GaInNAsSb cell with a
three-band solar simulator set to AM1.5D in the temperature
range of 30°C to 150°C. Reproduced from Ref. [215].   94
Figure 4.22   Structure of a triple-junction photovoltaic cell. Adapted
from Ref. [216].   95
Figure 4.23   Measured EQE spectra for GaInP, GaAs, and GaInNAsSb
sub-cells at room temperature and 80°C. Reproduced from
Ref. [215].   96

Figure 5.1   Solar module, string, and array at the outdoor test facility
                (Doha, Qatar)   112
Figure 5.2   Record cell and module efficiencies with a CTM gap [2]   116
Figure 5.3   Mechanisms responsible for the loss/gain by upscaling
                from cell to module under STC   117
Figure 5.4   Detailed loss analysis involved for evaluating CTM   117
Figure 5.5   (a) Pseudo-square, full square, or round wafers used for
                making solar cells and (b) assembly of a module showing
                the effect of wafer type on active and inactive areas   118
Figure 5.6   Effect of cell spacing on the aperture module efficiency
                for full square and pseudo-square wafers [25]   119
Figure 5.7   Power losses from different resistive components include
                contact resistance between emitter and tabs, base and tabs,
                interconnection between adjacent cells, string connections
                and cable losses   121
Figure 5.8   Ribbon-based interconnection and shingled-based
                interconnection technology   122
Figure 5.9   Current mismatching due to non-uniform radiation in a
                60-cell module connected with three bypass diodes   123
Figure 5.10   IV and the PV curve of the module (left) without bypass
                diodes and (right) with bypass diodes   123
Figure 5.11   Different tab design for current collection: standard
                rectangular, round tabs, and textured tabs   125
Figure 5.12   Performance parameters for mono-facial and bifacial
                PV module installation   126
Figure 5.13   Effect of mounting height on mono-facial and bifacial
                modules annual energy yield   127
Figure 5.14   Effect of ground albedo on mono-facial and bifacial
                modules annual energy yield   128
Figure 5.15   Effect of tilt angle on mono-facial and bifacial modules   129
Figure 5.16   Module temperature of glass–glass mono-facial and bifacial
                PV modules   130
Figure 5.17   (Left) Ground coverage ratio (GCR) and its impact on
                bifacial gain (BG) with a tilt angle and (right) mounting
                height   131
Figure 5.18   (Left) Schematic showing the azimuth angle and (right)
                typical power profiles for E–W and N–S mounted bifacial
                and mono-facial modules   131
Figure 5.19   PV test systems at the outdoor test facility in Doha, Qatar 132
Figure 5.20   Average daily SY versus nameplate PTC of 26 PV test
                systems, for one year (2017). Adapted from our Ref. [12]. 134
Figure 5.21   An example to highlight the importance of temperature
                masking of IR images with an appropriate temperature scale.
                Adapted from Ref. [50].   137

Figure 5.22   For the PV silicon technology under study, the AC power,
               normalized energy and the calculated degradation rate for:
               Multi_D, Multi_E, Multi_F, Mono_G and SHJ_H,
               respectively. Adapted from our Ref. [12].   141
Figure 5.23   Annual PV module degradation of the PV modules maximum
               power (P_{max}) of different silicon PV technologies installed in
               desert climate between 2014 and 2018. Each point is an
               average of PV modules measured (see number of module
               in each array, Table 5.3).   142
Figure 5.24   Electrical parameters temperature dependence of different
               silicon PV module technologies. The temperature coefficient
               of (a) P_{max}, (b) I_{sc}, (c) V_{oc}, and (d) FF were determined
               from the linear fitting of the normalized measurement points.
               Adapted from our Ref. [12].   143
Figure 5.25   (a) Visual inspection photographs showing yellowing
               observed over large area at the front multi-crystalline
               silicon array (module Multi_F). (b) Visual inspection
               photographs PV module with polyamide-based back
               sheet cracking (indicated by arrows) after exposure for
               5-years in desert climate (Multi_B). Not all PV module in
               this array showed back sheet cracking. Adapted from our
               Ref. [12].   143
Figure 5.26   (a) Indoor EL image of a silicon heterojunction PV module
               (SHJ_H) after 5-years exposure in the field. (b) and
               (d) The magnified image of the cracked cell shown in (a).
               (c) The rear side of the PV module where an indent with
               a sharp object (indicated by a red circle). (e) Comparing the
               IV curve of the reference module and the module with
               cracked cells. Adapted from our Ref. [12].   144
Figure 6.1   Difference between monofacial and bifacial PV at cell
               and module levels. (a) Passivated emitter rear cell (PERC)
               monofacial with full aluminum coverage from rear side.
               (b) Bifacial PERC+ solar cell with aluminum fingers at
               rear side allowing radiation to pass. Ref. [12]. (c) Schematic
               for monofacial PV module with opaque back sheet not
               allowing albedo reflections. (d) Bifacial with a transparent
               back sheet for collecting albedo radiation.   154
Figure 6.2   Number of refereed publications on bifacial solar cell
               publications per year, along with their associated distribution
               of the published papers per discipline. Data were collected
               from 1991 to December 2022, inclusively from Scopus[®].   156
Figure 6.3   Spectral albedo radiation spectrum for various materials.
               Spectral albedo data collected from Ref. [46].   159

Figure 6.4   Typical specific power profiles of bifacial and monofacial
               PV installed with vertical east–west (E–W) and north–south
               (N–S) directions   160
Figure 6.5   The general device structure of solar cells   162
Figure 6.6   (a) TRPL signals arising from recombination pathways.
               (b) Conventional fitting method of TRPL measurements for
               fitting constants (in logarithmic scale). PL curves are often
               fitted by one-, two-, and stretched exponential functions.
               Here τ_1, τ_2, and β show lifetimes and stretched constant,
               respectively. (c) The proposed charge dynamics method
               predicts physical parameters by solving rate equation
               with appropriate boundary conditions based
               on the geometry.   168
Figure 6.7   Record cell and module PCE with a cell to module gap [7] 170
Figure 6.8   Mechanisms responsible for the losses by upscaling from
               cell to module to field. STC deviation shows the difference
               between STC and operating conditions.   171
Figure 6.9   Detailed loss analysis involved in evaluating CTM
               performance   172
Figure 6.10   (a) Ambient temperature histogram using an annual data for
               the location of OTF, Qatar. (b) Histogram of albedo values
               monitored over a year for local sand at OTF site. (c) Possible
               radiation sources for bifacial power harnessing from different
               orientations. Note the difference between the rear-side
               measurements at the bottom and top pyranometers.
               (d) Typical specific power profile of bifacial and
               monofacial PV installed in Qatar with vertical E–W and
               N–S directions.   174
Figure 6.11   (a) Comparison of average daily specific yield for three
               in-array modules with 2 months cleaning frequency taken
               over 1 year. Two bifacial PV with different bifaciality (bi)
               factors of 65% and 90% were tested under natural albedo
               with standard 22° tilt configuration facing south. (b) Relative
               loss from a standalone system to realistic in-array bifacial
               PV modules for a Bi90 module. The graph shows the typical
               daily power profile for the month of May 2018 with an
               inset showing the average annual percentage losses from
               standalone to in-array mounting.   176
Figure 6.12   (a) Effect of installation height from the ground on the SEY
               of 22° tilt N–S and vertical E–W panel for Bi90 module and
               (b) I–V curves showing kinks due to nonhomogenous
               irradiation from the rear side of bifacial PV at
               different heights   177

Figure 6.13   (a) Albedo effect on power output, corrected to STC, of
               bifacial PV (Bi90) modules installed in vertical and 22° tilt
               position under standalone mounting at OTF, Qatar.
               The dashed line shows the monofacial front-side rating
               at 270 W, which is used for comparison, and (b) bifacial
               gain from monofacial PV module under different ground
               reflective materials.   178

Figure 6.14   (a) Temperature effect of glass/glass monofacial (mono) and
               bifacial (Bi90) PV module for a range of front-side POA
               irradiation under 22° tilt N–S module with both mounted
               in-array. Module temperature relative to ambient temperature
               is plotted against front-side irradiation with a standard
               deviation of 1.37°C for two temperature sensors' readings
               installed at the rear side and (b) difference between bifacial
               module temperature and monofacial module temperature as a
               function of front-side irradiation. The line at $T_{bi} - T_{mono} = 0$
               separates two regimes $T_{bi} > T_{mono}$ and $T_{mono} > T_{bi}$.   180

Figure 6.15   (a) Combined effect of the tilt angle, height, and albedo
               coefficient on the annual energy yield performance of bifacial
               modules using simulations and (b) optimal tilt angle for
               maximum annual energy yield as a function of height and
               albedo for bifacial PV modules installed at OTF obtained
               from simulations.   180

Figure 6.16   (a) Schematic showing the three studied configurations for
               azimuth effect, (b) simulation results for azimuth angle
               effect on annual energy yield of vertical E–W and tilted
               N–S modules, and (c) seasonal variation for vertical E–W
               and tilted N–S to highlight the difference in performance
               during each month.   182

Figure 6.17   Simulation results for solar cell device-level parameters.
               (a) Effect of a bifaciality factor on annual energy yield of
               vertical E–W and tilted N–S modules. Vertical bifacial gains
               more than tilted modules with higher slope and (b) effect
               of the power temperature coefficient (TC) on the annual
               energy yield of bifacial modules.   183

Figure 7.1   (a) Global dust intensity, the darker colors represent a higher
               $\mu g/m^3$ PM10. (b) Daily output PV power loss per day in
               selected countries/cities. Adapted from [1,2] and historical
               soiling data from Outdoor Test Facility (OTF).   198

Figure 7.2   Example of the field measurement of the daily soiling ratio
               percentage in Qatar from October to December 2021   199

Figure 7.3   Recent refereed publications related to the field of
               PV-dust/soiling, together with their corresponding
               distribution of the employed keywords vocabulary and
               histogram of the type of documents, in the inset. Statistics are

available from 1999 to January 2022 inclusively. Data
were collected from Scopus® Expertly curated abstract
and citation database information service. The arrow shows
the year when QEERI started working on PV soiling.   200

Figure 7.4   Uneven soiling on a PV array following a sandstorm in Doha,
Qatar. Modules are tilted south; dominant wind direction
was from the North West; PV arrays facing south.   201

Figure 7.5   Photo of the QEERI OTF's area   202

Figure 7.6   Soiling degree of a pyranometer operating in Doha, Qatar,
showing the decrease of the measured global horizontal
radiation per day due to sensor soiling   205

Figure 7.7   Global map of modeled annual median airborne $PM_{2.5}$ dust
concentration (World Health Organization 2016) (modified),
selected regions for analysis: Middle East (ME), East- and
Southeast Asia (E&SE-As.), North- and West Africa
(N&W Afr.), South Asia (S-As.), Europe (Eur.),
North America (N-Amr.), South America (S-Amr.). c [20],
modified and reprinted with permission from (World Health
Organization 2016). Reprinted with permission from
Elsevier [18].   206

Figure 7.8   Left: Maximum soiling rates (power/efficiency/transmission
loss per day) for different regions of the world, deducted
from a review of data from 63 publications on soiling [19].
Right: annual mean airborne dust concentrations of PM_{10} for
different regions of the world deducted from the Ambient Air
Pollution Database, WHO, May 2016 [20]. Reprinted with
permission from Elsevier [18].   206

Figure 7.9   Mineral composition of resuspensions of dust collected from
over 60 different ground sites throughout the world [17].
Reprinted with permission from Elsevier [18].   208

Figure 7.10   Particle size distribution of dust samples collected from PV
module surfaces and analyzed by laser scattering diffraction (wet)   209

Figure 7.11   Overview of particle size dependency of forces relevant for
dust adhesion and removal thresholds for rolling, sliding,
or direct lift-off by blowing wind. Reprinted with permission
from Elsevier [18].   210

Figure 7.12   Transmittance of atmosphere and black body radiation
indicating radiative losses in the atmospheric window   212

Figure 7.13   Mean RH levels at morning hours (4:30–8:00) for different
locations throughout the world (compiled from Hutchinson
World Weather Guide (Pearce und Smith 2000)). For Qatar
2016, monthly mean values at 6:00, monthly maximum
values at 6:00, and monthly total average values were
deducted from weather station at the OTF, Doha. Reprinted
with permission from Elsevier [18].   213

Figure 7.14   Schematic illustration of soiling mechanisms caused by dew
               formation and subsequent evaporation. Reprinted with
               permission from Elsevier [18].   214
Figure 7.15   Relative rebound as a function of wind speeds   218
Figure 7.16   Relative rebound as a function of RH % for different wind
               speeds   219
Figure 7.17   Relative resuspension per minute as a function of exposure
               time in hours   220
Figure 7.18   Dust accumulation as a function of the WS   221
Figure 7.19   Effect of surface tilt angle on deposition and resuspension.
               Left axis: coefficient of deposition of magnetite particles,
               solid blue circles: 1 m s^{-1} flow speed, open circles: 3 m s^{-1}
               flow speed. Right axis: detachment number of glass
               particles from steel surface.   222
Figure 7.20   Cleaned and un-cleaned PV modules at the OTF in Doha,
               Qatar   223
Figure 7.21   Apparent soiling vs. AOI with respect to time   228
Figure 7.22   Soiling mitigation technologies (a) manual cleaning,
               (b) automated cleaning, (c) dew mitigation for cementation
               control using heating, (d) anti-soiling coating, and
               (e) electrodynamic screen or shields for repelling dust particles 230
Figure 7.23   Overview of different cleaning technologies sorted by
               category: manual, semi-automatic (including truck-mounted
               solutions and portable robots), and fully automatic.
               Reproduced from [94].   230
Figure 7.24   Economic gain in $/kWp year as a function of various
               cleaning rates   232
Figure 7.25   Schematic illustration of Soiling Mitigation Technologies.
               Reproduced from [137]. (A) Important soiling mechanisms
               which could be addressed by ASCs. (B) Single-axis tracking
               and optimization of night stowing position. (C) Working
               Principle of EDS (standing wave version). (D) Dew mitigation
               by low-ε coatings and active and passive heating. (E) PV
               module design approaches for soiling loss reduction: the red
               overlay indicates lost cell strings dew to soiling. (F) Site
               adaption.   234
Figure 7.26   Optical images of glass substrate samples (a) uncoated and
               (b) coated with TiO_2 after soiling for 7 days in Doha for all
               samples (reproduced from [91])   236
Figure 7.27   Impact of soiling on solar power generation. Reproduced
               from [94]. (A) PV capacity installed by 2018 and medium
               estimate for 2023, sorted by country for the top 22, and
               global CSP capacity. (B) Corresponding soiling rates
               reported in literature. (C) Reported cleaning costs per
               cleaning and square meter. (D) Typical energy yield in

	kWh/kWp for representative locations. (E) Calculated range of optimal number of yearly cleaning cycles (bars) and actual range of typical yearly cleaning cycles reported in literature (blue lines, see Model Validation). The arrow indicates that for CSP, the numbers are out of range (up to 85 in 2018 and 55 in 2023). (F) Minimum expected financial losses due to soiling calculated from optimum cleaning cycles. (G) Overview of locations used for estimation of average specific yield.	240
Figure 7.28	(a) Price of utility-scale electricity generation technologies in the GCC. (b) Sustainable energy targets. (c) Installed renewable energy capacity in GCC countries as a share of the total by 2018. (d) Renewable power planned additions by country. Reproduced with permission from [154].	242
Figure 8.1	Daily average solar radiation – horizontal at five different locations in Qatar [4]	260
Figure 8.2	Average air temperature at five different locations in Qatar [4]	260
Figure 8.3	Average wind speed at five different locations in Qatar [4]	260
Figure 8.4	A clear day in winter	261
Figure 8.5	Clear day with haze/dust	261
Figure 8.6	A clear day in summer with typical max radiation	262
Figure 8.7	A day in summer with dust	262
Figure 8.8	A day with clouds and rain	263
Figure 8.9	A day with cloud, haze, and dust	263
Figure 8.10	Day in winter with a dust storm	263
Figure 8.11	A day with dust, haze, and clouds	264
Figure 8.12	Partial shading impact on $I–V$ curve of serially connected PV cells	265
Figure 8.13	Impact of partial shading on PV modules with bypassing diodes. (a) PV module with partial shading. (b) $I–V$ and $P–V$ curves for PV module with partial shading.	265
Figure 8.14	An aerial view of a typical desert farm	268
Figure 8.15	Load profile of typical day for two-bedroom apartment	272
Figure 8.16	Load profile of a typical day for four-bedrooms villa	272
Figure 8.17	The PV-powered water pumping system	273
Figure 8.18	PV-powered lighting system	273
Figure 8.19	Off grid PV-hybrid system	274
Figure 8.20	Off-grid PV-battery system	274
Figure 8.21	Grid connected PV-battery system	275
Figure 8.22	Load demand and potential PV generation	276
Figure 8.23	Commercial farm energy system architecture	278
Figure 8.24	Daily load profile for the farm	278
Figure 8.25	Average monthly load demand for the farm	279
Figure 8.26	The monthly average solar global horizontal irradiance	279

Figure 8.27	NPV between the base case and lowest cost system	281
Figure 9.1	Learning curve for module price as a function of cumulative shipments. Source: International Technology Roadmap for Photovoltaic, Ninth Edition, September 2018.	286
Figure 9.2	Sydney Olympic Village PV. Source: IEA-PVPS-Task 10 Community-Scale PV: real examples of PV-based housing and public developments.	288
Figure 9.3	Average module prices in Australia over the last two and a half decades	289
Figure 9.4	PV installations in Australia since 2001 by system size	289
Figure 9.5	Utility scale solar installations	290
Figure 9.6	Australia climate zones based on a modified Köppen classification system using a standard 30-year climatology dataset (1961–1990). Source: Australian Bureau of Meteorology (BOM).	297
Figure 9.7	Average daily solar exposure in MJ/m². Source: Australian Bureau of Meteorology (BOM).	297
Figure 9.8	Average daily maximum temperature in degrees Celsius. Source: Australian Bureau of Meteorology (BOM).	298
Figure 9.9	Australia's electricity infrastructure. Source: Australian Energy Resource Assessment, Geoscience Australia and BREE, 2014, 2nd Ed.	299
Figure 9.10	Historical weighted average spot prices. Source: Australia Energy Regulator (AER).	300
Figure 9.11	Installed PV capacity per state by system size. Source: APVI.	301
Figure 9.12	Australia PV installations per postcode based on (a) PV density and (b) PV capacity. Source: APVI.	302
Figure 9.13	(a) Schematic of the DKASC demonstration facility. Source: http://dkasolarcentre.com.au/locations/alice-springs and (b) aerial image of the site. *Source*: Google Earth.	304
Figure 9.14	Average monthly temperatures and daily irradiance for Alice Springs. Source: BOM.	305
Figure 9.15	Performance of CdTe systems from the DKASC	308
Figure 9.16	Average performance of crystalline silicon systems	309

List of tables

Table 1.1   ISO 9060:1990 classification for solar irradiance sensors;
            see text for descriptions of the specifications   8
Table 3.1   The commonly used parameters for the materials considered
            cells   48
Table 4.1   Relative TCs at AM 1.5G irradiance of 1,000 W/m^2 of the
            devices shown in Figure 4.8, derived from linear fitting between
            25°C and 75°C of the temperature-dependent J(V) parameters
            shown in Figure 4.8.   71
Table 4.2   Data of commercial modules taken from module data sheets.   71
Table 4.3   Record power conversion efficiencies obtained for CIGS and
            CdTe PV technologies at the solar cell and module levels   79
Table 4.4   Examples of degradation rates (assuming a linear decrease
            of the output) of CdTe and CIGS PV module arrays installed
            in different field locations   83
Table 4.5   Experimentally measured various temperature coefficients   96
Table 5.1   Module technologies of PV test systems used in the study   133
Table 5.2   Various defects types with their associated power losses   135
Table 5.3   Electrical parameters and temperature coefficient (TC) of the
            various PV modules in OTF   140
Table 5.4   Measured temperature coefficient of the silicon PV module
            technologies   144
Table 6.1   A range of band-to-band radiative recombination constant
            and product of trap density and cross section   164
Table 6.2   Summary of key performance parameters studied for
            bifacial modules   184
Table 7.1   Systems in OTF measured continuously since 2013   203
Table 7.2   Soiling reduction potential and costs for selected soiling
            mitigation technologies   231
Table 7.3   PV soiling research and market—state-of-the-art   237
Table 7.4   Predicted renewable energy capacity in GCC countries
            by 2030.   243
Table 8.1   Temperature, GHI, and wind speed at Abu Samrah   257
Table 8.2   Temperature, GHI, and wind speed at Ar-Ruwais   258
Table 8.3   Temperature, GHI, and wind speed at Doha International
            Airport   258

Table 8.4	Temperature, GHI, and wind speed at Dukhan	259
Table 8.5	Temperature, GHI, and wind speed at Messaeid	259
Table 8.6	Typical energy consumption of different desalination techniques	270
Table 8.7	Typical load in the farm and its energy consumption	271
Table 8.8	Techno-economic analysis of PV-diesel-battery	280
Table 9.1	Average specific yield for small-scale PV systems (<100 kW) in 2018	303
Table 9.2	Sample of systems installed in DKASC	305
Table 9.3	Average PR per cell type	306
Table 9.4	Average SY per cell type	307
Table 9.5	Average annual degradation (%) calculated using linear regression for systems with at least 7 years of data	307

Preface

The book "Photovoltaics in Desert Environments" explores the unique challenges and opportunities offered by the deployment of solar energy systems in arid, desert regions. Deserts are often seen as inhospitable, but they are usually located in highly sunny regions with offer some of the most constant, stable, and intense sunlight in the world. This makes them an ideal location for harnessing solar power, which is a clean and renewable source of energy that is becoming increasingly important as the world looks for ways to reduce our reliance on fossil fuels to contribute to the reduction of greenhouse gases emissions from the power sector. These amazing opportunities face several challenges associated with deploying photovoltaic (PV) systems in desert environments. These challenges include the harsh and extreme environmental conditions that can affect the performance and lifespan of PV panels, the efficient operation of power plants, as well as the difficulty of transporting and installing PV systems in remote and often rugged terrain. In addition to the technical challenges of soiling, high UV radiation, high level of humidity, high temperature and high temperature oscillation cycles between day and night, there are also cultural and economic considerations to take into account when implementing PV projects in desert regions. These can include working with local communities and addressing issues of land ownership and access, while at the same time complying with environmental challenges. Despite these challenges, there are many benefits to be gained and opportunities for innovation, from implementing PV systems in desert environments. These benefits include the potential for large-scale energy production, developing new methods to develop solutions to reduce construction and operation costs, between others, as well as the economic and social benefits that can come from creating jobs and bringing electricity to remote areas.

Based on part of research undertaken at Qatar Environment and Energy Research Institute (QEERI), and with contributions from chapter authors from a range of international institutions, this book addresses recent trends and developments in its field. Solar energy resources and harvesting technologies are introduced and detailed in Chapter 1. Chapters 2 and 3 are dedicated to the solar cell fundamentals and the thermodynamics of solar energy conversion, while the different solar cell technologies are reviewed in Chapter 4. Chapter 5 highlights the module technology and energy yield under desert environment conditions, while Chapters 6 and 7 revolve around the bifacial solar technology and module installation and the associated PV soiling characterizing desert environments. In Chapters 8 and 9, we have put the focus on two different case studies representing two different desert climates, namely that of Qatar and Australia. Chapter 10 concludes with an outlook into the future. The book is amended with an extensive and updated survey of the published articles and conference reports. It is hoped that the book will be of interest to those involved in the investigation of PV energy at large, ranging from novice students to the most experienced end-users. The intention is also to stimulate debate and reinforce the importance of a multidisciplinary approach in this exciting field.

Whether you are an engineer, a policy maker, or simply someone interested in the potential of solar energy, "Photovoltaics in Desert Environments" is an essential read that will provide you with a comprehensive understanding of the unique challenges and opportunities of using PV to harness the power of the sun in desert regions.

Acknowledgment

This book would not have been possible without the tremendous support of many people. As authors and editors of "Photovoltaic Technology for Hot and Arid Environments," we would like to express our sincere gratitude to all of the scientists who have contributed to the creation of this project.

First and foremost, we would like to thank our colleagues from the Qatar Environment and Energy Research Institute (QEERI), The University of New South Wales (UNSW) and mentors in the field of photovoltaics, who have shared their knowledge and expertise with us over the years. We are grateful for the opportunity to learn from and collaborate with such talented and dedicated individuals.

We would also like to thank the reviewers who generously provided their time and insights to help improve the quality of this book. Their comments and suggestions were invaluable and greatly appreciated.

We wish to underline the valuable help provided by the staff of The Institution of Engineering and Technology (IET) publishing group, with a special mention to Sarah Lynch, Olivia Wilkins, Christoph von Friedeburg, and Paul Deards for their availability and willingness to provide guidance and advice during the preparation of the manuscript. Thank you for having worked closely with us to bring this book to fruition. Your professionalism and dedication made the process of writing and publishing this book a real pleasure.

Finally, we would like to thank our families and friends, who have supported and encouraged us throughout this process. Your love and encouragement has meant the world to us, and we could not have achieved it without you.

It is our hope that this book will be a valuable resource for those interested in the unique challenges and opportunities of deploying photovoltaic systems in desert environments. We are grateful for the opportunity to share our knowledge and experience with you, and we hope that it will inspire others to pursue their own interests in this exciting and important field. Thus, this book is a collaborative effort of many scientists, and we are deeply thankful to everyone who has contributed to its creation.

We are thankful to all people who have created the particular magic environment, where the business objectives are realized through the cultivation of innovation and by harnessing scientific curiosity.

Chapter 1

Solar energy resources and harvesting technologies

Daniel Perez Astudillo[1] and Nouar Tabet[2]

1.1 Introduction

The sun is a huge nuclear reactor where the fusion process of hydrogen atoms occurs continuously. It releases at its surface a power of about 3.8×10^{26} W. The planet earth being at an average distance of 150 million km away from the sun, receives only 1.7×10^{17} W, which corresponds to an irradiance of 1,367 W/m^2. Part of this power is reflected back into space while the rest has to travel through the earth atmosphere where part of it is absorbed. About 60% reaches the surface of earth (10^{17} W). This average incident power is not uniformly distributed over the globe. The desert regions of North Africa and the Middle East known as the *Sun Belt* and the north of Chile have an average daily insolation exceeding 6 kWh/ m^2. Photovoltaics (PV) and concentrated solar power (CSP) are the main solar technologies that have been deployed at large scale in many regions of the world. However, the deployment of these technologies in the desert regions face some major challenges, among them the dusty atmosphere, high ambient temperatures, high UV component of the light spectrum, and high humidity in the regions close to the sea. These characteristics induce a reduction of the energy yield of the solar installations and a faster degradation over time which lead to an increased levelized cost of energy (LCOE). The impact of these environmental factors is discussed in detail in various chapters of this book.

In this introductory chapter, we review the basic characteristics of the solar light and the techniques to assess the insolation in a given place and harvest the solar energy.

1.2 Solar resources

1.2.1 Sun–earth system

Our sun is the ultimate source of practically all of the energy available on earth, not only in the form of direct sunlight but also in other different forms, mainly driven

[1]Qatar Environment and Energy Research Institute (QEERI), Hamad Bin Khalifa University (HBKU), Qatar Foundation, Qatar
[2]Deptartment of Applied Physics and Astronomy, University of Sharjah, United Arab Emirates

by its gravity. Solar energy, however, commonly refers to the radiant forms of energy, light and heat, from the sun, and this book focuses on the conversion of sunlight to electricity for use on earth. It is thus important to understand how this light is generated, transmitted to our planet, and affected by our atmosphere until it reaches ground level, where it is used.

The sun is a star that continuously emits electromagnetic radiation in all directions due to the nuclear reactions occurring in its interior. Seen as a black body, the total radiant power, or radiant flux, emitted at the surface of the sun per unit surface area, depends on its absolute temperature according to the Stefan–Boltzmann law: $P = \sigma T^4$ where $\sigma = 5.67 \times 10^{-8}$ W m^{-2} K^{-4} is the Stefan–Boltzmann constant and T is the temperature in K, around 5,778 K for the surface of the sun. This results in an emission of around 60 MW/m^2 at the sun's surface. This value, however, is not constant, due to variations in solar activity with a nearly constant period of 11 years, the so-called solar cycles. On the other hand, there are other long-term changes in solar activity that occur at much larger timescales, not of interest for the purposes of current solar power applications.

The earth revolves around the sun in an elliptical orbit with a mean distance of about 150 million km; although the orbit is almost circular due to its low eccentricity of 0.0167, this distance varies between a minimum of 146 million km at the perihelion in early January and a maximum of 152 million km at the aphelion in early July.

After travelling across space, the amount of solar radiation at the sun–earth distance, called the solar constant, has an average value of 1366 W/m^2 measured on a plane normal to the sunlight rays' travel direction, and varies by about ±3 per cent through the year due to the changes in distance. This radiation is also called top-of-atmosphere radiation, or beam extraterrestrial (ET) radiation, denoted I_{0b}, and determines the maximum possible value of available solar radiation on earth. I_{0b} is also commonly projected to a horizontal plane at the top of the atmosphere, resulting in the horizontal ET radiation, H_0.

As mentioned, I_{0b} is measured on a plane perpendicular to the sun rays, which are for most purposes parallel for all points on earth, given our distance to the sun. Due to our planet's curvature, however, the projection to a horizontal plane at any (coincident) time is different for different locations on the planet. It is therefore necessary to know the location of the sun in the local sky of a given observer, at any given time, in order to obtain H_0 and other related quantities.

Using a spherical coordinate system with the observer on the surface of the earth at the origin, the solar position is given by three coordinates, see Figure 1.1: the sun–earth distance, the sun's azimuth angle, and the sun's zenith angle. The azimuth angle is commonly measured using either the north or the south as origin, and in the range [0°,360°] or [−180°,180°], so when choosing any set of equations for its calculation, care must be taken to note which convention is used. The zenith angle θ_Z is measured from the vertical overhead, with possible values from 0° to 180°; the solar elevation angle is also frequently used during daytime, and is given, in degrees, by elevation = 90° − θ_Z.

There exist various sets of formulas to obtain the solar position for any given place on earth at any given time, with the algorithms differing in complexity,

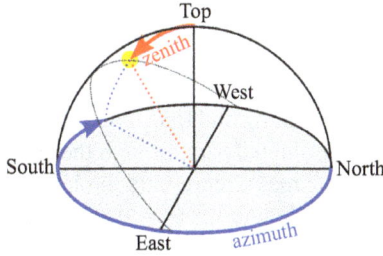

Figure 1.1   The zenith and azimuth angles commonly used to describe the sun's position in the sky dome of an observer; in this case, azimuth is measured clockwise from north, from 0° to 360°. The finest dotted line illustrates the sun's path through a random day, from sunrise to sunset, in the northern hemisphere.

required inputs (some even need actual measurements of some parameters), accuracy, and temporal validity. A simple, easy to implement set of equations is the one provided by NOAA, the USA's National Oceanographic and Atmospheric Administration [1], accurate within one minute. NREL's SolPos v.2 algorithm [2] can provide more output calculations including, for example, the intersection angle between the solar rays and a tilted plane and is accurate to a fraction of a degree. The equations can be implemented in the user's code, but some are available through online interfaces; for example, NREL has an online calculator that uses the even more accurate Solar Position Algorithm (SPA), on https://midcdmz.nrel.gov/solpos/spa.html.

1.2.2   Effects of earth's atmosphere on sunlight

Before reaching the earth's surface, extraterrestrial radiation must traverse the atmosphere, and along this path, part of the incoming sunlight can pass through unaffected, but other part is absorbed, and the rest is scattered.

Scattering depends mainly on the relative difference between the size of the atmospheric particles and the wavelength of incoming light. A more detailed discussion can be found in Chapter 7 of [3], but in general, three regimes are usually considered, of which the two main ones are Rayleigh and Mie scattering. Rayleigh scattering is relevant for atomic-sized atmospheric components, where the particle size is much smaller than the light's wavelength; thus, larger wavelengths are scattered more than smaller ones. Mie scattering occurs when the scattered particles are of sizes comparable to the light wavelength; this is the case of molecules and micro-sized particles. The angular distribution of Mie scattering varies widely depending on size and other factors, but in general, the forward scattering is dominant. For large scatterers, such as water droplets or dust particles, geometric optics is used to model reflections off their surfaces.

For a cloudless sky, the amount of transmitted light as a function of traversed distance follows an exponential extinction function, given by the Beer–Lambert law, which can be written as a function of wavelength (λ) as:

$$I(\lambda) = I_0(\lambda) \cdot \exp(-m \cdot t(\lambda)) \tag{1.1}$$

where $I_0(\lambda)$ and $I(\lambda)$ are the incoming and transmitted light intensities, respectively; $m \cdot \tau$ is the optical depth for the slanted path through the atmosphere traversed by the light; m, called the relative air mass, represents the length of the path through the atmosphere relative to the path length for $\theta_Z = 0$ at sea level and is approximately equal to $1/\cos(\theta_Z)$ for small θ_Z but with larger differences at low solar elevations, so a commonly used correction is the one proposed by Kasten and Young [4]; $\tau(\lambda)$ is the sum of the optical depths of all absorption contributors, mainly aerosols, uniformly mixed gases (O_2 and CO_2), water vapor (which affects light at several wavelengths, but mainly around 900 nm), ozone and Rayleigh scattering, plus other attenuators for specific wavelengths.

1.2.3   Standard solar spectrum

Although several of the atmospheric components are highly variable both in time and in space, a number of standard models have been developed to describe different atmospheric properties (temperature, pressure, density, etc.) for different altitudes. Some models provide formulas for calculation of the parameters as a function of height but, usually, tables of the values in bins of altitude are given. Some common examples are: the International Standard Atmosphere from the International Standards Organization; the ICAO Standard Atmosphere from the International Civil Aviation Organization; the WMO Standard Atmosphere; and the US Standard Atmosphere. More recent models include NRLMSISE-00 by the US Naval Research Laboratory and JB2008 by the US Air Force Space Command and Space Environment Technologies.

Sunlight arriving to the earth surface depends not only on the specific atmospheric composition but also on the time of day and location; so standard reference spectra have been defined too, which allow the comparison of the performance of different photovoltaic, solar thermal, and other devices that use direct or the hemispherical solar irradiance (these irradiances are defined in Section 1.4). The International Standard ISO 9845-1: 1992 [5] defines two standards, known as ASTM G-173 (ASTM stands for American Society for Testing and Materials), both for an air mass of 1.5 using the US Standard Atmosphere; one standard spectrum is for direct normal radiation, with a field of view (FOV) of 5.8° and an integrated power density of 900 W/m^2; the other standard is for hemispherical radiation on an equator-facing plane at a tilt of 37° and an albedo of 0.2, with an integrated power density of 1,000 W/m^2.

A number of standard spectra can be generated using Gueymard's SMARTS software [6]. As an example, Figure 1.2 shows a comparison of the spectral distributions of incoming extraterrestrial radiation and solar radiation at sea level for a US Standard Atmosphere and an air mass of 1.5, obtained using SMARTS 2.9.5.

1.2.4   Solar light components: GHI, DHI, DNI

Although, as will be discussed in Chapter 2, the spectral distribution of sunlight is relevant to the performance of PV systems, for resource assessment and other

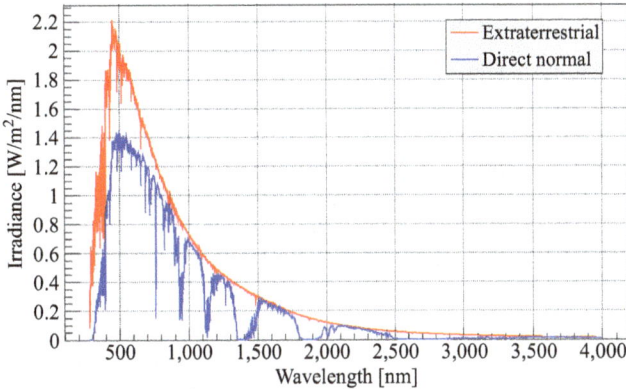

Figure 1.2   Spectral distributions of the solar radiation before (extraterrestrial) and after (direct normal irradiance at ground level) crossing the earth's atmosphere at an air mass of 1.5; both were generated with SMARTS 2.9.5 using a US Standard Atmosphere in the "test for July" example in SMARTS

general purposes the solar resources are most commonly determined in the so-called broadband radiation, namely the integrated solar radiation in the range from 200 to 4,000 nm. In the rest of this chapter, solar radiation refers to this broadband radiation, unless explicitly noted.

As described in Section 1.2, while crossing the earth's atmosphere, solar radiation undergoes different interactions with the atmospheric components (gases, water vapor, aerosols), and can be absorbed, scattered, or transmitted unaffected.

Regardless of the details of scattering and absorption processes, the resulting solar light incident on a surface at ground level can come from three sources or directions: from the direct line to the sun, from the sky dome excluding direct sunlight, and from surface reflections.

- From the direct line to the sun. This is sunlight that was not affected by scattering or absorption along its path, and on clear sky conditions can amount to up to 70% of the total ET radiation.
- From the sky dome, excluding direct sunlight. This is light that was not originally travelling in the direction of the surface of interest but arrives to it after being scattered one or more times in the atmosphere from any other direction.
- Surface reflections. Ground-reflected light can reach the surface of interest if it is not facing horizontally upwards, that is, if it is tilted at some angle such that the surface has at least a partial view of the ground. Sunlight can also be reflected by other surfaces such as nearby walls or other structures; however, for solar power applications, these extra reflections are generally avoided.

Accordingly, solar radiation is measured through three parameters:

- GHI, global horizontal irradiance, is the total amount of downwards sunlight, per unit area, reaching a horizontal surface that looks upwards. Thus, GHI includes direct and scattered or reflected light. The ratio GHI/H_0, called the global clearness index, provides an indication of the transparency of the atmosphere to sunlight, and is mainly sensitive to the presence of clouds, while being less affected by atmospheric aerosols.
- DNI, direct or beam normal irradiance, is the light coming from the sun disk that was not affected by scattering or absorption along its path, although as will be shown in Section 1.5, the measurement of DNI actually includes a small region around the solar disk. DNI is measured per unit area on a plane perpendicular to the sun rays, and since the sun is continually moving across the observer's sky, the orientation of this plane is also continuously changing through the day. Unlike GHI, DNI is very sensitive not only to clouds but also to aerosols. Under ideal clear sky conditions, the beam clearness index DNI/I_{0b} can reach up to 0.75.
- DHI, diffuse horizontal irradiance, is the light falling on an upwards-looking horizontal plane, per unit area, coming from the hemispheric sky dome excluding direct sunlight; in other words, excluding from GHI the horizontal projection of DNI.

These three quantities are related through a simple equation:

$$GHI = DNI * \cos(\theta_Z) + DHI \qquad (1.2)$$

where DNI $\times \cos(\theta_Z)$, sometimes called the beam horizontal radiation, is the projection of DNI on the horizontal plane. When the three irradiances are measured, the points that deviate largely from this equality are one indication of wrong measurements (see Figure 1.3).

1.2.5   Estimating insolation

As previously discussed, the amount of solar radiation reaching the top of the Earth's atmosphere can be calculated, but this radiation is affected by the atmospheric components before reaching ground level. While the effects of a clear, ideal atmosphere can also be calculated, the real composition of an observer's local atmosphere is constantly changing, and so are its effects on sunlight; these changes depend not only on climatic conditions, e.g. fog, cloudiness or airborne desert dust, but also on human-induced factors, mainly in the form of aerosols which, apart from their direct influence on solar radiation, can affect cloud formation. Therefore, available solar resources have to be separately determined for any given location and time, either by models (physical, empirical or hybrid) or by observations (direct or indirect). The two most commonly used methods, namely direct measurements and satellite-based observations, are discussed here.

1.2.6   Ground measurements, instruments

The most precise estimations of solar radiation at ground level can be achieved through on-site measurements. The instruments used to this end are called

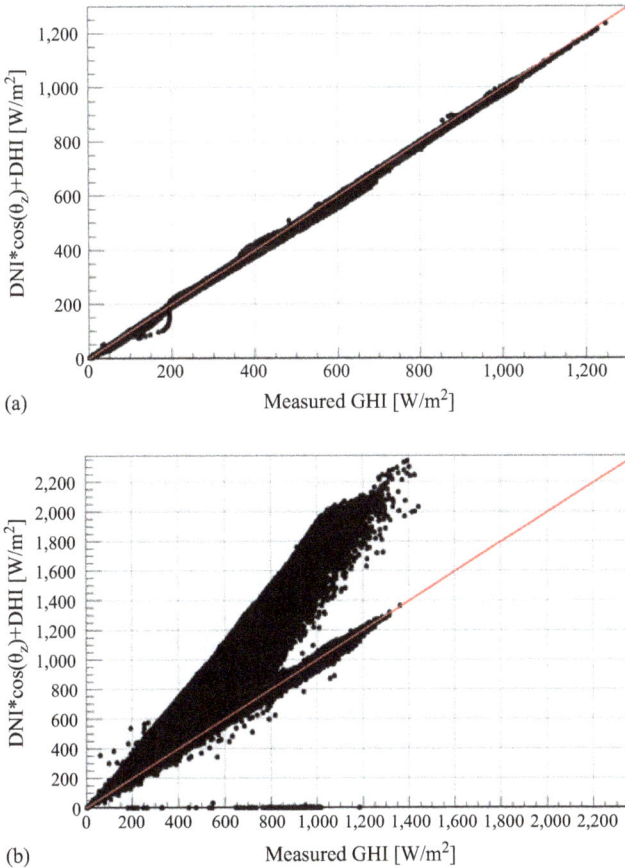

Figure 1.3 (a) and (b) Samples of calculated vs. measured GHI at one-minute resolution for two sites [7]. Large deviations from the one-to-one line are an indication of measurement issues.

radiometers, of which different types exist, with different advantages and disadvantages, but currently the most common use either thermoelectric or photoelectric disc-shaped sensors with a flat surface (since the measurements are done on planes) that is exposed to the incoming solar radiation. The radiometers used to measure DNI are generically called pyrheliometers, while the ones for GHI and DHI are called pyranometers. Although the sensors have to be exposed to solar radiation in order to measure it, they should also be protected from the elements; thus, in solar radiometers the sensor is placed inside enclosures, with windows or domes (see next paragraphs) that only allow solar radiation to reach the sensor. Independently of the type of sensor, radiometers for solar resource assessment are classified by their response and characteristics as shown in Table 1.1, according to the ISO 9060:1990 classification [8].

Table 1.1   ISO 9060:1990 classification for solar irradiance sensors; see text for descriptions of the specifications

Specifications	Secondary standard	First class	Second class
Response time	< 15 s	< 30 s	< 60 s
Zero offset A	± 7 W/m^2	± 15 W/m^2	± 30 W/m^2
Zero offset B	± 2 W/m^2	± 4 W/m^2	± 8 W/m^2
Non-stability	± 0.8 %	± 1.5 %	± 3.0 %
Non-linearity	± 0.5 %	± 1 %	± 3 %
Directional response	± 10 W/m^2	± 20 W/m^2	± 30 W/m^2
Spectral selectivity	± 3 %	± 5 %	± 10 %
Temperature response	2%	4%	8%
Tilt response	± 0.5 W/m^2	± 2 W/m^2	± 5 W/m^2

The response time is the time required to reach a 95% response in the signal. Zero offset type A is the response to 200 W/m^2 net thermal radiation, ventilated. Zero offset type B is the response to 5 K/h change in ambient temperature. Non-stability is the change in responsivity per year, as percentage of the full scale. Non-linearity is the deviation from responsivity at 500 W/m^2 due to any change in irradiance within the range from 100 to 1,000 W/m^2. Directional response, for beam irradiance, is the range of errors caused by assuming that the normal incidence responsivity is valid for all directions when measuring, from any direction, a beam radiation whose normal incidence irradiance is 1,000 W/m^2. Spectral selectivity is the deviation of the product of spectral absorbance and spectral transmittance from the corresponding mean, within the range from 350 to 1,500 nm. Temperature response is the deviation due to any change in ambient temperature within an interval of 50 K. Tilt response is the deviation in responsivity relative to 0° tilt (horizontal) due to any change in tilt from 0 to 90° at 1,000 W/m^2 beam irradiance.

This ISO 9060 classification was updated in November 2018. The ISO: 9060:2018 specifications [9] define three instrument classes, namely A, B, and C, equivalent to the previous secondary standard, first class, and second class, respectively, but with some changes. Each of the new classes has two types: either "spectrally flat" or "fast response"; this means that, for instance, rotating shadow band radiometers (see next paragraphs), which did not have an ISO 9060:1990 class, can now be classified with the new ISO 9060. For class A pyranometers, there are new additional requirements of individual tests of temperature response and directional response.

1.2.6.1   DNI

By definition (see Section 1.4), DNI should exclude any light outside the solar disk, which has an apparent angular diameter of about 0.5° as seen from earth; however, standard pyrheliometers are designed with an FOV of five degrees; thus, a small region of sky around the sun is included in the measurement of DNI. To achieve this, the basic pyrheliometer design consists of a tube with the sensor at one end, a protective window at the other end, and collimating disks along its inner length. As the sun continuously moves across the sky, pyrheliometers are usually mounted on a sun-tracking device that follows the solar path from sunrise to sunset, all year, so that the sensor is always normal to the sun rays. This design with moving parts adds complexity to the measuring system, so sometimes DNI is not directly measured, but derived from GHI and DHI measurements by using (1.2), if these two are measured, which can be achieved by two pyranometers as will be described next, or with devices know as rotating shadow band radiometers (RSRs), which use only one sensor measuring GHI, and a shading arm or band blocks the sensor's view of the sun at certain intervals, during which DHI is measured, and then DNI is calculated as per (1.2).

1.2.6.2 GHI

To measure GHI, no tracking is necessary, the sensor needs only to be placed on a horizontal and with a view of the full sky hemisphere. To protect the sensor, domes are installed covering the full hemispheric view; some designs use a single dome, and some use two concentric domes to help reduce thermal effects. In RSRs, the sensor is directly encapsulated inside a disk-shaped white light-diffusing material, so no domes are used.

1.2.6.3 DHI

Since the only difference between DHI and GHI is the exclusion of direct sunlight, both are measured with the same kind of pyranometer. To remove the direct contribution, the sun disk has to be covered from the DHI sensor view. When the solar radiation components are measured separately, there are two common approaches to measure DHI: the simplest one is placing a shading band oriented along the sun's path through the day, based on the sun's declination angle, so that it projects a shadow fully covering the sensor at all times of the day without tracking needed; as the solar declination angle changes through the year, the shading band has to be adjusted, which can be done manually every few days. As the shading arm also covers part of the sky, some corrections are needed to account for this, which introduce additional uncertainties. The second approach is using a ball or a disk that only covers the sun; this, however, requires that the shading ball/disk be mounted on a sun tracking device; if DNI is being measured with a pyrheliometer, the same tracker is used, only adding

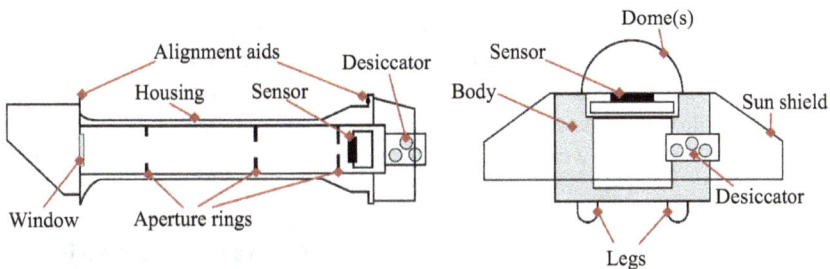

Figure 1.4 Main components of standard thermoelectric (left) pyrheliometers to measure DNI and (right) pyranometers for GHI and DHI

a support arm for the shading ball/disk. In RSRs, as previously mentioned, the shading band is not fixed, but stays most of the time outside the sensor's FOV so that GHI is measured, and rotates frequently (e.g. once or twice per minute) to measure DHI and calculate DNI at the moments when the sensor is shaded.

1.2.7   Satellite-based data

A properly set up and maintained solar radiation monitoring station will provide the most accurate solar data at a given site. However, the associated costs and resources needed may impede the deployment of such instrumentation. Even when this is not an obstacle, the station will only be able to provide current measurements, going as far back as its date of start of operation, and spatially the measured values gradually lose their validity as distance to its point of origin increases; additionally, any down times due to malfunctions, repairs, maintenance, etc. of the equipment result in gaps in the data, which could last from few minutes to weeks or months, until normal operation resumes.

An alternative method of obtaining solar resource information is through satellite observations. Satellites, either geostationary or circling in orbit around the earth, have the advantage of looking at large areas of the planet at a time, providing wide spatial coverage, and some satellites have been in operation for decades, which makes longer term data available. The basic idea followed by some of the most common methods to derive ground-level solar irradiance from satellite observations is to take visible spectrum images of the earth and somehow correlate pixel brightness to the received solar radiation at that pixel. This requires accounting for other factors that affect pixel brightness, such as the natural brightness of the ground in that pixel, as well as its albedo, and for any cloudiness or other atmospheric components – factors that change at different time scales, even from seconds or minutes as in the case of clouds or fog. On top of these factors, parallax effects can be significant, for example at areas near the edges of the satellite's FOV and can result in clouds appearing to be above a given area when they are, in fact, over another.

Light captured by the satellites has been reflected from the earth's surface, i.e. this light has traversed the atmosphere for a second time, often through a different path from the incoming sunlight, when the sun is not directly behind the satellite. Therefore, additional information is frequently added to satellite-based models, from weather observations to model predictions or long-term averages, resulting in different accuracies, and satellite-derived databases are continuously validated and improved with the help of comparisons with data from ground measurements from as many locations as possible. Although aerosols affect mainly DNI, their effects can be noticeable in GHI as well, especially in regions with high aerosol loads in the atmosphere. Introducing an explicit, advanced treatment of aerosols to simulate GHI with the WRF numerical weather prediction model (see [10] and references therein), the relative root-mean-square difference with measured values was cut almost by half in a study conducted in Qatar, a desertic region with hot, arid climate. A comparison of average diurnal profiles of measured and predicted GHI values with ("WRF/Chem-QEERI") and without ("WRF") the improved aerosol treatment is shown in Figure 1.5, for the month of August 2015; in both predicted cases, the rapid radiative transfer model parameterization [11] was used.

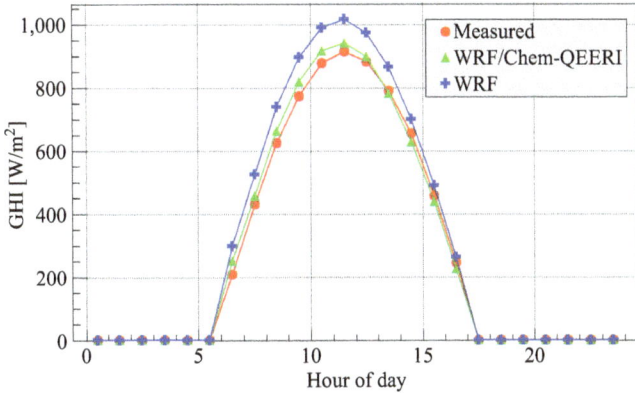

Figure 1.5   Average profiles of measured and predicted GHI for Doha, Qatar, for August 2015, in local time (UTC+3) [10]

1.2.8   Insolation maps and data sets

Although ground-level measurements of solar radiation are at present still limited in both spatial and temporal coverages around the world, the use of satellite observations and models, combined or not with ground measurements, has allowed for a practically worldwide coverage of available solar resource data.

Given that the primary source for satellite-based data are raster images consisting of pixels, the first outputs come at the spatial resolution of those pixels, which currently are on the order of kilometers. Some datasets are kept in that spatial resolution, while others are downscaled by using interpolation techniques and ancillary information such as other parameters obtained at higher resolution, ground elevation profiles, etc.

The temporal resolution of satellite images depends on the frequency at which they acquire the images. For example, MeteoSat, a geostationary satellite, captures its full FOV (about one third of the earth's surface) every 15 min; however, the images are taken by scanning that FOV through taking consecutive images of smaller areas, which are finally combined into a large image. Similar to the case of spatial resolution, higher temporal resolution can be delivered by interpolation and by adding inputs from additional sources.

Solar radiation information for one or many given locations, either in the form of data time series or as maps with current or long-term values, can be obtained from a number of existing providers. Some sources provide the data for free and some at a cost. The delivery options are also varied, ranging from manual downloads via online interfaces, to e-mails or scripted requests to servers. The World Bank Group (partnership of the World Bank and the International Finance Corporation), for example, created a Global Solar Atlas funded by the Energy Sector Management Assistance Program (ESMAP) of the World Bank, freely accessible (at https://globalsolaratlas. info/) to provide maps of long-term averages of solar irradiance, potential PV output and other data at global, regional and country levels. Figure 1.6 shows a GHI map for

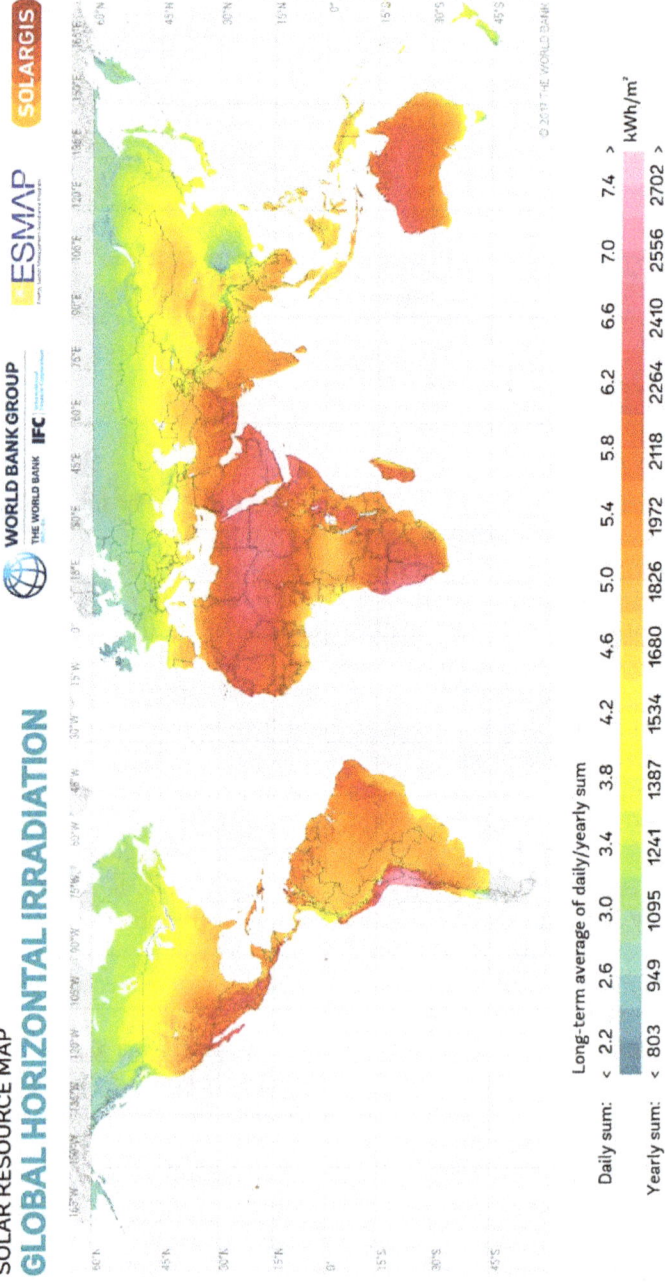

Figure 1.6 World map of long-term GHI, obtained from the Global Solar Atlas, owned by the World Bank Group and provided by Solargis

the world, from the Global Solar Atlas online application (at https://globalsolar-atlas.info), provided under the Creative Commons Attribution license (CC BY 3.0 IGO). Although datasets at higher temporal resolution are usually sold by commercial entities, some efforts have been made to fund and provide free datasets, such as the Photovoltaic Geographical Information System (PVGIS) of the European Commission's Joint Research Centre, which currently provides maps and data for Europe, Africa and Asia through the PVGIS website (http://re.jrc.ec.europa.eu/pvgis/) and the CM SAF web user interface (https://wui.cmsaf.eu/safira/action/viewProduktSearch/); the PVGIS data are provided by the CM SAF collaboration, the EUMETSAT Satellite Application Facility dedicated to Climate Monitoring.

1.3   Harvesting technologies

1.3.1   Solar energy conversion

The most straightforward way to harvest solar energy is to convert it into heat and use it for heating purposes such as heating a building, making hot water, cooking or generating steam that drives a turbine to generate electricity. One of such interesting applications is the purification of water using solar energy in remote areas. Figure 1.7 shows a schematic diagram of a basic working device. Dark substances with strong absorption of light such as carbon fibers are used to maximize the absorption. Sponge-like carbon material has been successfully tested to convert up to 85% of solar energy into heat and used to evaporate water under one sun exposure [12].

However, in most of our daily applications, energy is used in the form of electricity. The world electricity production in 2018 was 26,700 TWh. More than 60% of it is generated from traditional fossil fuels: coal (38%), natural gas (23%), and oil (3%). A tiny fraction only is generated from the solar energy (2%) (Figure 1.8). However, the share of renewable energies in the world

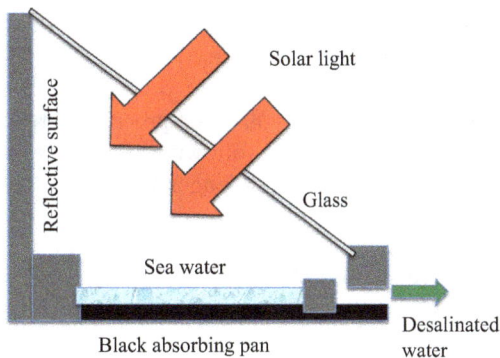

Figure 1.7   Conversion of sun light to heat for water desalination

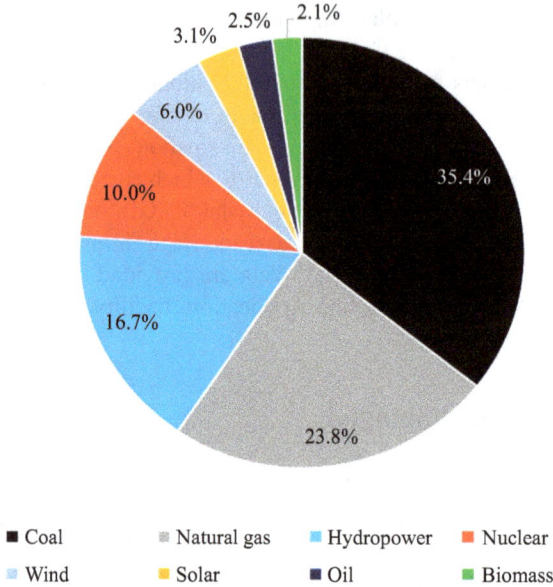

Figure 1.8   Electricity production by sources: data adapted from Ref. [13]

energy mix is growing very fast (7% for wind and 19.8% for solar over 2017–2018) [13].

Two major technologies have been developed and implemented to convert solar energy into electricity at a large scale: PV and CSP. We give in the following section a brief description of each technology and discuss the major challenges to be addressed to ensure further development of these technologies at large scale, more specifically in the desert environment.

1.3.2   Photovoltaics

PV technology enables a direct conversion of solar energy into electricity. This is achieved through solar cells which are made of semiconducting materials that absorb strongly part of the solar spectrum. The free charge carriers that are generated as a result of the light absorption produce a photocurrent flowing across the device. The generation and collection processes of the photocurrent are discussed in details in Chapter 2. In the following section, we review the current status of the technology and the key challenges facing its deployment in desert climates.

1.3.2.1   PV technology from cell to panel

Today, the PV technology is essentially based on the use of silicon material for the conversion of solar light into electricity. In 2018, silicon-based PV modules made about 95% of the PV market shares. Monocrystalline (c-Si) and multicrystalline silicon (mc-Si) had similar shares. Thin film technologies share the remaining 4.5%

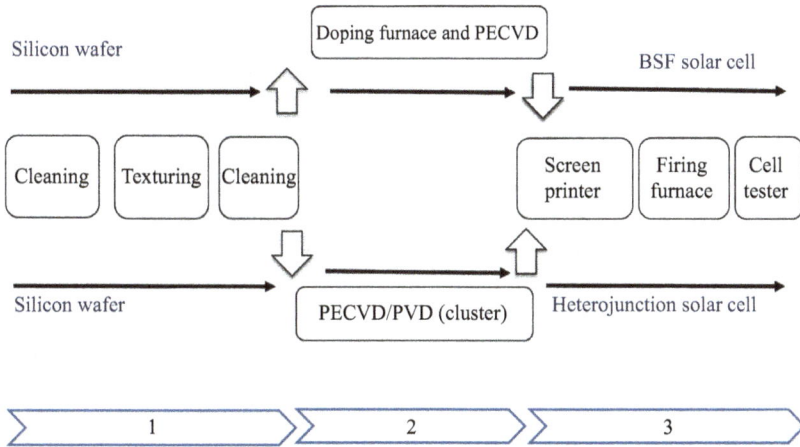

Figure 1.9   Fabrication process of conventional solar cells and heterojunction solar cells: (1) Wafer preparation. (2) Cell fabrication. (3) Electric contacts and testing.

of the market: cadmium telluride (CdTe): 2.3%, copper indium (gallium) sulfides (CI(G)S): 1.9%, and amorphous Si (a-Si): 0.3%. [14]

The fabrication of a silicon PV panel includes five major steps:

1. Reduction of silica via a reaction with carbon at high temperature (up to 2,000°C) to obtain metallurgical silicon (MG-Si) with 98% purity.
2. Silicon refinement to produce pure silicon powder ($<10^{-9}$ impurity content, i.e. "nine 9" purity). This level of purity of silicon is not actually required for the production of solar cells of high efficiency but is essential for other devices in microelectronics such as metal oxide field effect transistors (MOSFET). This is usually achieved using Siemens process [15].
3. Production of silicon wafer: Multicrystalline silicon wafers (mc-Si) are produced by casting process [16] while monocrystalline wafers (c-Si) are commonly produced using Czochralski (Cz) growth process [17,18]. More recently high-performance multicrystalline silicon wafers (HPmc-Si) were developed through a careful control of the grain size during growth [19,20].
4. Fabrication of solar cells: Different processes including many steps are used. These steps depend on the architecture of the solar cells to be produced. They are described in detail in Chapter 3. All silicon solar cells are mainly PN junctions (diodes). However, one can distinguish two basic processes (Figure 1.9):
   (i) High-temperature process for the fabrication of conventional solar cells called back surface field (BSF) solar cells.
   (ii) Low-temperature process used for the fabrication of a heterojunction solar cells (HJSC).

In both processes, the preparation of the wafers is the same: it includes texturization and chemical cleaning of the wafers. In the case of conventional cells, the

Monofacial   Bifacial

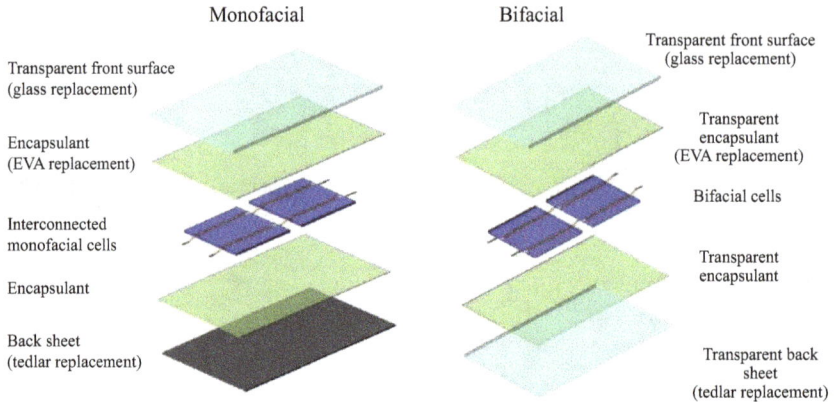

Transparent front surface
(glass replacement)

Transparent front surface
(glass replacement)

Encapsulant
(EVA replacement)

Transparent
encapsulant
(EVA replacement)

Bifacial cells

Interconnected
monofacial cells

Transparent
encapsulant

Encapsulant

Back sheet
(tedlar replacement)

Transparent back
sheet
(tedlar replacement)

Figure 1.10   Configuration of PV module layers for monofacial and bifacial module technologies

pn junction is obtained by phosphorous diffusion at high temperature in p-type wafer. In the case of HJSC cells, plasma enhanced chemical vapor deposition is used to deposit a phosphorous-doped thin silicon layer on the p-type wafer. The electric contacts are deposited by thermal evaporation. Next, the cells are placed in sandwich between two transparent layers of encapsulant, usually made of ethyl vinyl acetate (EVA). In the common configuration of a conventional PV panel, the back of the panel is made of a polymer sheet (Tedlar) and the front of a glass sheet. Recently, bifacial PV technology gained a great deal of attention as it enables harvesting the albedo (light reflected from the ground) leading to the enhancement of the energy yield by up to 30%. The typical structure of the bifacial panel is as follows: glass/EVA/cell/EVA/glass (Figure 1.10).

The 2019 international PV roadmap predicts a growth of the market share of the bifacial technology to reach up to 60% in 2023 [21].

A key driver for the development of PV technology is the reduction of LCOE produced by the PV system. The LCOE is defined as the ratio of the total life cycle cost to the life cycle energy produced:

$$LCOE = \frac{\text{Total life cycle cost}}{\text{Life time energy generation}} \tag{1.3}$$

The total life cycle cost includes the system cost (capital cost, operating and maintenance cost and others). The total energy produced over the life cycle of the system depends on the local insolation and the power conversion efficiency (PCE) of the system. It was found that the sensitivity of the LCOE to changes of any factor affecting it is maximum in the case of PCE change [22]. This justifies and explains the race for efficiency in the photovoltaics industry.

It should be pointed out the LCOE is system location dependent. This is due to (i) non-uniform distribution of the insolation over the earth surface, and (ii) the dependence of the PCE upon the weather conditions, and more specifically the temperature of the module as explained in Chapters 2 and 4. The PCE of the PV module decreases as the temperature increases leading to the reduction of the energy yield and the enhancement of the LCOE. In addition to high temperatures, the harsh and dusty environment of desert climates reduces the energy yield and increases the maintenance cost of the PV systems including inverters and some components of the balance of system (BOS). The effect of dust accumulation and its mitigation will be analyzed in details in Chapter 6.

The cost of PV has been decreasing exponentially over the last few decades as it was reduced by un factor 250 since 1977. This drastic decrease is essentially related to a massive investment of China in the PV industry leading to an excess of supply in the market.

1.3.2.2   Concentrated photovoltaic

One of the important issues hindering the deployment of PV farms for electricity generation is the need for large areas of land. One way to reduce the use of land is to concentrate the incident light through optical systems on a smaller area covered by solar cells of high efficiency although more costly. Systems with concentrations exceeding 400 suns are called high concentrated photovoltaic (CPV) (HCPV). Low concentration systems (LCPV) correspond to concentrations lower than 3 suns. Multi-junction solar cells based on III–V compounds are used for both space and terrestrial concentrator systems. The most common one is the triple junction $Ga_{0.50}In_{0.50}P/Ga_{0.99}In_{0.01}As/Ge$. Cells with 10 mm × 10 mm size and efficiency exceeding 40% under 500× sun concentration are available on the market [23]. Figure 1.11 shows 1 MWp CPV system using Fresnel lenses for light concentration installed in Morocco in 2016.

The light concentration has another advantage as it increases slightly the PCE of the cell. However, this gain is reduced due to the simultaneous increase of the cell temperature which has the opposite effect.

Various sun tracking systems are used in CPV installations to increase the energy yield by up to 30%. However, this adds to the initial cost of the system as well as to the maintenance cost due to the moving parts.

The total installed CPV capacity in 2017 was about 350 MWp. This represents a tiny fraction of the total PV installed capacity in 2017 which was 300 GWp and reached 500 GWp in 2019 [24].

1.3.3   CSP

The CSP technologies concentrate sun light by using different types of mirrors: parabolic trough, plan mirrors, Fresnel or dish reflectors. In the case of parabolic trough, the concentrated light heats up a liquid (oil) that is pumped through a pipe placed in the focal line of the reflector (Figure 1.12). The oil is heated up to about 400°C and used to produce higher pressure steam through a heat exchanger. The steam drives a turbine to generate electricity.

The parabolic troughs are usually mounted on tracking systems to enhance the thermal efficiency up to 60–80% [25].

Figure 1.11   Sumimoto CPV installation of 1 MWp in Morocco, using two axis tracking system and Fresnel lenses, 2016 https://global-sei.com/company/press/2016/05/prs034.html

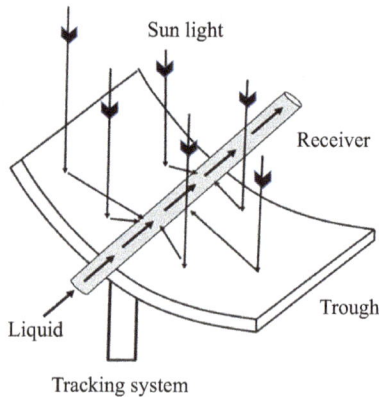

Figure 1.12   Parabolic trough solar concentrator

Plane mirrors equipped with tracking systems (heliostats) are used in CSP tower plants. The sun light is concentrated onto a receiver atop a central tower surrounded by the mirrors (Figure 1.13).

The concentrated light heats up a salt mix, such as sodium nitrate and potassium nitrate, up to about 565°C. The hot molten salt flows from the top of the tower

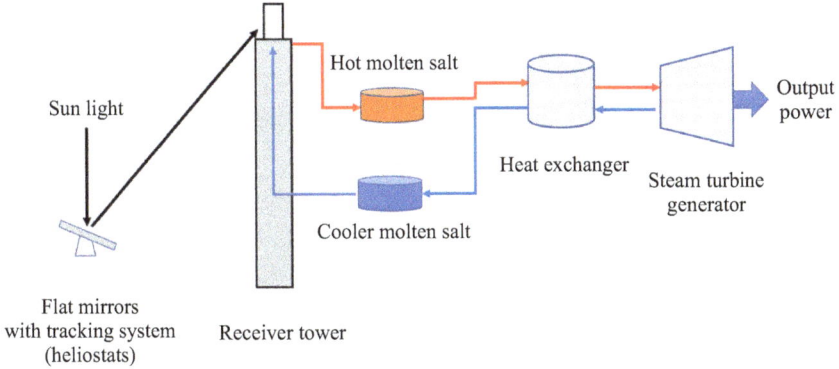

Figure 1.13   CSP with molten salts for energy storage during the day and electricity supply overnight

to a storage tank or it is directed immediately to a heat exchanger to produce steam and generate electricity via a turbine. The cooler salt that comes out of the heat exchanger at about 260°C is stored in a second tank or cycled to the power tower to be heated again.

CSP offers the important advantage to produce electricity even in the absence of the sun or in case of significant decrease of the incident solar power due to the weather conditions.

However, the efficiency of CSP technology is limited because of its inability to harvest the diffuse light as it is continuously tracking the direct light only. CSP is more feasible in regions of high annual direct normal radiation (DNI $\sim$ 2,000 kWh/m^2 per year).

A second major challenge facing the large-scale deployment of CSP technology is the high LCOE as compared to PV and other sources of energy. The most recent record (2019) for the installation of 900 MW PV solar power plant in Dubai, United Arab Emirates (UAE) was 1.7 cents per kWh. This is much lower than the cost of electricity from gas which is about 4 cents per kWh in the region. It is also more than four times lower than the record achieved in 2019 for the construction of the world's first advanced hybridization of (CSP) and (PV) with 800 MW capacity in Morroco (7 cents per kWh) .

Despite its high cost, CSP remains an important technology to ensure the autonomy of solar plants after sunset and in case of drastic reduction of the PV plants output caused by weather conditions. Consequently, the CSP–PV hybridization is emerging as an attractive alternative for the design of the large-scale solar plants of the future [26]. PV technology is used to generate electricity needed at low cost during the days and CSP installations of optimized capacity are added to provide high-quality power and autonomy over night or under weather conditions that reduce the output power of the PV modules.

References

[1]   NOAA Global Monitoring Division, "NOAA General Solar Position Calculations," 2005. http://www.esrl.noaa.gov/gmd/grad/solcalc/solareqns. PDF. [Accessed 29 November 2018].

[2]   NREL, "SolPos 2.0." http://rredc.nrel.gov/solar/codesandalgorithms/solpos/. [Accessed 29 November 2018].

[3]   R. M. Goody and Y. L. Yung, *Atmospheric Radiation: Theoretical Basis*, Oxford: Oxford University Press, 1995.

[4]   F. Kasten and A. T. Young, "Revised optical air mass tables and approximation formula," *Applied Optics*, vol. 28, no. 22, pp. 4735–4738, 1989.

[5]   International Organization for Standardization, ISO 9845-1:1992 Solar energy – Reference Solar Spectral Irradiance at the Ground at Different Receiving Conditions – Part 1: Direct Normal and Hemispherical Solar Irradiance for Air Mass 1.5, Geneva, Switzerland.

[6]   C. A. Gueymard, "Interdisciplinary applications of a versatile spectral solar irradiance model: a review," *Energy*, vol. 30, no. 9, pp. 1551–1576, 2005.

[7]   D. Perez-Astudillo, D. Bachour, and L. Martin-Pomares, "Improved quality control protocols on solar radiation measurements," *Solar Energy*, vol. 169, pp. 425–433, 2018.

[8]   International Organization for Standardization, ISO 9060:1990 Solar Energy – Specification and Classification of Instruments for Measuring Hemispherical Solar and Direct Solar Radiation, Geneva, Switzerland.

[9]   International Organization for Standardization, ISO 9060:2018(E) Solar Energy - Specification and Classification of Instruments for Measuring Hemispherical Solar and Direct Solar Radiation, Geneva, Switzerland.

[10] C. Fountoukis, L. Martin-Pomares, D. Perez-Astudillo, D. Bachour, and I. Gladich, "Simulating global horizontal irradiance in the Arabian Peninsula: sensitivity to explicit treatment of aerosols," *Solar Energy*, vol. 163, pp. 347–355, 2018.

[11] E. J. Mlawer, S. J. Taubman, P. D. Brown, M. J. Iacono, and S. A. Clough, "Radiative transfer for inhomogeneous atmospheres: RRTM, a validated correlated-k model for the longwave," *Journal of Geophysical Research: Atmospheres*, vol. 102, no. D14, pp. 16663–16682, 1997.

[12] G. Ni, G. Li, S. V. Boriskina, *et al.*, Steam generation under one sun enabled by a floating structure with thermal concentration, *Nature Energy*, vol. 1, Article number: 16126, 2016.

[13] Electricity Statistics, IEA Report, https://www.iea.org/data-and-statistics/data-tools/energy-statistics-data-rowser?country=WORLD&fuel=Electricity%20and%20heat&indicator=ElecGenByFuel.

[14] Photovoltaics Report, Published by Fraunhofer Institute ISE, Freiburg, 14 March 2019.

[15] J. A. Kilner, S. J. Skinner, S. J. C. Irvine, P. P. Edwards (eds.), *Functional Materials for Sustainable Energy Applications*. Cambridge, UK: Woodhead Publishing Limited; 2012.

[16] J. Friedrich, "Methods for bulk growth of inorganic crystals," in *Reference Module in Materials Science and Materials Engineering*, New York, NY: Elsevier, 2016.

[17] J. A. Kilner, S. J. Skinner, and P. P. Edwards (eds.), *Functional Materials for Sustainable Energy Applications*, Sawston: Woodhead Publishing, 2012.

[18] G. Müller and P. Rudolph, "Crystal growth from the melt", in *Encyclopedia of Materials: Science and Technology* (2nd ed.), 2001, pp. 1866–1872.

[19] Y. M. Yang, A. Yu, B. Hsu, W. C. Hsu, A. Yang, and C. W. Lan, "Development of high-performance multicrystalline silicon for photovoltaic industry," *Progress in Photovoltaics: Research and Applications*, vol. 23, no. 3, pp. 340–351, 2015.

[20] Y. Yu, J. Ding, W. Chen, *et al.*, "Growth of high-quality multicrystalline silicon ingot through the cristobalite seeded method," *Journal of Materials Science and Engineering*, vol. B6, nos. 11-12, pp. 304–310, 2016.

[21] International Technology Roadmap for Photovoltaics (ITRPV), 10th Edition, March 2019, https://itrpv.vdma.org/en/ueber-uns.

[22] D. M. Powell, M. T. Winkler, and A. Goodrich, and T. Buonassisi, "Modeling the cost and minimum sustainable price of crystalline silicon photovoltaic manufacturing in the United States," *IEEE Journal of Photovoltaics*, vol. 3, no. 2, pp. 662–668, 2013.

[23] Fullsuns Website, https://www.gaas.fullsuns.com.

[24] Report IEA-PVPS T1-35-2019, ISBN 978-3-906042-83-1.

[25] P. G. Jordan, An analysis of the global solar industry," in *Solar Energy Markets*, New York, NY: Elsevier, 2014, https://doi.org/10.1016/C2011-0-07234-4.

[26] X. Ju, Y. Hu, C. Xu, X. Huan, G. Wei, and X. Du, "A review on the development of photovoltaic/concentrated solar power (PVCSP) hybrid systems," *Solar Energy Materials & Solar Cells*, vol. 161, pp. 305–327, 2017.

Chapter 2

Solar cell fundamentals

Brahim Aïssa[1], Fahhad Alharbi[2] and Nouar Tabet[3]

2.1 Introduction

The first silicon solar cell was produced at Bell Lab in 1954 [1]. It consisted of a silicon p–n junction. In principle, the operational concept of such a cell is quite simple [2–5]. We describe first the physical process of light conversion into electricity and the working principle of a solar cell. We define the three key parameters that are commonly used to characterize solar cells, namely the short-circuit current I_{sc}, the open circuit voltage V_{oc}, and the fill factor FF. Then we will describe how materials properties and environmental factors such as the temperature and insolation affect the cell performance and discuss approaches to mitigate cell efficiency deterioration under hot climates.

Finally, we will discuss briefly the theoretical limit of cell efficiency and introduce the recent computational approach to expand the materials space for the fabrication of solar cells of high performance.

2.2 Cell structure and light conversion

The basic structure of a conventional solar cell includes the following components (Figure 2.1):

- An absorbing layer, called base, which can be p- or n-doped semiconductor.
- A second layer called emitter which is n-type in the case of a p-type base and vice versa.
- An antireflection layer on the front of the cell to minimize the loss of light by reflection.
- Front and back metallic ohmic contacts to connect the cell to the external circuit.

[1]Qatar Environment and Energy Research Institute (QEERI), Hamad Bin Khalifa University (HBKU), Qatar Foundation, Qatar
[2]Department of Electrical Engineering, King Fahd University of Petroleum and Minerals, Saudi Arabia
[3]Department of Applied Physics and Astronomy, College of Sciences, University of Sharjah, United Arab Emirates

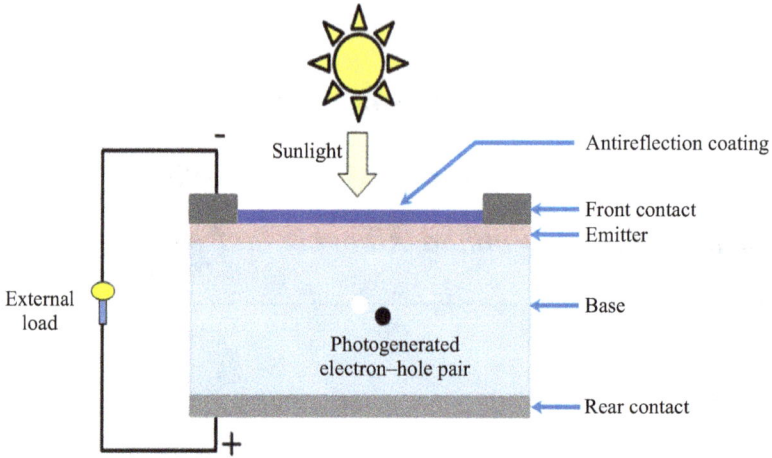

Figure 2.1   Schematic illustrating the cross section of a p–n junction solar cell

A variety of materials can be used for solar energy conversion [5,6], but in practice most solar cells use semiconductor materials in the form of a $p-n$ junction [3,7–9].

Basically, photons of energy equal or larger than the energy gap of the used semiconducting material are absorbed in both the p-type and n-type regions and generate quasi-free electrons in the conduction band and holes (which are vacant electron states) in the valence band (Figure 2.1).

Carriers generated in the space charge region of the pn junction are immediately separated by the built in electric field present in this region and produce a drift current. The quasi-free charge carriers that are generated outside the space charge region diffuse randomly until reaching the edge of the space charge region where they get separated by the built in electric field and give rise to a photocurrent. The total photocurrent is collected through an external load and generates electric power (i.e. voltage times current).

Two important observations are worth noting:

(i)   Photons whose energy is below the gap energy are not absorbed, therefore do not contribute to the photocurrent.

(ii) Photons of energy higher than the energy gap produce photocarriers with an excess of energy in the bands. This excess of energy is lost by thermal relaxation within a few tens of picoseconds towards the minimum of the conduction band for electrons and the maximum of the valence band for holes (Figure 2.2) [10]. This process generates heat that increases the device temperature and reduces the efficiency of the conversion process as described below.

(iii) Photocarriers that are generated out of the space charge region may *recombine* with diffusing randomly with holes before they reach the edge of the

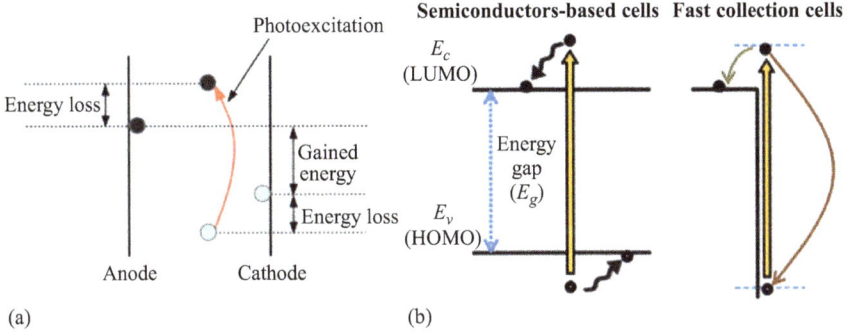

Figure 2.2   The basic operational concept of solar cells

space charge region. If this happens these carriers are "lost" and do not contribute to the photocurrent.

The above observations indicate that only a fraction of the incident photons generate free carriers, only a fraction of the generated carriers are collected and produce a current and finally, part of the energy of absorbed photons is lost as heat and not converted into electric energy. These are the main sources of losses that limit the power conversion efficiency of the cell as discussed in Section 2.3.

2.3   Current–voltage (I–V) characteristics

Under *short circuit* conditions, the collected photocarriers produce a current (I_{sc}) directed from the n-side to the p-side of the junction (due to the direction of the built in electric field) without generating any bias on the cell ($V = 0$). In the absence of current losses through carrier recombination, the collected current is maximum. However, the output power is zero.

If a load resistance is connected to the cell, a *forward bias* is produced on the device under light which weakens the built in electric field (i.e. reduces the barrier height) for both electrons and holes. As a result, electrons are injected from the n-side to the p-side and holes from the p-side to the n-side. Under steady-state conditions, these injected carriers produce a non-uniform distribution of the excess carriers on both sides of the junction (Figure 2.3). Consequently, a *diffusion current* I_D is produced. This current is directed from the p-side to the n-side, which is opposite to the direction of I_{sc} and leads to the reduction of the total collected current given by the following equation [11]:

$$I = I_{sc} - I_0\left(e^{\frac{eV}{kT}} - 1\right) \tag{2.1}$$

where the first term corresponds to the diffusion current I_D. I_0 is the saturation current of the diode, V the voltage induced by the load, k Boltzmann constant, and T the temperature in (K).

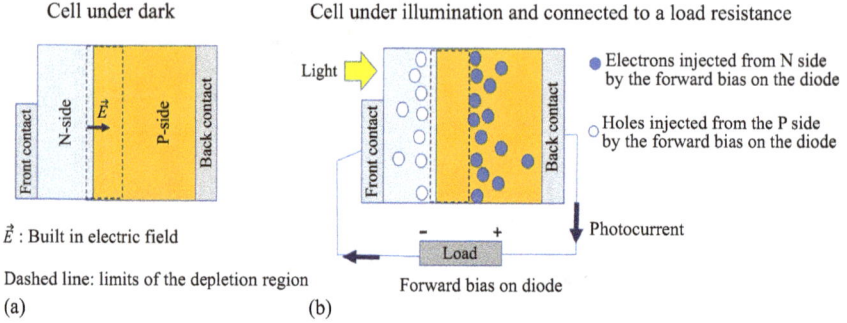

Cell under dark — Cell under illumination and connected to a load resistance

$\vec{E}$: Built in electric field

Dashed line: limits of the depletion region
(a)   (b)

Figure 2.3 *(a) p–n junction at equilibrium under dark, (b) photocurrent and carrier storage in both sides of the junction under illumination. Electrons and holes are injected from the n-side and p-side, respectively, due to the forward bias which weakens the built in electric field (i.e. reduces the barrier height).*

On the other hand, under *open circuit* conditions, the forward bias of the junction grows up in such a way that the photogenerated current is balanced totally by the forward bias diffusion current I_D, and the net resulting current is zero. The "open-circuit voltage" can be obtained by setting $I = 0$:

$$V_{oc} = \frac{kT}{e} Ln\left(\frac{I_{sc}}{I_0} - 1\right) \qquad (2.2)$$

A typical shape of the I–V characteristics is shown in Figure 2.3.

In practice, solar cells operate under the load conditions that maximize the output power (maximum power point: MPP). The MPP is obtained by solving numerically the following equation:

$$\frac{\partial P}{\partial V} = 0 \qquad (2.3)$$

where P is given by

$$P = \left[I_{sc} - I_0\left(e^{\frac{eV}{kT}} - 1\right)\right]V \qquad (2.4)$$

By using (2.3), one obtains

$$I_{sc} - I_0\left(e^{\frac{eV}{kT}} - 1\right) - \frac{qI_0}{kT}V_m e^{\frac{qV_m}{kT}} = 0 \qquad (2.5)$$

Equation (2.5) can be solved numerically to determine V_m. The corresponding current I_m is then obtained by replacing V_m in (2.1).

The optimum load resistance R_m that maximizes the collected output power is given by:

$$R_m = \frac{V_m}{I_m} \tag{2.6}$$

The power conversion efficiency (PCE) η of the cell is defined as the ratio of the output power P_{out} to the incident power P_{in}:

$$\eta = \frac{P_{out}}{P_{in}} \tag{2.7}$$

Using the *I–V* characteristics as shown in Figure 2.3, the PCE can be expressed as follows:

$$\eta = \frac{I_m V_m}{P_{in}} \tag{2.8}$$

The fill factor FF of the cell is defined as

$$FF = \frac{I_m V_m}{I_{sc} V_{oc}} \tag{2.9}$$

Then the PCE can be expressed as follows:

$$\eta = FF \frac{I_{sc} V_{oc}}{P_{in}} \tag{2.10}$$

The following are typical values for silicon solar cell illuminated under standard conditions, i.e. AM1.5 spectrum, and $T = 25\,°C$, $I_{sc} = 40$ mA/cm^2, and $V_{oc} = 0.76$ V, FF = 0.77, PCE = 23.4%.

Note that the short circuit current is directly proportional to the area of the cell because the number of absorbed photons is proportional to the area as well. However, it appears clearly from (2.2) that the open circuit is independent upon the cell area.

We may also think that the PCE is independent upon the incident power as the output power is expected to be proportional to incident power. In reality, this is not the case because as the open circuit increases logarithmically upon the short circuit current as shown by (2.2).

Furthermore, the increase of the intensity of the incident light also affects the recombination rate of the photocarriers and leads to a complex dependence of the PCE versus incident power as discussed in the section below.

2.4 Factors limiting the cell performance

There are many factors that limit the PCE of solar cells. For example, the reflection of the incident photons on the top surface of the device can be a major source of performance limitation. For instance, silicon surface reflects about 30% of incident light in the visible region. However, different techniques have been developed over the years to reduce significantly the light loss by reflection (down to few %). These techniques include the following:

(i)   Thin anti-reflecting coating (ARC) commonly made of silicon nitride (SiN_x) or/and $SiOx/TiO_2$ of appropriate thickness [12].
(ii) Texturing the surface using wet chemical etching [13] or reaction etching (RIE) [14].

The current density of photoelectrons J_n has a drift component due to the built in electric field E in the space charge region and a diffusion component due to the non-uniform distribution of the carriers and can be expressed as follows:

$$J_n = ne\mu_n E + eD_n \frac{\partial n}{\partial x} \tag{2.11}$$

where D_n is the diffusion constant of the electron, n the photoelectron density, E the electric field within the space charge region, and μ_n the electron mobility. Note that the diffusion constant D is related to the mobility by the Einstein equation:

$$D = \frac{kT}{e} \tag{2.12}$$

Similar equation can be written for holes,

$$J_p = pe\mu_p E - eD_p \frac{\partial p}{\partial x} \tag{2.13}$$

where p is the hole density, D_p and μ_p are the diffusion constant and mobility of holes, respectively.

The built in electric field in the space charge region is given by the following equation:

$$\frac{dE}{dx} = \frac{q}{\varepsilon}(p - n + N_D - N_A) \tag{2.14}$$

where ε is the permittivity of the material, N_D and N_A the donor and acceptor densities in the n-type and p-type regions, respectively.

The continuity equations under light are the following:

$$\frac{1}{e}\frac{dJ_n}{dx} = U(x) - G(x). \tag{2.15}$$

and

$$\frac{1}{e}\frac{dJ_p}{dx} = -U(x) + G(x). \tag{2.16}$$

where $U(x)$ and $G(x)$ are the recombination and generation rates, respectively (cm^{-3}/s).

The mathematical expression of $U(x)$ depends upon the recombination mechanisms involved: band to band recombination, defect assisted recombination or Auger recombination. $G(x)$ is usually taken as an exponential function:

$$G(x) = G_0 e^{-ax} \qquad (2.17)$$

where a (cm^{-1}) is the linear coefficient of absorption of light in the semiconductor and G_0 is a constant which depends on the intensity of the incident light.

The calculation of the photocurrent requires to solve the system of (2.11)–(2.17) along with appropriate boundary conditions [11].

Various codes have been developed to solve numerically the system of (2.11)–(2.17) and compute the collected photocurrent. Among these, one can list PC1D, AMPS, and SCAPS. The codes offer the possibility to investigate the effect of any material property and device characteristics on the performance of the cell [15].

The above equations reveal clearly some of the material properties that affect the collected photocurrent such as the electron mobility, the recombination, and generation rates.

The nature of the gap (direct/indirect) and the presence of point defects such as impurities and structural defects (dislocations and grain boundaries) can affect drastically the carrier mobilities and the current loss through recombination [16].

These material properties affect the three key parameters, namely the short circuit current Jsc, the open circuit voltage V_{oc}, and the fill factor (FF). Optimizing all three factors simultaneously is somehow a challenging task, and trade-offs are often necessary [17–20].

We review in the following section the effect of the energy gap and the temperature on the temperature.

2.4.1 Effect of the energy gap

The energy gap is a key characteristics of the absorbing material that impacts the performance of solar cell in a complex way. The light absorption is much stronger in direct gap semiconductors but so are the current losses through band to band recombination.

The correlation between the value of the band gap of the absorber and the short circuit current is straightforward. Smaller is the gap, more photons are absorbed, thus more photocurrent is produced. Therefore, as long as the illumination level is not too high, the photocurrent is proportional to incident power.

However, the energy gap does not affect the photocurrent only but the output voltage as well. This can be observed by having a closer look at the expression of the open circuit voltage (2.2). One can observe that V_{oc} depends on both I_{sc} and the saturation current (I_0) of the p–n junction. The current I_0 is proportional to n_i^2, n_i being the carrier intrinsic density. In the case of one side $p-n$ junction (i.e. heavily doped emitter), I_0 is given by [21]:

$$I_0 = qA\,\frac{Dn_i^2}{LN_D} \qquad (2.18)$$

where n_i is given by

$$n_i = AT^{3/2}e^{-\frac{E_g}{2kT}} \tag{2.19}$$

A is a constant, T the temperature, k the Boltzmann constant, and E_g is the energy gap.

It appears clearly that larger the gap, smaller are n_i and I_0, thus larger is V_{oc}. Lowering the gap has opposite effects on the photocurrent and the output voltage. Furthermore, the excess energy $(E_{ph}-E_g)$ of high energy photons is not collected but rapidly lost by thermalization as mentioned above. Consequently, there is an optimal bandgap that gives the maximum output power for the spectrum of the sun, which in fact is close to that of c-Si [2].

Based on a detailed balance approach, Shockley and Queisser [17] showed that the trade-off between current and voltage leads to a theoretical limit of the PCE of silicon cell exceeding 30% (around 33% for a spectrum of AM1.5 [22] if only radiative recombination is considered). Nevertheless, as the silicon has an indirect gap, other recombination processes have to be taken into account. The Shockley–Queisser limit is therefore an overestimation. Assuming a perfect absorber, Tiedje *et al.* [23] extended the calculations for the AM1.5 spectrum, and included non-radiative recombination processes, namely Auger recombination. Doing so, they obtained 29.8% as the new device limit. More recently, the Auger coefficients have been updated leading to the most recent efficiency limit of 29.43% for c-Si-based devices (110 μm-thick, un-doped wafer) [24].

2.4.2   Current losses through defects

Band-to-band or radiative recombination prevails in direct bandgap semiconductors such as gallium arsenide (GaAs), which dominates solar cell applications in space and concentrator. However, in terrestrial applications, solar cells are mainly made from silicon, which is an indirect bandgap semiconductor, and hence radiative recombination is extremely low and usually neglected.

Recombination through defects, called Shockley–Read–Hall (SRH) recombination, usually dominates under the working conditions of cells. It is a two-step process (Figure 2.4) [25,26].

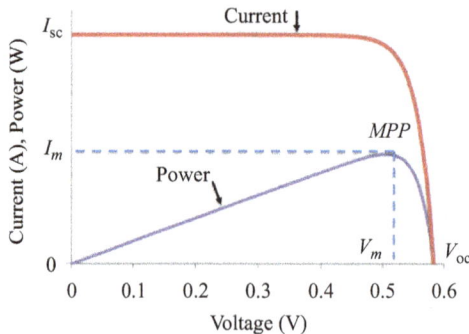

Figure 2.4   I–V characteristics and power curve under illumination

- An electron (or hole) is first trapped by an energy state located in the forbidden region. This energy step, also called *recombination-center* or *trap-state*, is introduced through defects in the crystal lattice of the semiconductor such as impurities or native defects in the crystalline structure (vacancies or interstitial atoms).
- If a hole (or an electron) moves up (or down) to the same energy state before the electron (or hole) is thermally re-emitted into the conduction or valence bands, then it recombines.

The SRH recombination rate depends drastically on the location of the defect level in the energy gap. Levels near mid-gap are called *Deep Levels* and are very effective recombination centers because a trapped electron has low probability to be reemitted at room temperature. However, trap levels close to the minimum of the conduction band or to the maximum of the valence band, called *shallow levels*, have no significant contribution to the recombination process at room temperature.

The SRH recombination rate depends upon many parameters, namely the trap density N_t, trap level E_t, temperature, doping level, N_D, and the injection level. Under low injection conditions, i.e., if the excess photocarrier density is lower than the doping concentration, the SRH recombination rate in N-type semiconductors can be expressed in the following simplified form:

$$U = \frac{\Delta_n}{\tau} \tag{2.20}$$

where Δ_n is the photocarrier carrier density, and τ is the electron hole pair lifetime. It is important to note that the recombination process under low injection level conditions is governed by the minority carriers. Therefore, in p-type semi-conductors, Δn is replaced by Δp in (2.20).

It is worth mentioning that the SRH process is less likely in pure crystalline materials with low structural defect concentration than the band to band recombination in direct gap semiconductors. However, in practice, it is the dominant mechanism under common operational conditions due to the presence of high density of growth defects. Lifetime measurements have been systematically carried out in silicon used for solar cells. The lifetime in boron doped silicon containing various concentration of interstitial iron atoms has been found varying from 10^{-4} s to 10^{-8} s as Fe concentration changed from 10^{10} cm^{-3} to 10^{14} cm^{-3}, respectively [27].

2.4.3 Auger recombination

Auger recombination is a *three particle process*, as illustrated in Figure 2.5. An electron and a hole recombine, but rather than emitting the energy as heat or as a photon, the energy is transferred to a third carrier, an electron in the conduction band. This electron then thermalizes back down to the conduction band edge.

Auger recombination is significant at high carrier concentrations caused by heavy doping or high level injection under concentrated sunlight. Under such conditions, it limits the lifetime and hence the photoconversion efficiency. The

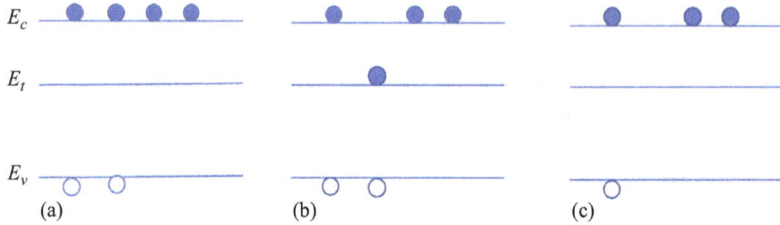

*Figure 2.5   Electrons in the conduction bands and holes in the valence band, trap
energy level (E_t) associated with a defect, (b) electron trapped from
the conduction band, (c) recombination of the trapped electron with a
hole from the valence band. The complete process leads to the
annihilation of one electron hole pair.*

more heavily doped the material is or higher is the light concentration, the shorter
the Auger recombination lifetime.

The total bulk lifetime is defined as follows:

$$\frac{1}{\tau} = \frac{1}{\tau_{Band}} + \frac{1}{\tau_{Auger}} + \frac{1}{\tau_{SRH}} \qquad (2.21)$$

where τ is the minority carrier lifetime.

Finally, the carrier diffusion length L defined by $L = \sqrt{\tau D}$ represents the
average distance crossed by the minority carriers before they recombine. $D\left(\frac{cm^2}{s}\right)$ is
the diffusion coefficient.

Typical values of the carrier lifetime range from *millisecond* in defect free single
crystals to *microsecond* or lower in as grown multicrystalline silicon ingots, respectively.

2.4.4   Surface recombination

SRH recombination in a solar cell can occur not only is bulk but also at the surface
and interfaces present in the device as a result of various defects such as *dangling
bonds* or impurities. In n-type semiconductors, the surface recombination rate R_s
can be expressed in the following form [28,29]:

$$R_s \approx v_{th}\sigma_p N_{sT}(p_s - p_0) \qquad (2.22)$$

where v_{th} is the thermal velocity in cm/s, N_{sT} is the surface trap density in cm^{-2},
and σ_p is the capture cross-section for holes in cm^2, p_s is the hole concentration at
the surface, and p_0 is the equilibrium hole concentration in the *n*-type semiconductor. For a *p*-type semiconductor, $\sigma_p p_0$ and p_s need to be replaced by σ_n, n_0
and n_s, respectively.

Note that the product $v_{th}\sigma_p N_{sT}$ has the unit of a velocity and is called *surface
recombination velocity*:

$$S_r = v_{th}\sigma N_{sT} \qquad (2.23)$$

For $v_{th} = 10^5 \text{cm/s}$, $\sigma = 10^{-15}\text{cm}^2$ and $N_{sT} = 10^{14}\text{cm}^{-2}$, the surface recombination velocity is about 10^4 cm/s. Above this typical value, the surface recombination affects significantly the performance of the device.

In contrast to radiative or Auger recombination, SRH recombination can be controlled by:

(i)   using high-quality materials [e.g. float-zone (FZ) wafers] with extremely low amounts of impurities or by

(ii) using some schemes for surface passivation, which rely on *chemical* or *field-effect passivation*. The chemical passivation relies on the application of a material which reduces N_{sT} and hence increases the photogenerated carrier lifetime τ. Typical example for this type of passivating is the use of a-Si:H and derivatives such as oxides (a-SiOx:H) in silicon heterojunction (SHJ) solar cells [30,31]. In contrast, the passivation by field-effect results from the energy band bending at the interfaces – induced by fixed charges – and prevents one type of charge carrier from reaching the interface which increases τ as well. Materials that are used for this type of passivation are e.g. thermally grown oxides, amorphous silicon nitrides (a-SiN$_x$) [32], and aluminum oxide (Al$_2$O$_3$) [33–35].

The various bulk and surface recombination processes described above lead to the reduction of the carrier lifetime and the diffusion length. The collection probability of a photocarrier that is generated at any point of the cell depends on the ratio of the diffusion length to the distance to the edge of the depletion region. Larger is the ratio, higher the probability. The optimum thickness of a cell is the result of a tradeoff between a maximum light absorption that requires enough thickness and a maximum carrier collection that requires thinner absorbing layer.

2.4.5   *Effect of temperature*

The temperature affects all properties of semiconducting materials, especially those that impact significantly the solar cells performance such as the light absorption, the carrier transport, and recombination processes.

The energy band gap of most semiconductors decreases as the temperature increases, due mainly to the thermal expansion. This leads to an increase of the light absorption range to low photon energy and an increase of the short circuit current (assuming that the carrier collection process is not affected) as shown in Figure 2.6.

However, it has been reported recently that the band gap of methyl ammonium lead triodide CH$_3$NH$_3$PbI$_3$ (MAPbI$_3$) decreases from 1.61 to 1.55 eV as the temperature decreases from 300 to 150 K [36,37].

The temperature variation of the energy band gap of the absorbing layer has an important impact on the open-circuit voltage V_{oc} of the cell because of the temperature dependence of the saturation current I_0 as shown in (2.18). If the band gap decreases as the temperature increases, then the saturation current increases because of the exponential increase of the intrinsic carrier density n_i. Consequently,

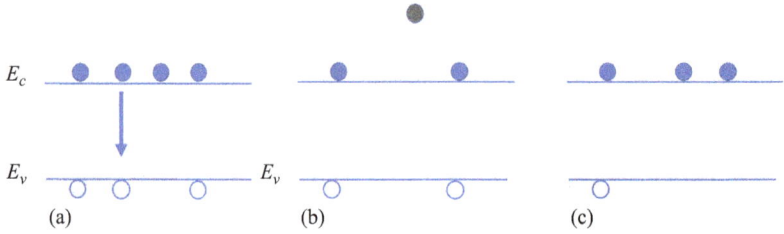

Figure 2.6 (a) Electrons in the conduction band and holes in the valence band, (b) electron hole pair recombination with excitation of an electron (Auger electron) to upper level in the conduction band, and (c) Auger recombination with excitation of a hole to upper level in the valence band

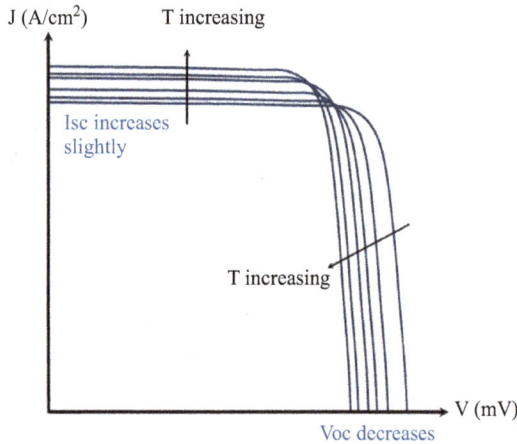

Figure 2.7   Schematic effect of temperature on the I−V characteristics of a solar cell

the open circuit decreases as indicated by (2.2) (Figure 2.6) leading to a deterioration of the cell performance.

The temperature dependence of the band gap can be derived using (2.24):

$$\frac{dV_{oc}}{dT} = -\frac{V_{go} - V_{oc} + \gamma\left(\frac{kT}{q}\right)}{T} \tag{2.24}$$

Varying the light intensity that is incident on a solar cell device impacts all the solar cell characteristics, including J_{sc}, V_{oc}, the FF, the photoconversion efficiency and the series and shunt resistances. The light intensity at which the solar cell is operating is called the number of Suns. One (1) Sun corresponds to standard

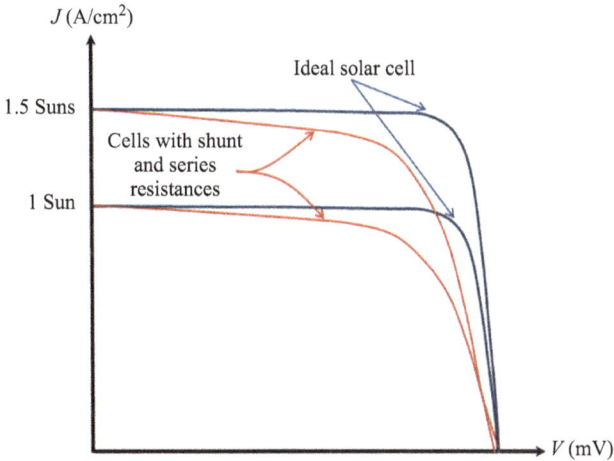

Figure 2.8   Effect of light intensity on the I–V characteristics of a solar cell

illumination at AM1.5, or 1 kW/m². For instance, a solar cell-based system with a power intensity of 10 kW/m² incident light would be operating at 10 Suns, or at 10×. A conventional PV module is usually designed to operate under 1 Sun condition and is called a "flat plate" module while those using concentrated sunlight are called "concentrators" [38–41].

PV panels undergo a daily variation in the light intensity, where the incident power from the Sun is changing between zero (0) and 1 kW/m². At low light levels, the effect of the shunt resistance becomes progressively crucial. As the light intensity diminishes, the photogenerated current – of the solar cell device – decreases as well, and the equivalent resistance of the solar cell may start to approach that of the shunt resistance. When these two resistances are equal, the fraction of the total current flowing through the shunt resistance becomes important, increasing thereby the fractional power loss due to shunt resistance. Consequently, under cloudy conditions, a solar cell with a high shunt resistance holds a higher fraction of its original power comparatively to a solar cell with a low shunt resistance. Figure 2.8 shows the effect of light intensity on the *I–V* characteristics of a solar cell.

2.5   Theoretical limit of silicon cell efficiency: Shockley–Queisser limit (SQ limit)

By plotting the *I–V* characteristic curve, it is possible to obtain the maximum power point for the current and voltage and then it is possible to obtain the limit efficiency of Shockley–Queisser. The following graph is the famous limit efficiency (%) of Shockley–Queisser for p–n junctions in function of the energy band gap (eV):

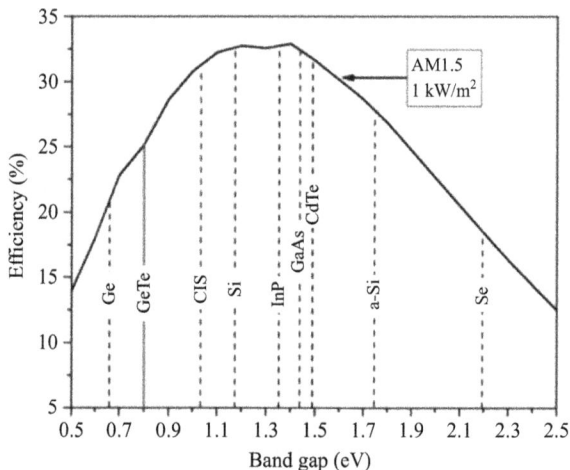

Figure 2.9   SQ efficiency limit in function of the energy band gap and the correspondence for different solar cells materials

As shown in Figure 2.9, the SQ efficiency limit is plotted as a function of the energy band gap and the of the most used solar cells materials, the maximum efficiency. The maximum efficiency is reached around 33.7% for a single pn-junction PV cell, assuming AM 1.5 solar spectrum and a temperature of the cell of 300 K and it occurs at a band gap of 1.34 eV, which is very close for silicon solar cells, which have a band gap of 1.12 eV.

References

[1]   April 25, 1954: "Bell Labs Demonstrates the First Practical Silicon Solar Cell," *APS News*, American Chemical Society, 18 (4), April 2009.
[2]   P. Würfel and U. Würfel, *Physics of Solar Cells: From Basic Principles to Advanced Concepts*, New York, NY: John Wiley & Sons, 2016.
[3]   S. Fonash, *Solar Cell Device Physics*, New York, NY: Elsevier, 2012.
[4]   A. Luque and S. Hegedus, *Handbook of Photovoltaic Science and Engineering*, New York, NY: John Wiley & Sons, 2011.
[5]   F. H. Alharbi and S. Kais, "Theoretical limits of photovoltaics efficiency and possible improvements by intuitive approaches learned from photosynthesis and quantum coherence," *Renewable and Sustainable Energy Reviews*, vol. 43, pp. 1073–1089, 2015.
[6]   A. Goetzberger, C. Hebling, and H.-W. Schock, "Photovoltaic materials, history, status and outlook," *Materials Science and Engineering: R: Reports*, vol. 40, pp. 1–46, 2003.
[7]   C. Kittel, P. McEuen, and P. McEuen, *Introduction to Solid State Physics*, vol. 8, New York, NY: Wiley, 1996.

[8] D. A. Neamen, *Semiconductor Physics and Devices: Basic Principles*, New York, NY: McGraw-Hill, 2012.

[9] S. M. Sze and K. K. Ng, *Physics of Semiconductor Devices*, New York, NY: John Wiley & Sons, 2006.

[10] M. El-Amine Madjet, A. Akimov, F. El-Mellouhi, *et al.*, "Enhancing the carrier thermalization time in organometallic perovskites by halide mixing," *Physical Chemistry Chemical Physics*, vol. 18, pp. 5219–5231, 2016, doi:10.1039/C5CP06603D.

[11] G. Martin, "Solar cells: operating principles, technology and system applications." In: *Prentice-Hall Series in Solid State Physical Electronics*, Hoboken, NJ: Prentice-Hall, 1981, ISBN-13: 978-0138222703.

[12] C. R. Xue, Y. Q. Gu, and M. L. Deng, "Investigation on antireflection coatings for silicon solar cells," *Applied Mechanics and Materials,* vol. 521, pp. 33–36, 2014.

[13] U. Kaiser, M. Kaiser, and R. Schindler. "Texture etching of multicrystalline silicon." In: A. Luque, G. Sala, W. Palz, G. Dos Santos, and P. Helm (eds), *Tenth E.C. Photovoltaic Solar Energy Conference*, Dordrecht: Springer, 1991.

[14] H. Dekkers, F. Duerinckx, F. Duerinck, J. Szlufcik, and J. Nijs, "Silicon surface texturing by reactive ion etching," *Opto-Electronics Review*, vol. 8, no. 4, pp. 311–316, 2000.

[15] M. I. Hossain, F. H. Alharbi, and N. Tabet, "Copper oxide as inorganic hole transport material for lead halide perovskite based solar cells of enhanced performance," *Solar Energy*, vol. 120, pp. 370–380, 2015.

[16] H. J. Queisser and E. E. Haller "Defects in semiconductors: some fatal, some vital", *Science, New Series*, vol. 281, no. 5379, pp. 945–950, 1998.

[17] W. Shockley and H. J. Queisser, "Detailed balance limit of efficiency of p-n junction solar cells," *Journal of Applied Physics*, vol. 32, pp. 510–519, 1961.

[18] F. H. Alharbi, S. N. Rashkeev, F. El-Mellouhi, H. P. Lüthi, N. Tabet, and S. Kais, "An efficient descriptor model for designing materials for solar cells," *NPJ Computational Materials*, vol. 1, p. 15003, 2015.

[19] D. Ferry, *Semiconductor Transport*, London: CRC Press, 2016.

[20] M. C. Hanna and A. J. Nozik, "Solar conversion efficiency of photovoltaic and photoelectrolysis cells with carrier multiplication absorbers," *Journal of Applied Physics*, vol. 100, p. 074510, 2006.

[21] H. Plagwitz and R. Brendel, "Analytical model for the diode saturation current of point-contacted solar cells," *Progress in Photovoltaics: Research and Applications*, vol. 14, pp. 1–12, 2006.

[22] R. E. Bird, R. L. Hulstrom, and L. J. Lewis, "Terrestrial solar spectral data sets," *Solar Energy*, vol. 30, pp. 563–573, 1983.

[23] T. O. M. Tiedje, E. L. I. Yablonovitch, G. D. Cody, and B. G. Brooks, "Limiting efficiency of silicon solar cells," *IEEE Transactions on Electron Devices*, vol. 31, pp. 711–716, 1984.

[24] A. Richter, M. Hermle, and S. W. Glunz, "Reassessment of the limiting efficiency for crystalline silicon solar cells," *IEEE Journal of Photovoltaics*, vol. 3, pp. 1184–1191, 2013.

[25]   W. T. R. W. Shockley and W. T. Read Jr, "Statistics of the recombinations of holes and electrons," *Physical Review*, vol. 87, p. 835, 1952.

[26]   R. N. Hall, "Electron-hole recombination in germanium," *Physical Review*, vol. 87, p. 387, 1952.

[27]   A. A. Istratov, H. Hieslmair, and E. R. Weber "Iron contamination in silicon technology," *Applied Physics A*, vol. 70, pp 489–534, 2000.

[28]   D. J. Fitzgerald and A. S. Grove, "Surface recombination in semiconductors," *Surface Science*, vol. 9, pp. 347–369, 1968.

[29]   D. E. Aspnes, "Recombination at semiconductor surfaces and interfaces," *Surface Science*, vol. 132, pp. 406–421, 1983.

[30]   K. Yoshikawa, H. Kawasaki, W. Yoshida, *et al.*, "Silicon heterojunction solar cell with interdigitated back contacts for a photoconversion efficiency over 26%," *Nature Energy*, vol. 2, p. 17032, 2017.

[31]   L. G. Gerling, S. Mahato, A. Morales-Vilches, *et al.*, "Transition metal oxides as hole-selective contacts in silicon heterojunctions solar cells," *Solar Energy Materials and Solar Cells*, vol. 145, pp. 109–115, 2016.

[32]   S. Gatz, H. Plagwitz, P. P. Altermatt, B. Terheiden, and R. Brendel, "Thermal stability of amorphous silicon/silicon nitride stacks for passivating crystalline silicon solar cells," *Applied Physics Letters*, vol. 93, p. 173502, 2008.

[33]   P. Saint-Cast, J. Benick, D. Kania, *et al.*, "High-efficiency c-Si solar cells passivated with ALD and PECVD aluminum oxide," *IEEE Electron Device Letters*, vol. 31, pp. 695–697, 2010.

[34]   R. Hezel and K. Jaeger, "Low-temperature surface passivation of silicon for solar cells," *Journal of the Electrochemical Society*, vol. 136, pp. 518–523, 1989.

[35]   T.-T. Li and A. Cuevas, "Effective surface passivation of crystalline silicon by rf sputtered aluminum oxide," *Physica Status Solidi (RRL) – Rapid Research Letters*, vol. 3, pp. 160–162, 2009.

[36]   A. Thankappan and S. Thomas, *Perovskite Photovoltaics: Basic to Advanced Concepts and Implementation*, London: Academic Press, 2018.

[37]   H. Kim, J. Hunger, E. Cánovas, *et al.*, "Direct observation of mode-specific phonon-band gap coupling in methylammonium lead halide perovskites," *Nature Communication*, vol. 8, p. 687, 2017.

[38]   B. Parida, S. Iniyan and R. Goic, "A review of solar photovoltaic technologies," *Renewable and Sustainable Energy Reviews*, vol. 15, pp. 1625–1636, 2011.

[39]   A. Luque and G. L. Araújo, *Solar Cells and Optics for Photovoltaic Concentration*, Bristol: A. Hilger, 1989.

[40]   M. I. Hossain, A. Bousselham, and F. H. Alharbi, "Optical concentration effects on conversion efficiency of a split-spectrum solar cell system," *Journal of Physics D: Applied Physics*, vol. 47, p. 075101, 2014.

[41]   A. Polman, M. Knight, E. C. Garnett, B. Ehrler, and W. C. Sinke, "Photovoltaic materials: present efficiencies and future challenges," *Science*, vol. 352, p. aad4424, 2016.

Chapter 3

Thermodynamics of solar energy conversion

Fahhad Alharbi[1], Fedwa EL Mellouhi[2], Brahim Aïssa[2] and Nouar Tabet[3]

3.1 Introduction

The main aim of any energy conversion model is to establish upper limits for the conversion efficiency. The more detailed the thermodynamic model is, the more realistic upper bounds are obtained. However, the increase in the model's complexity is accompanied by more involved calculations. A generalized introduction to the fundamental principles of future solar energy systems, based on consistent physics is presented. In describing the various conversions, we make use of endo reversible thermodynamics – a subset of irreversible thermodynamics. In this way, readers are supplied with the information to enable them to calculate the explicit values for a broad class of processes. Throughout, general principles are illustrated using idealized models, and end-of-chapter Materials for Solar Cells are described and practical examples are merely presented so as to compare reality with theory.

3.2 The thermodynamics of solar energy conversion

To estimate the maximum possible energy conversion efficiencies of solar cells, various theoretical models have been used. For tracking, they can be categorized in two general classes. The first class is phenomenological. It is based on detailed balance of radiations between two-extended-level system. This accounts for absorption and recombination. To estimate the maximum possible efficiency, only the inevitable radiative recombination is considered. Originally, this was introduced by Shockley and Queisser in 1961 [1] and then followed by many others [2–7]. In this section, a model following this class is adopted [6,7]. The second class is more rigorous and fundamental. The analyses of this class are based on maintaining the balance of both energy and entropy fluxes [8,9].

[1]Department of Electrical Engineering, King Fahd University of Petroleum and Minerals, Saudi Arabia
[2]Qatar Environment and Energy Research Institute (QEERI), Hamad Bin Khalifa University (HBKU), Qatar Foundation, Qatar
[3]Department of Applied Physics and Astronomy, College of Sciences, University of Sharjah, United Arab Emirates

In this section, detailed balance model is used where the focus on single junction solar cells. Hereunder listed are the main assumptions considered in the analyses:

- Solar radiation intensity and spectrum is not uniform and vary from a place to another. Actually, the variations on the spectrum are minimal as discussed in Chapter 1. However, there is a considerable variation in the intensity mainly based on the Air Mass coefficient. Besides the intensity, the coefficient accounts for the optical path length through the Earth's atmosphere. The intensity is more for the more direct radiation and shorter optical paths. The reference density is set by the American Society for Testing and Materials standard (ASTM G173-03) [10]. Here, we use AM1.5g, which is the commonly used spectrum, as the base reference. For hot areas, the main difference is on the directedness of the radiation with very small effect on the spectrum. The intensity is increased by some factor X. X is usually between 1.2 and 1.3 in the Arabic countries and Australia. Theoretically, the maximum value is 1.5. In the first subsection, AM1.5g is used while the factor X (between 1.0 and 1.4) is used in the second subsection.
- It is assumed that any photon above the energy gap E_g is absorbed.
- Carrier multiplication is not considered. Thus, any absorbed photon can only produce one exciton.
- As known, there are a lot of possible recombination mechanisms. Unfortunately, the most practically dominant ones are material quality dependent and not necessarily intrinsic. In principle, such nonfundamental limitations do not determine the theoretical limit. The only inevitable mechanism is radiative recombination. Hence, it is the only considered one in the analysis. However, in the last subsection, we consider non-radiative recombination for practical analyses.

At the beginning, the thermodynamic limit of solar energy conversion for AM1.5g is presented and discussed. Then, the effect of hot climate is analyzed by considering the increase of solar radiation intensity. Finally, the effect of temperature is incorporated as well to study the effect of hot climate on the commonly used solar cells technologies: namely Si, CdTe, and GaAs. The effect of the temperature is heavily material dependent and thus it is considered for specific technologies.

3.2.1 *Thermodynamics limit of solar energy conversion: AM1.5g*

As aforementioned, a detailed balance model is used to estimate the theoretical limit [1]. The balance is applied between the excitation and the radiative recombination in two-extended-level system. As AM1.5g radiation flux $\phi(E)$ reach the cell, the resulted photogenerated current density as a function of E_g is then

$$J_g\left(E_g\right) = q \int_{E_g}^{\infty} \phi(E)\, dE \qquad (3.1)$$

where q is the electron charge and E is the photon energy. Concerning radiative recombination, it is assumed to be only due to spontaneous emission, which follows the generalized black body radiation. Thus, the recombination current density can be calculated consequently. So,

$$J_r\left(E_g, V, T\right) = \frac{2\pi q^3}{c^2 h^3} q \int_{E_g}^{\infty} \frac{E^2}{\exp\left(\frac{E-V}{k_B T}\right)} dE \tag{3.2}$$

where V is the photogenerated voltage across the cell, T is the temperature in Kelvin (K), c is the speed of light, h is Planck's constant, and k_B is Boltzmann's constant. Then, the net current density is

$$J\left(E_g, V, T\right) = J_g\left(E_g\right) - J_r\left(E_g, V, T\right) \tag{3.3}$$

Thus, the efficiency of energy conversion is then the ratio between the gained power and the input power P_{in}:

$$\eta\left(E_g, V, T\right) = \frac{P_{out}}{P_{in}} = \frac{VJ\left(E_g, V, T\right)}{P_{in}} \tag{3.4}$$

where:

$$P_{in} = q \int_0^{\infty} E\phi(E)\, dE \tag{3.5}$$

The upper limit of η is then obtained by maximizing it vs. V for a particular E_g at a given temperature T, i.e.:

$$\eta_{max}\left(E_g, T\right) = \max_V \eta\left(E_g, V, T\right) \tag{3.6}$$

Based on this model, there are three main causes of losses as listed below:

- Unabsorbed photons:
  Photons with energies below E_g are not absorbed. So,

$$L_{Ph} = \frac{q}{P_{in}} \int_0^{E_g} E\phi(E)\, dE, \tag{3.7}$$

- Thermalization:
  This loss is due to the fact that the absorbed energy (i.e. E) above the maximum obtainable energy (i.e. E_g) is lost rapidly to heat:

$$L_{Th} = \frac{q}{P_{in}} \int_{E_g}^{\infty} \left(E - E_g\right)\phi(E)\, dE, \tag{3.8}$$

- Recombination:

  Any system that can absorb can emit. Depending on the system and environment, an equilibrium shall be reached. At this point, it is inevitable to lose part of the gained energy due to re-emission. This process losses are two-fold. The first is due to the direct loss by recombination. The second is the further reduction in the averaged gained energy below its initial separation energy after thermalization (i.e. E_g). The overall loss due to the two contributions of recombination losses are:

$$L_{RR} = \frac{V J_r}{P_{in}} + \frac{q}{P_{in}} \int_{E_g}^{\infty} (E_g - V)\phi(E)dE. \tag{3.9}$$

The efficiency η and the associated losses L_{Ph}, L_{Th}, and L_{RR} at temperatures 300 K, 200 K, 100 K, and at 1 K are shown in Figure 3.1. Clearly, the main effect of the temperature is on η and L_{RR} (as can be clearly observed as well from Figure 3.2). As the temperature increases, radiation from the cell obviously increases (Black body radiation) and hence the associated recombination increases. This is on account of η. η_{max} decreases from around 49% for 1.13 eV gap at 1 K to 33.5% for 1.16 eV gap at 300 K. Nonetheless, at room temperature around 300 K, there is a wide window of E_g between 0.9 eV and 1.6 eV that has η_{max} around 30%. Many materials have

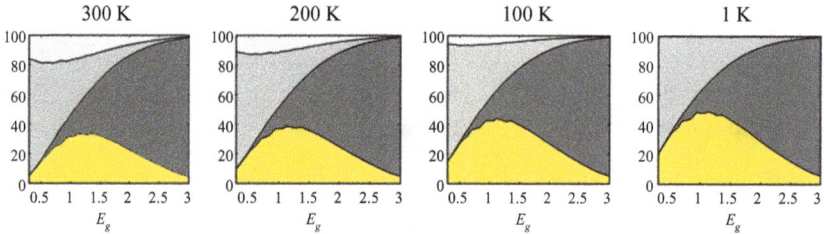

Figure 3.1   The theoretical limit of η for single junction solar cells with the associated losses vs. E_g at temperatures 300 K, 200 K, 100 K, and at 1 K. (η is in yellow and the losses in gray. The darkest is for L_{Ph}, the middle is for L_{Th}, and the lightest is for L_{RR}).

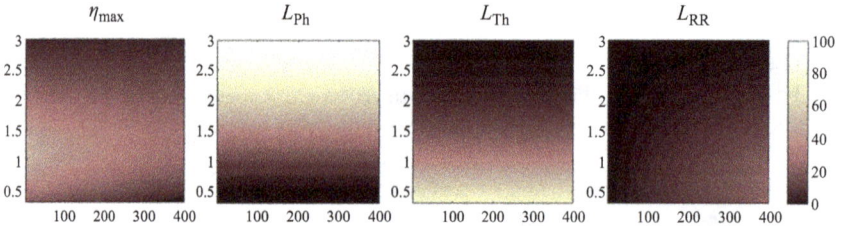

Figure 3.2   The contours of η, L_{Ph}, L_{Th}, and L_{RR} vs. E_g and temperature

gaps in this range and can conceptually be used to develop relatively efficient solar cells (providing that their transport is good).

3.2.2   Single junction solar cells

Furthermore, it is clear from Figure 3.1 that more that 50% of energy is lost either due to thermalization or for not absorbed photons over the whole range of E_g. Actually, this fact is the driver for seeking alternative device concepts to increase conversion efficiency like multi-junction, split-spectrum, hot carrier, and carrier multiplication cells. Seemingly, the losses L_{Ph} and L_{Th} are in principle independent of temperature and the balance between absorption and re-emission. Additionally, they are mostly not affected by any uniform increase of the intensity of solar radiation (assuming no variation in the spectrum).

In this subsection, the thermodynamic analysis focuses on AM1.5g radiation, which is definitely not suitable for hot areas where rays are more direct. Furthermore, there are small variations on the spectrum as shown in Chapter 1. However, these spectral variations are very small and will be ignored. The only effect to be considered on η is the increase of the intensity due to the more direct rays. Also and as aforementioned, optoelectronic materials' properties are temperature dependent. However, the quantitative dependences are materials specific and thus the limit cannot be generalized. Nonetheless, in the last subsection, this temperature effect on the main three solar cell technologies (i.e. Si, CdTe, and GaAs) is discussed.

3.2.3   Thermodynamics limit of solar energy conversion: the effect of increased intensity in hot areas

AM1.5g assumes that the average inclination angle is around 48.19° such that the Air Mass coefficient $1/\cos(\theta)$ is 1.5 which represent the increase of the exposed area (i.e. reduction in intensity) due to the angle. Closer to the equator, this explosion angle is getting less and hence the intensity increases. At most, the increase of the intensity over AM1.5g is 1.5. However, due to the tilted Earth axis, this is averaged to be between 1.2 and 1.3. In this section, we consider an intensity increase (X) between 1.0 and 1.5.

At the beginning, let us identify the quantities that will be affected by the increase of the intensity (i.e. with X). The first parameter is obviously the radiation flux $\phi(E)$ which almost increase linearly with X. This J_g, P_{in}, L_{Ph}, and L_{Th} are changed linearly according to (3.1), (3.5), (3.7), and (3.8). However, J_r (3.2) is independent of X as it is merely Black Body Radiation, which is only temperature dependent. So, the influence of X on η and L_{RR} is somewhere in the middle. Since J_g increases linearly with X while J_r is independent of it. η increases slightly and sublinearly with X as shown in Figure 3.3. For 1.2 and 1.4 eV gaps, η_{max} increases from 33.3% with $X = 1$ to around 33.75% when $X = 1.5$. However, this is for the theoretical limit. The effect of X on the materials (which is associated with an increase in the temperature) is completely ignored in this analysis.

Although there is a slight increase in the theoretical limit of the efficiency, the most important part is the increase in the maximum possible power yield which can

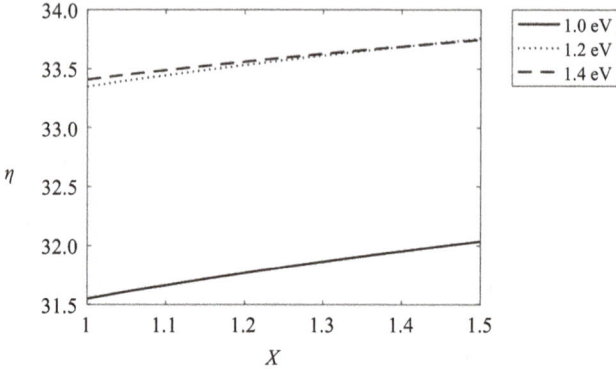

Figure 3.3 η_{max} vs. X for single junction solar cells with 1.0, 1.2, and 1.4 eV energy gaps

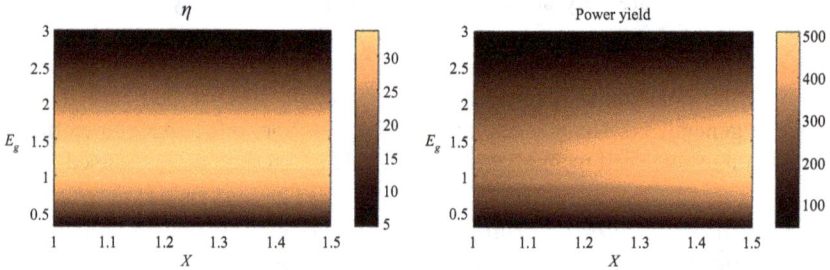

Figure 3.4 η_{max} and maximum power yield vs. X and E_g for single junction solar cells

be generated. This increases with X due to the relevant increase of P_{in}. Figure 3.3 shows contours representing η_{max} and maximum power yield vs. X and E_g for single junction solar cells.

So far, we only consider the effect of X on the theoretical limit. It could be useful to consider its practical effect. In practice, the efficiency of solar cell is:

$$\eta = \frac{J_{sc}V_{oc}FF}{P_{in}} \tag{3.10}$$

J_{sc} is related linearly to photo-generation. In the above analysis, we just considered the radiation recombination. But the main practical recombination is non-radiative and it is due to defect-assisted recombination known as Shockley–Read–Hall (SRH). We can empirically combine the non-radiative recombination in a single associated J_{nr} current density. So, J_{sc} is approximately

$$J_{sc} = J_g - J_r - J_{nr} \tag{3.11}$$

Among the three terms, only J_g depends on X. So, as a function of X, J_{sc} becomes

$$J_{sc}^{(X)} = XJ_g - J_r - J_{nr} = J_{sc}^{(1)} + (X - 1)J_g \tag{3.12}$$

where the subscripts stand for the value of X. In the limit that $J_g \gg J_r + J_{nr}$:

$$J_{sc}^{(X)} \approx XJ_{sc}^{(1)} \tag{3.13}$$

As for V_{oc}, according to Shockley diode equation, it is:

$$V_{oc} = \frac{k_B T}{q} \ln\left(\frac{J_{sc}}{J_0} + 1\right) \approx \frac{k_B T}{q} \ln\left(\frac{J_{sc}}{J_0}\right) \tag{3.14}$$

where J_0 is the device saturation current and it is extremely small compared to J_{sc}. It depends on the used materials and the design of the device and completely independent of any variation in the spectrum. As a function of X and using (3.14), V_{oc} becomes

$$V_{oc}^{(X)} \approx \frac{k_B T}{q} \ln\left(\frac{J_{sc}^{(X)}}{J_0}\right) =\approx \frac{k_B T}{q} \ln\left(\frac{X J_{sc}^{(1)}}{J_0}\right) \approx V_{oc}^{(1)} + \frac{k_B T}{q} \ln(X) \tag{3.15}$$

At room temperature and even in hot areas, $k_B T/q$ is small (just above 25 mV). Furthermore, as X is slightly more than 1, $\ln(X) \approx X - 1$. So, the change on V_{oc} with respect to X is logarithmic and very small and not comparable to that linear dependence of $J_{sc}^{(X)}$.

 FF, which is the last parameter, depends empirically on V_{oc} as

$$FF = \frac{\frac{q}{nk_B T} V_{oc} - \ln\left(\frac{q}{nk_B T} V_{oc} + a\right)}{\frac{q}{nk_B T} V_{oc} + b} \tag{3.16}$$

where n is the identity factor and a and b are fitting parameters. All of them are material dependent. Usually, a and b are small when compared to the reduced voltage $v = V_{oc}^{(1)} q/k_B T$ and can be ignored. So, the above equation can be approximated using this fact and (3.16) by

$$FF^{(X)} \approx FF^{(1)} + \frac{n}{v^2} \ln(X) \left[\ln\left(\frac{v}{n}\right) - 1\right] \tag{3.17}$$

Hence, clearly, FF increases slightly and logarithmically with X.

 In term of efficiency, the resulted improvement is:

$$\frac{\eta^{(X)}}{\eta^{(1)}} = \frac{J_{sc}^{(X)}}{X J_{sc}^{(1)}} \frac{V_{oc}^{(X)}}{V_{oc}^{(1)}} \frac{FF^{(X)}}{FF^{(1)}} \tag{3.18}$$

It can be shown easily from the above equations that the efficiency certainly increases slightly and sublinearly with X. Figure 3.4 shows the maximum power conversion efficiency and and maximum power yield as a function of X and bandgap energy for single junction solar cells.

3.2.4   Hot climate implications on common solar cell technologies

So far, only the effects of increased intensity in hot areas on the theoretical limit of solar power conversion are considered. However, it is obvious that this increase in the intensity is accompanied by an increase in the temperature. Though the effects of the temperature follow qualitatively some patterns, they are quantitatively material dependent. It is well established that η decreases – almost linearly – with temperature in normal operation ranges (between 0°C and 100°C) for many of the conventional inorganic materials such as Si, CdTe, CIGS, and GaAs, while $In_xGa_{1-x}N$ cells exhibit a different behavior where η increases slightly until 100°C and then decreases [11].

Detailed studies illustrated that temperature effect on the efficiency is generally a balance between the increase in J_{sc} and the decrease in V_{oc}. Interestingly, in many cases, these effects are due to common causes. For example, E_g in most cases decreases with temperature as will be discussed in more details shortly. This permits absorbing more photons and hence increases J_{sc}. On the other side, V_{oc} is proportional to the gap and hence it decreases with temperature. Yet, absorbing more photons results in increasing the photo-generated carrier density, proportionally affecting the built-in potential and hence can increase V_{oc}.

To practically illustrate the effect of increased temperature in hot areas on the cell performance, we study the three of the mostly used single junction solar cell technologies, namely Si, CdTe, and GaAs where the operational temperature changed from 300 K and 350 K. This covers the range between normal operation temperature in cold areas and hottest regions in the globe.

At the beginning, the main models used to estimate the temperature effects on the properties of semiconductors materials are listed and discussed briefly.

- **Energy gap (E_g):**
  Many thermal mechanisms have influence on E_g. The first one is due to thermal expansion. The lattice volume increases with the temperature and hence the overlapping and interactions of atomic orbitals decrease. This results in reducing interaction splitting and hence the gaps. However, this is not general; but it is the case for elemental and binary semiconductors and for a wide range of structures of ternary semiconductors. The second important temperature effect on the gap is due to lattice vibration. It results in reducing – slightly – the gap. The reduction increases with the temperature. It starts quadratically for small T and then shifts till it become linear for large T. At room temperature, the effect is almost linear. Both these effects are captured very well my many suggested models like those by Varshni [12], Vina *et al.* [13], Passler [14], and Manoogian and Leclerc [15]. Here we consider Varshni [12] model where:

$$E_g(T) = E_0 - \frac{\alpha \; T^2}{T + \beta} \tag{3.19}$$

  where E_0 is the energy gap at 0 K.

- **Carriers mobilities (μ):**
  As known, efficient transport of carriers is essential for proper operation of solar cells. It depends heavily on materials quality; the better the growth quality, the better is the transport. If the quality is perfect, then, the intrinsic features of a material dominate the limit of the transport properties. Here, we just consider them and assume that the growth quality is perfect. The main intrinsic features that influence μ are carrier scattering mechanisms and the doping concentration. Concerning the scattering, the main mechanisms are lattice scattering, ionized donor–acceptor scattering, and electron–hole scattering. At low temperature, lattice scattering dominates over others, where their contributions increase with the time. Most of the suggested models (like Klassen [16], Reggiani [17], and Sah [18]) to estimate the temperature dependence of carrier mobilities are based on empirical forms starting from some reference point where the mobility is known. This reference point is usually at room temperature (300 K). For the analysis, Klassen's model [16] is used where

$$\mu = \mu_0 \left(\frac{300}{T}\right)^\theta \tag{3.20}$$

  where μ_0 is the reference mobility of a material at the reference temperature of 300 K, T is the temperature in Kelvin, and θ is a fitting parameter.

- **Effective masses (m^*):**
  The basic temperature dependence of m^* is due to the temperature dependence of both conduction- and valance-band edges. Furthermore, many other mechanisms alter it as well. In return, m^* affect many other semiconductor properties implicitly where the dependence of many optoelectronic properties on the effective masses are well established. Conceptually, due to the reduced interactions between atomic orbitals due to lattice volume increase with the temperature, the dispersion of the bands is altered in a way that reduce the edge curvatures. In term of m^*, this means that it decreases with the temperature. This point is the starting point for most of the used models for T-dependence of m^* as they depend on identifying a reducing factor with T starting from a reference point at the reference T of 300 K. The most commonly used models are those by Sharma *et al.* [19], Ravich *et al.* [20], Adachi [21], and Vasileff [22]. Sharma's model [19] is used in the following analysis:

$$\frac{m_0^*}{m^*} = 1 + C\left(\frac{2}{E_g(T)} + \frac{1}{E_g(T) + \Delta}\right) \tag{3.21}$$

- **Electron affinity (χ):**
  Proper band alignment is crucial for solar cell operation. It is governed by χ and quisi-Fermi levels. Unfortunately, there are few T-depenence model for χ. The one used in this work is proposed by Hossain *et al.*, which is a rough approximation. It is assumed that the reduction of E_g is split evenly between

the conduction and valance bands [11]. So,

$$\chi(T) = \chi_0 - \frac{1}{2}\left(E_0 - E_g(T)\right) \tag{3.22}$$

where χ_0 is the affinity at 0 K. This illustrates as well that E_g has further implicit effect on the performance of the cell.

- Densities of states: For three-dimensional semiconductors, the temperature dependence of density of states is well established. With parabolic band edge model,

$$N_i = \left(\frac{2\pi m_i^* k_B T}{h^2}\right)^{1.5} \tag{3.23}$$

where i stands for either valance or the conduction band and h is Planck' constants. Few further "empirical" simplifications were suggested by setting a reference point at the reference T of 300 K. Since, N_i depends on effective masses, Sharma's model [19, Eq. (2.24)] can be used to approximate N_i as well. The commonly used model is

$$N_i(T) \approx N_i(300)\left(\frac{1 + C\left(\frac{2}{E_g(300)} + \frac{1}{E_g(300)+\Delta}\right)}{1 + C\left(\frac{2}{E_g(T)} + \frac{1}{E_g(T)+\Delta}\right)}\right)^{1.5} \tag{3.24}$$

As aforementioned, temperature dependences are highly materials dependent. So, to illustrate the effect, only Si, CdTe, and GaAs solar cells are considered. Table 3.1 lists the commonly used parameters for the materials considered cells

Table 3.1   The commonly used parameters for the materials considered cells

Materials property	Unit	Parameter	Si	GaAs	CdTe	CdS
Energy gap	eV	E_0	1.1557	1.152	1.61	2.58
		α	7.02×10^{-4}	8.87×10^{-4}	3.10×10^{-4}	4.02×10^{-4}
		β	1,108	572	108	147
Electron mobility	cm^2/Vs	μ_{n0}	1,500	8,500	320	100
		θ_n	2.4	1.0	1.178	1.1765
Hole mobility	cm^2/Vs	μ_{p0}	450	400	40	25
		θ_n	2.2	2.1	0.714	2.33
Effective masses		C	7.0396	7.0834	7.0847	7.7700
Elec. eff. density	cm^{-3}	$N_c(300)$	2.8×10^{19}	4.7×10^{17}	8.65×10^{17}	2.2×10^{18}
		$E_g(300)$	1.120	1.425	1.513	2.51
		Δ	0.044	0.34	0.92	0.06
Hole eff. density	cm^{-3}	$N_v(300)$	1.04×10^{19}	7.0×10^{17}	1.27×10^{19}	1.8×10^{18}
Electron affinity	eV	χ_0	4.05	4.07	4.28	4.445

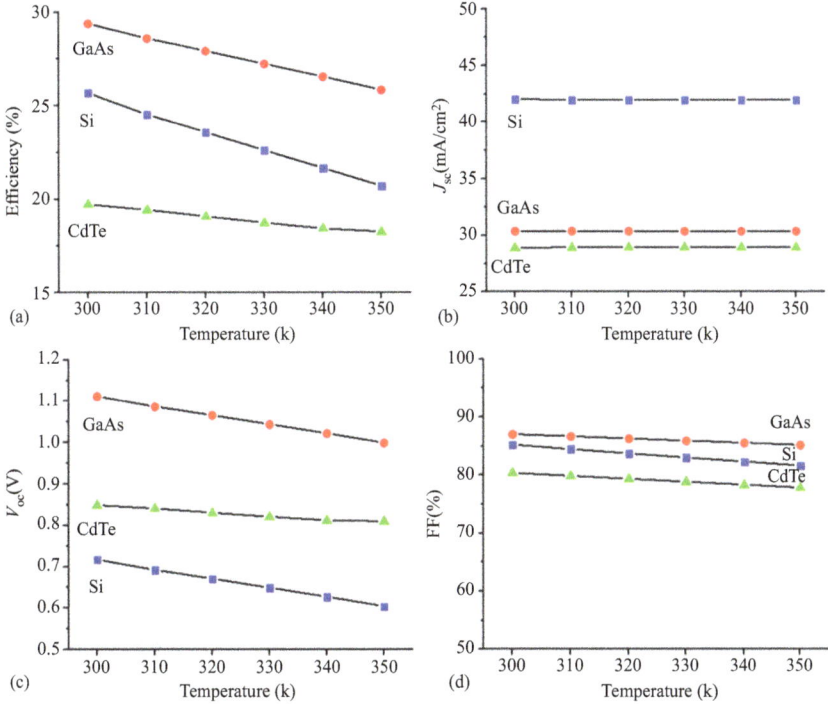

Figure 3.5   Si, GaAs, and CdTe solar cells performances vs. temperature in the range of 300–350 K: (a) η, (b) J_{sc}, (c) V_{oc}, and (d) FF

(collected from Ref. [11] and the references within). The cells' performances are calculated using the Solar Cell Capacitance Simulator SCAPS.

Figure 3.5 presents the simulated performances of the cells. Clearly, the conversion efficiency decreases with the temperate (but not at the same rate). By proper correlation, it was found that the drop in V_{oc} is the main cause of performance drop with the temperature. This is due to the inverse proportionality between E_g and T. As E_g, V_{oc} decreases accordingly. This reduction in the gap results in a slight increase in J_{sc} because the cell can then absorb photons with longer wavelengths and hence the generated photocurrent increases. To have a better quantitative comparison, the calculated performances are scaled relatively as depicted in Figure 3.6. Clearly, V_{oc} is the most affected performance indicator by the temperature. In term of relative efficiency drop, the used models agree very well with the experimentally obtained values as shown in Figure 3.7. The usual relative efficiency drop ranges are:

- for Si: 0.069–0.080 per degree,
- for GaAs: 0.069–0.099 per degree,
- for Si: 0.034–0.047 per degree.

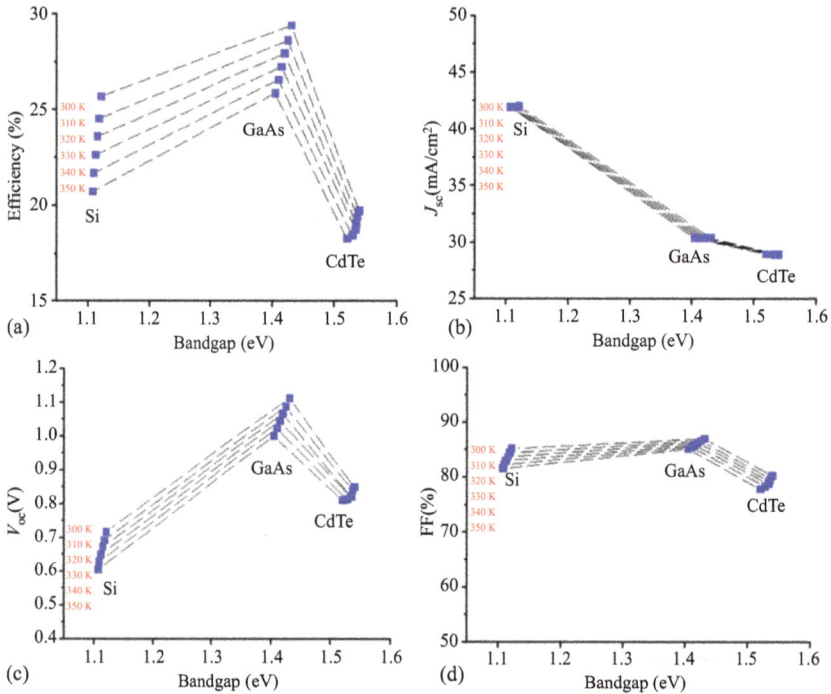

Figure 3.6   Si, GaAs, and CdTe solar cells performances vs. E_g at different temperatures: (a) η, (b) J_{sc}, (c) V_{oc}, and (d) FF

Clearly, CdTe is less affected by temperature when compared to Si and GaAs. However, due to their higher efficiencies at room temperature, both Si and GaAs are better than CdTe within the normal operation temperature ranges below 80°C.

3.3   Materials for solar cells

3.3.1   Conventional materials

The recent past years have witnessed a very persistent dedication to improve the performance of absorbers and to develop novel solar cell materials. It is expected that the field of photovoltaics, especially as it relates to alternative and renewable energy, will grow further in the coming decades as the conventional fossil-fuel-based energy resources are depleted. Today, world record single-junction solar cell efficiencies are close to their practical and theoretical efficiency limits. Record efficiencies of c-Si and TF solar cells [23] are about 25% and 23%, respectively, and hence approaching their theoretical limit Shockley–Queisser (SQ), which takes into account Auger recombination) of 29.4% under standard illumination conditions (AM 1.5G = 1 Sun). As the gap between the laboratory and industrial scale efficiencies becomes narrower, further cost reduction will only be possible through

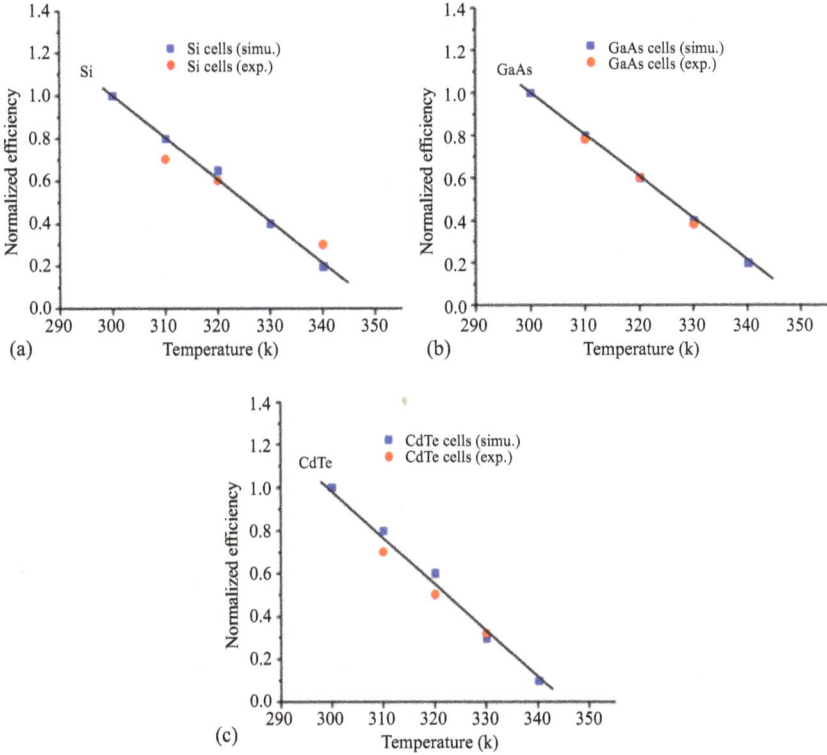

Figure 3.7   *Relative efficiency for Si, GaAs, and CdTe solar cells temperature in the range of 300–350 (red dots are for experimental data and the blue ones are by simulations)*

more economical manufacturing, unless a practical solution can be found to push efficiency to 30% and beyond. The fundamental loss mechanisms for all single-junction cells, such as thermalization and transmission losses (about 35% and 20% of the incident energy, respectively), cannot be easily avoided, and so to reach efficiency ranges beyond the theoretical limit of the state of the art single junction solar cell, alternative approaches have to be used.

Figure 3.8 presents the current market share of solar cell technologies. Clearly, silicon-based solar cell technology dominates the market. Furthermore, the market share of monocrystalline silicon has increased drastically over the last years increasing from 25% in 2016 to more than 80% in 2020 [24].

3.3.2   Emerging materials

Since 1970s, many other materials have been explored for solar cells. In recent days, the most notable achievement arising from these efforts is the emergence of a new generation of hybrid organic–inorganic perovskites as solar energy conversion

PV production by technology
Percentage of global annual production

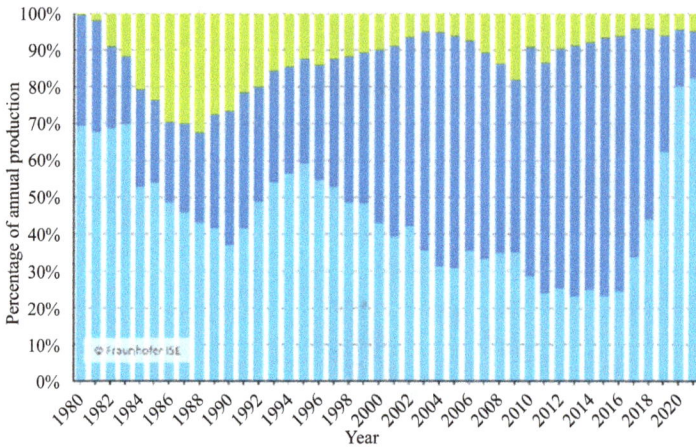

Figure 3.8   Market share of solar cell technologies, adapted from [24].

materials [25,26]. Over just a few years, a variety of perovskite solar cells (PSCs) with different device designs (see Figure 3.9) have been developed with power conversion efficiencies (PCEs) already exceeding 25.7% [23]. From a practical perspective, it is impressive how simple it is to fabricate the PSCs and how many efficient cells made with various hybrid perovskites absorbers and with dierent device designs. Among the hybrid halide ABX_3 perovskites developed for PV, methyl-ammonium lead triiodide ($MAPbI_3$) is the most widely employed compound, its absorption onset is placed at $1 = 780$ nm ($E_g = 1.6$ eV) and its absorption coefficient is high enough to absorb 100% of the photons provided by the sun with energies higher than the bandgap with only 1-mm thick material. Resulting excitons perform as low as 6 meV binding energy at room temperature, which allow them to be separated in free electrons and holes very quickly. Nevertheless, three major obstacles before large-scale commercialization of the cells are:

(i)   degradation of PCE due to absorber instability;
(ii) the presence of toxic element lead, and;
(iii) scaling leading to the degradation of efficiencies. All the above parameters for the time being compromise the competitiveness of PSCs. This has motivated lead universities and companies in the world to work on various PSCs technologies (refer to Figure 3.10).

Figure 3.10 illustrates a summary of different solar cell absorbers and their PCEs achieved in solar cells for halide and non-halide perovskites. Major institutions holding the record efficiencies are listed.

(a)

FTO, ITO, IZO, AgNW, DMDs, graphene, In_2O_3:H
Ti$_2$, SnO$_3$, C60
If necessary, Ti$_2$, SiO$_2$, Al$_2$O$_3$
Perovskite
Spiro-OMeTAD, TaTm, PEDOT, PTAA
Au, Ag, C

(b)

(c)

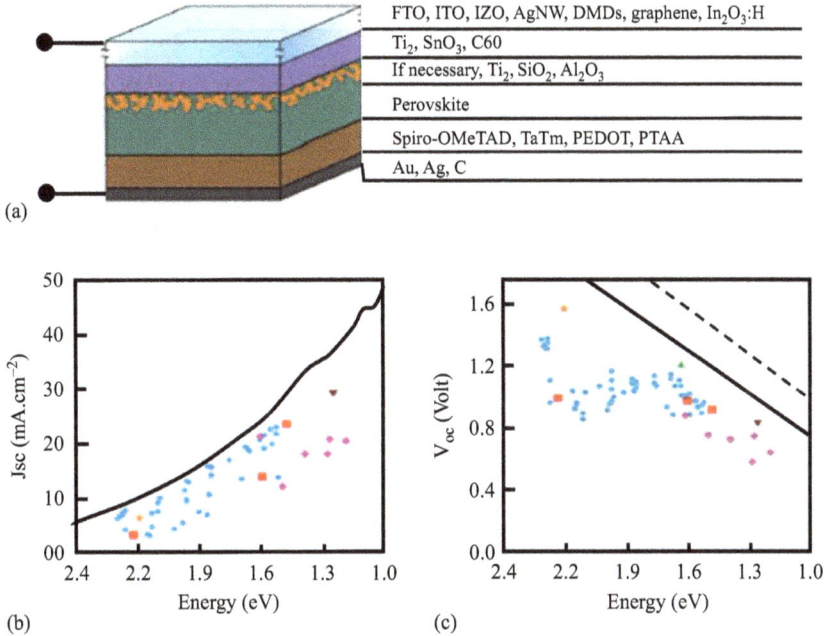

Figure 3.9   *State-of-the-art of single-junction hybrid halide perovskite solar cells.*
(A) Scheme of the standard architecture of a perovskite solar cell. (B
and C) Calculated ideal short-circuit current. (B) Open-circuit voltage
(C) as a function of the bandgap energy of the absorber material
according to Shockley–Queisser theory. A selection of the best
experimental parameters found in the literature is also plotted [27].

3.3.3   Computational materials design for solar cells

In principle, there is nothing special about Si or MAPbI$_3$. They just have the required properties to make efficient solar cells. An efficient solar cell can be made of any semiconductor with E_g ranging between 1.0 and 1.7 eV and with adequate transport allowing the photogenerated carrier to be collected. Many semiconductors have been explored and truly used to make solar cells. Previously, the assortment was mostly based on known materials as experimental data were the main "catalogue" for materials selection for solar cells [29]. Despite the vast data, this is definitely short and confines the screening space. However, the sophisticated and continuously growing computational capabilities have provided an alternative route to explore new materials for solar cells much beyond the rich experimental data. This has become one of the main trends in materials science and engineering. Currently, there are many "enormous" state-sponsored initiatives in this regard worldwide.

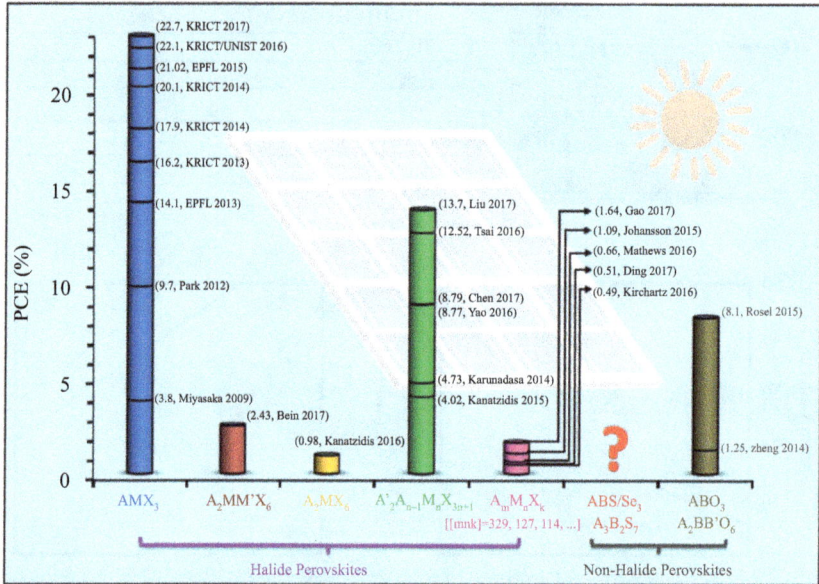

Figure 3.10   Summary of different solar cell absorbers and their PCEs achieved in solar cells for halide and non-halide perovskites. Major institutions holding the record efficiencies are listed. Adapted from [28].

For faster screening and obviously cost reduction, computational materials discovery has become a practical routine alternative way to discover new materials before their experimental realization. Yet, the field as a whole is still in its beginning and needs lot of further efforts. For example, there is a need for standardization, related data-mining, database construction, intelligent database search, and experimental certainly validation.

Another important aspect that should be stated is the influence of the hybrid perovskites solar cells. Previously, both the experimental and computational materials screening have focused on certain families of materials and usually either organic of fully inorganic. However, the hybrid nature of MAPbI$_3$ has enforced considering the extremely large hybrid possibilities. As a matter of fact, considering the extremely vast compositional space of such hybrid materials, it should be believed that the witnessed progress could be just the start and the tip of the iceberg.

Usually, computational materials design is performed in multistep [29–31]. Initially, a very large material space of materials is defined in some predefined material subspace. Then, the pool of potential materials is reduced at each step. The criterion for the selection of the remaining candidates after each step is called the descriptor. It is a simple quantity that can be calculated from the electronic band structure to connect the microscopic properties to some device observables. For example, one such descriptor is the band gap. An example workflow for the screening process that can be used is shown in Figure 3.11.

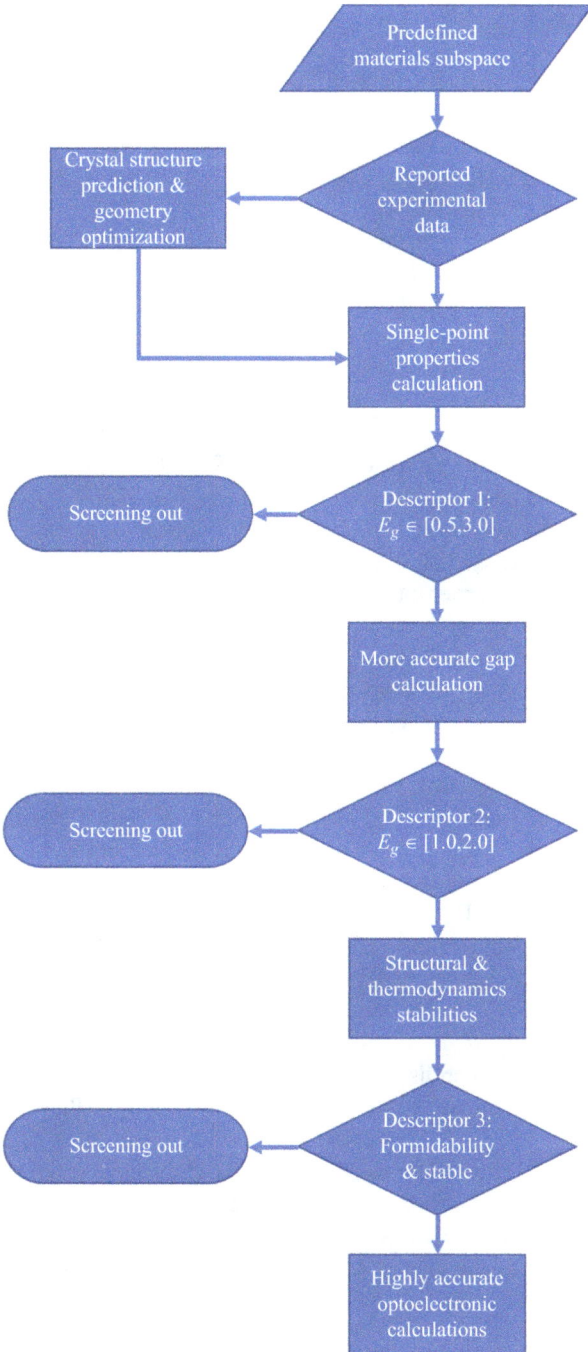

Figure 3.11   An example of a workflow of materials screening for solar cells

References

[1]   W. Shockley and H. J. Queisser, "Detailed balance limit of efficiency of p-n junction solar cells," *Journal of Applied Physics*, vol. 32, pp. 510–519, 1961.

[2]   B. P. Fingerhut, W. Zinth, and R. Vivie-Riedle, "The detailed balance limit of photochemical energy conversion," *Physical Chemistry Chemical Physics*, vol. 12, pp. 422–432, 2010.

[3]   A. De Vos, "Detailed balance limit of the efficiency of tandem solar cells," *Journal of Physics D: Applied Physics*, vol. 13, p. 839, 1980.

[4]   S. P. Bremner, R. Corkish, and C. B. Honsberg, "Detailed balance efficiency limits with quasi-Fermi level variations [QW solar cell]," *IEEE Transactions on Electron Devices*, vol. 46, pp. 1932–1939, 1999.

[5]   A. S. Brown and M. A. Green, "Detailed balance limit for the series constrained two terminal tandem solar cell," *Physica E: Low-dimensional Systems and Nanostructures*, vol. 14, pp. 96–100, 2002.

[6]   M. C. Hanna and A. J. Nozik, "Solar conversion efficiency of photovoltaic and photoelectrolysis cells with carrier multiplication absorbers," *Journal of Applied Physics*, vol. 100, p. 074510, 2006.

[7]   F. H. Alharbi, "Carrier multiplication applicability for photovoltaics: a critical analysis," *Journal of Physics D: Applied Physics*, vol. 46, p. 125102, 2013.

[8]   P. T. Landsberg and P.Baruch, "The thermodynamics of the conversion of radiation energy for photovoltaics," *Journal of Physics A: Mathematical and General*, vol. 22, p. 1911, 1989.

[9]   A. Luque and A. Martí, "Entropy production in photovoltaic conversion," *Physical Review B*, vol. 55, p. 6994, 1997.

[10]   A. S. T. M. Standard, "G173-03 (reapproved 2012), standard tables for reference solar spectral irradiances: Direct normal and hemispherical on 37 tilted surface," West Conshohocken, PA: ASTM International, 2012.

[11]   M. I. Hossain, A. Bousselham, F. H. Alharbi and N. Tabet, "Computational analysis of temperature effects on solar cell efficiency," *Journal of Computational Electronics*, vol. 16, pp. 776–786, 2017.

[12]   Y. P. Varshni, "Temperature dependence of the energy gap in semiconductors," *Physica*, vol. 34, pp. 149–154, 1967.

[13]   L. Vina, S. Logothetidis, and M. Cardona, "Temperature dependence of the dielectric function of germanium," *Physical Review B*, vol. 30, p. 1979, 1984.

[14]   R. Pässler, "Temperature dependence of exciton peak energies in multiple quantum wells," *Journal of Applied Physics*, vol. 83, pp. 3356–3359, 1998.

[15]   A. Manoogian and A. Leclerc, "Determination of the dilation and vibrational contributions to the energy band gaps in germanium and silicon," *Physica Status Solidi (b)*, vol. 92, pp. K23–K27, 1979.

[16]   D. B. M. Klaassen, "A unified mobility model for device simulation—II. Temperature dependence of carrier mobility and lifetime," *Solid-State Electronics*, vol. 35, pp. 961–967, 1992.

[17] S. Reggiani, M. Valdinoci, L. Colalongo, M. Rudan, and G. Baccarani, "An analytical, temperature-dependent model for majority- and minority-carrier mobility in silicon devices," *VLSI Design*, vol. 10, pp. 467–483, 2000.

[18] C.-T. Sah, *Fundamentals of Solid State Electronics*, Singapore: World Scientific Publishing Company, 1991.

[19] A. C. Sharma, N. M. Ravindra, S. Auluck, and V. K. Srivastava, "Temperature-dependent effective masses in III–V compound semi-conductors," *Physica Status Solidi (B)*, vol. 120, pp. 715–721, 1983.

[20] Y. I. Ravich, B. A. Efimova, and V. I. Tamarchenko, "Scattering of current carriers and transport phenomena in lead chalcogenides," *Physica Status Solidi (B)*, vol. 43, pp. 11–33, 1971.

[21] S. Adachi, *Properties of Semiconductor Alloys: Group-IV, III–V and II–VI Semiconductors*, vol. 28, New York, NY: John Wiley & Sons, 2009.

[22] H. D. Vasileff, "Electron self-energy and temperature-dependent effective masses in semiconductors: n-type Ge and Si," *Physical Review*, vol. 105, p. 441, 1957.

[23] Z. He, C. Xu, L. Li, A. Liu, T. Ma, and L. Gao, "Highly efficient and stable perovskite solar cells induced by novel bulk organosulfur ammo-nium," *Materials Today Energy*, vol. 26, p. 101004, 2022, https://doi.org/10.1016/j.mtener.2022.101004.

[24] Photovoltaic Report, Fraunhofer Institute for Solar Energy Systems, ISE with support of PSE Projects GmbH , Freiburg, 6 December 2022. www.ise.fraunhofer.de.

[25] M. M. Lee, J. Teuscher, T. Miyasaka, T. N. Murakami, and H. J. Snaith, "Efficient hybrid solar cells based on meso-superstructured organometal halide perovskites," *Science*, vol. 338, pp. 643–647, 2012.

[26] H. J. Snaith, "Perovskites: the emergence of a new era for low-cost, high-efficiency solar cells," *The Journal of Physical Chemistry Letters*, vol. 4, pp. 3623–3630, 2013.

[27] M. Anaya, G. Lozano, M. E. Calvo, and H. Míguez, "ABX3 perovskites for tandem solar cells," *Joule*, vol. 1, pp. 769–793, 2017.

[28] Q. Xu, D. Yang, J. Lv, Y.-Y. Sun, and L. Zhang, "Perovskite solar absorbers: materials by design," *Small Methods*, vol. 2, p. 1700316, 2018.

[29] F. El-Mellouhi, A. Akande, C. Motta, *et al.*, "Solar cell materials by design: hybrid pyroxene corner-sharing VO_4 tetrahedral chains," *ChemSusChem*, vol. 10, pp. 1931–1942, 2017.

[30] H. Park, F. H. Alharbi, S. Sanvito, N. Tabet, and F. El-Mellouhi, "Searching for photoactive polymorphs of CsNbQ3 (Q = O, S, Se, Te) with enhanced optical properties and intrinsic thermodynamic stabilities," *The Journal of Physical Chemistry C*, vol. 122, pp. 8814–8821, 2018.

[31] H. Park, R. Mall, F. H. Alharbi, *et al.*, "Exploring new approaches towards the formability of mixed-ion perovskites by DFT and machine learning," *Physical Chemistry Chemical Physics*, vol. 21, pp. 1078–1088, 2019.

Chapter 4

Solar cell technologies

Brahim Aïssa[1], Marie Buffiere[1] and Mohammad I. Hossain[1]

4.1 Introduction

Bell Laboratories developed the first silicon solar cell in 1954 with an efficiency of 6%. The earliest commercial silicon traditional solar cells are made from silicon, are currently the most efficient solar cells available for residential use, and account for around above 80% of all the solar panels sold around the world. Silicon solar cells are the most efficient in terms of single cell photovoltaic devices, and silicon is the most abundant element on earth, only second to oxygen. It is a semiconductor material suitable for PV applications; with energy band gap of 1.1 eV crystalline, silicon cells are classified into three main types depending on how the Si wafers are made.

The types are based on the type of silicon used, specifically: monocrystalline (Mono c-Si), polycrystalline (Poly c-Si), and amorphous silicon cells the oldest solar cell technology and still the most popular and efficient are solar cells made from thin wafers of silicon. These are called mono-crystalline solar cells. Commercial production of c-Si modules began in 1963 when Sharp Corporation of Japan started producing commercial PV modules and installed a 242-Watt (W) PV.

Compared to the other types of solar PV, they have a higher efficiency (up to 27.6%), meaning you will obtain more electricity from a given area of panel. Single crystal wafers are made by Czochralski process, as in silicon electronics. It comprises about 30% of the market. The cost of fabricating single crystalline silicon solar cells is due to the purification process of bulk. Polycrystalline silicon and amorphous silicon are much less pure than the single crystalline silicon, and most common because they are least expensive. The reason polycrystalline solar panels are less expensive than monocrystalline solar panels is because of the way the silicon is made. Basically, the molten silicon is poured into a cast instead of being made into a single crystal. The highest recorded efficiency for polycrystalline silicon cell is 23.3%, silicon solar cells typically have two layers: a positive layer (p-type) and a negative layer (n-type). The positive layer is usually made by doping

[1]Qatar Environment and Energy Research Institute (QEERI), Hamad Bin Khalifa University (HBKU), Qatar Foundation, Qatar

silicon with boron to create extra holes in the silicon lattice, and the negative layer is usually made by doping silicon with phosphorus to have extra electrons available in the silicon lattice.

4.2   Discussion about the main parameters affecting the silicon solar cell device performance under desert conditions (V_{oc}, temperature coefficient TC_η)

The certification of solar cells is usually performed under standard testing conditions (STC), namely, 1,000 W/m^2, 25°C, AM1.5g spectrum. However, when operating under desert conditions, this certification process could be however questioned. Indeed, PV modules deployed in the desert field can reach operating temperatures (T) as high as 90°C [1].

Typically, photovoltaic devices show significant performance degradation with respect to increasing T. Consequently, the temperature coefficient of the photoconversion efficiency (TC_η) is a representative figure-of-merit for the photovoltaic energy yield of a given PV technology. As a matter of fact, compared to conventional c-Si homojunction technologies ($TC_\eta = -0.45\%$/°C for standard homojunction and -0.35%/°C for homojunctions with passivating contacts [2]), silicon heterojunction (SHJ) solar cells technology is the less sensitive to increasing operating temperatures among all the others ($TC_\eta = -0.23\%$/°C [3] to -0.10%/°C [4]). To highlight the relevance of an adequate TC_η, for instance, starting from a 20%-efficient-cell (at 25°C), a difference in TC_η of 0.1 or 0.3%/°C results in ~ 4.5 or 13.5% relative difference in PCE at a T of 70°C.

Panasonic which is one of the main fabricants and supplier of SHJ modules claims that compared to conventional c-Si-based solar modules, their technology yields up to 8% more in the cumulative power output throughout the year, owing to the superior temperature performance of the SHJ devices. This represents a difference of up to 65.3 kWh/kWp per year [5]. The unit Wp (Watt peak) refers to the maximum power output under standard testing conditions (25°C, AM1.5g spectrum, 0.1 W/cm^2).

As will be detailed later, SHJ technology shows the highest V_{oc} among the other silicon-based ones. For an ideal solar cell, the diode equation directly yields an inverse dependence of the absolute value of $TC_{V_{oc}}$ on its V_{oc} [6,7].

Green *et al.* [6] proposed the following equation for the temperature dependence of the V_{oc}:

$$\frac{\partial V_{oc}}{\partial T} = -\frac{V_{go} - V_{oc} + \gamma k_B T/q}{T} \tag{4.1}$$

where T is the absolute temperature in Kelvin, V_{go} is the voltage equivalent of the semiconductor bandgap extrapolated to $T = 0K$, $k_B T/q$ is the thermal voltage and γ is a parameter that determines the sensitivity of the V_{oc} towards variations in T (smaller value equals lower sensitivity).

The superior resistance to high-temperature degradation behavior of this SHJ technology is often attributed to its high V_{oc} [8]. As will be detailed in the sections below, the high V_{oc} values are a direct result of their excellent surface passivation achieved by the intrinsic hydrogenated amorphous silicon [a-Si:H(i)] layers deposited on both sides of the wafer.

4.3   Silicon-based photovoltaic technologies

Figure 4.1 shows the record conversion efficiencies of photovoltaic devices over the past 40 years. Efficiencies as high as 46% were obtained with a four-junction solar cell based on III–V materials under a concentration of 508 suns [10]. Without concentration, the highest efficiency of 38.8% is reached with a similar structure. However, due to their complex and expensive processing, such technologies are mainly used for space applications or in combination with a concentrator.

For terrestrial application, silicon-wafer-based single-junction solar cells are still the major technology with a market share of over 90% [11]. For silicon, the highest non-concentrated efficiency was obtained with a 200-μm-thick rear-contacted heterojunction solar cell with an impressive 26.7% [10,12] which is less than $3\%_{abs}$ below the theoretical limit of 29.4% for a 110 μm-thick cell [9,13]. Recently, a rear-contacted cell with oxide-based passivating contacts reached 26.1% [14]. The best both-side-contacted device—one side an oxide-based passivating contact and the other a traditional diffused contact—reached 25.8% efficiency [15].

Before the recent boost with oxide-based passivating contacts [16], most solar cells were contacted directly with metals in geometries of varying sophistication. A relatively simple design was the passivated emitter solar cell (PESC) of the early 1980s whose front surface was partially passivated. In the late 1980s the same concept was refined by localizing the area of the metal contacts at the rear to small areas in passivated emitter and rear cells (PERC). A further refinement was introduced by restricting the highly doped regions to the areas underneath the contacts in the passivated emitter, rear locally diffused (PERL) and the passivated emitter, rear totally diffused (PERT) solar cells. In industry, the most common cell type is essentially a PESC cell, slightly modified by the introduction of a back-surface field (BSF). At the time of writing, it represents about 65% of industrial production: ca. 30% are produced with PERC technology. The remaining 5% are silicon heterojunction cells (SHJ). Another group is thin-film solar cells that have the advantage that they use less material. The most industrially common are CIGS and CdTe; cells on the basis of these materials reach efficiencies of 23.4% and 22.1%, respectively. Recently, perovskites have gained a lot of interest with a record efficiency of 25.7% even though their high sensitivity to humidity and oxygen make the commercialization of this material challenging [17,18]. Additionally, thin-film silicon (TF-Si) solar cells based on amorphous silicon (a-Si) or microcrystalline silicon (μc-Si) mostly in a multi-junction configuration reached efficiencies of 14% [12]. Despite the low cell and module production cost for this technology, they continuously lost market share because of their low efficiencies which require more surface area and a higher balance of system (BOS) cost.

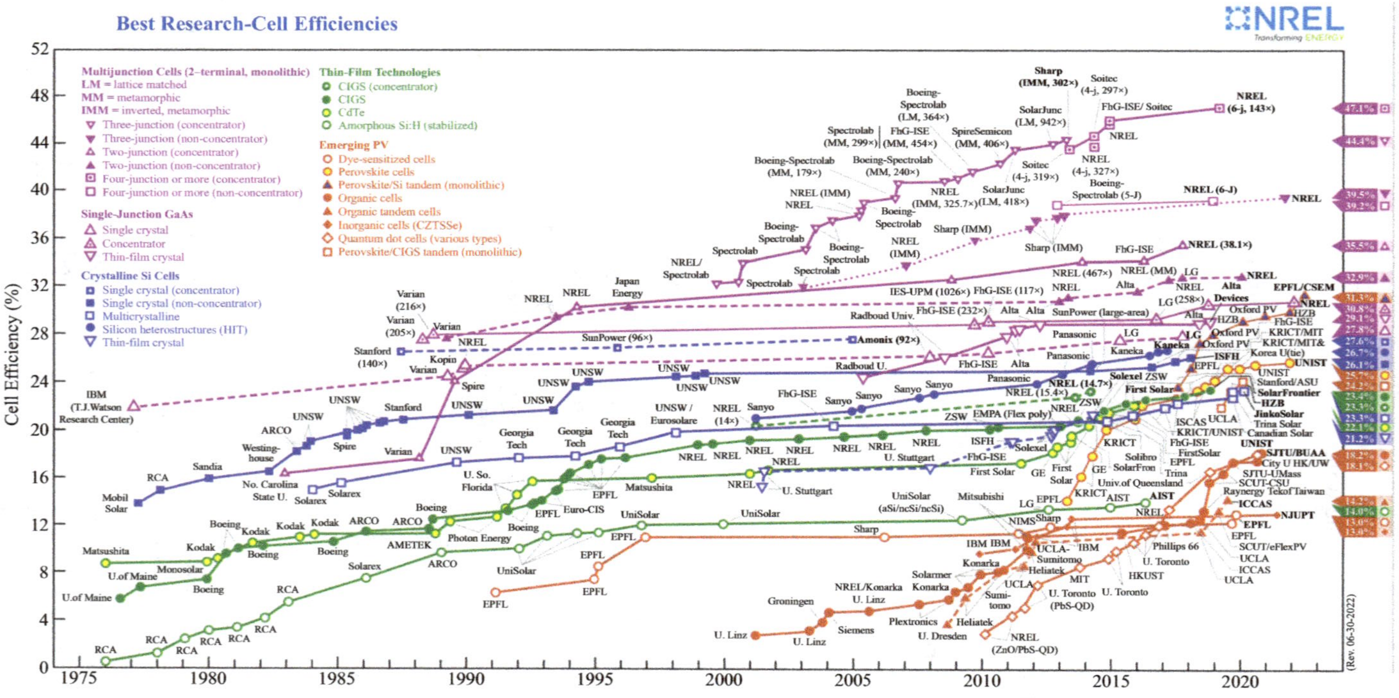

Figure 4.1 Record efficiencies for different photovoltaic technologies over the last 50 years. This plot is courtesy of the National Renewable Energy Laboratory, Golden, CO, USA [9].

4.3.1   *Aluminum back surface field (Al-BSF) cells*

The photovoltaic industry was dominated for years by the Al-BSF technology due to its simplicity of fabrication [19]. The structure is sketched in Figure 4.2(a) using either a mono-or multicrystalline silicon p-type wafer with a resistivity of ca. 1 Ωcm (acceptor density $N_A \sim 1 \times 10^{16}$ cm^{-3}). To scatter the incoming light and increase the path way of photons inside the absorber, the front side of the solar cell is textured to a depth of a few micrometers either by an alkaline wet etching in the case of mono-crystalline wafers or by acidic etching in the case of multi-crystalline wafers. Next, the wafers are annealed at 800°C–900°C in a tube furnace in an atmosphere of phosphorus oxychloride (POCl$_3$) and oxygen O$_2$. For this, nitrogen as the carrier gas is flowing through a bubbler with liquid POCl$_3$ before it is mixed with oxygen (O$_2$) and conducted directly to the quartz tube [21]. Phosphorous oxide P$_2$O$_5$ is deposited on the wafers and forms a two-layer stack of a phosphosilicate glass (PSG) and a silicon oxide (SiO$_2$) as well as a diffused region within the wafer forming the selective electron collector [22]. The released Cl$_2$ removes metal impurities while the diffused phosphorus reduces the concentration of impurities by gettering, both of which improve the quality of the bulk material [21]. The diffused region typically has a donor concentration N_D above 1×10^{20} cm^{-3} at the wafer surface which falls steeply below N_A within 1 μm. Typically, the diffused region has a sheet resistance of around 75Ω/□ [19].

The surface concentration and the depth of the diffusion profile can be optimized by a two-step annealing in which the POCl$_3$ supply is switched off during the second phase for a drive-in diffusion [23]. After removing the PSG/SiO$_2$ layer stack by hydrofluoric acid (HF), a silicon-rich silicon nitride SiN$_X$ layer with a refractive index of ~ 2.1 [24] and thickness of ~ 75 nm is deposited by plasma-enhanced chemical vapor deposition (PECVD) to form an anti-reflection coating (ARC). At the rear, an aluminum-based metallic paste is screen-printed on the full area and dried prior to the screen-printing of the silver front grid. Finally, both contacts are co-fired (a thermal treatment at around 800°C for a few seconds with fast heating and cooling). During firing, the aluminum back surface field is formed at the rear, while at the front, the silver paste locally penetrates the silicon nitride to contact the

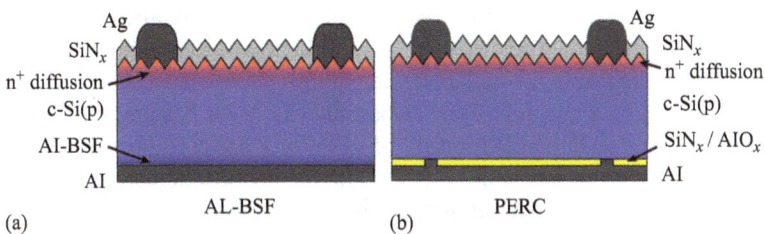

Figure 4.2   Sketch of a cross section of an (a) aluminum back surface field solar cell in comparison with (b) a PERC cell. Reproduced from Ref. [20] with permission from The Royal Society of Chemistry.

front n-type diffusion [25]. At the same time, hydrogen is released from the SiN_x layer to passivate defects at the wafer surface and in the bulk.

The enormous success of this structure relies on several main drivers: the simplicity of the production, a high tolerance against variations in wafer quality and perhaps most important—the main structure elements or process sequences are not severely protected by patents. On the other side, the main drawbacks are the direct metal–semiconductor interfaces at the front and the rear that lead to recombination losses. The diffused region at the front as well as the Al-BSF at the rear reduces the recombination by the reduced concentration of minority charge carriers, but the devices are still limited by recombination losses. There is little margin to further increase the doping levels since this would introduce more Auger recombination. The saturation current density J_0 at the front and the rear is in the range of 100–1,000 fA/cm^2 limiting the V_{OC} to around 660 mV [26].

4.3.2 PERC

To overcome the discussed recombination losses, the device structure was modified as illustrated in Figure 4.2(b). At the rear side of the cell (hole contact), a passivation layer is introduced and opened only locally for the highly recombinative metal contact. This adds several steps to the fabrication process. Thermally grown silicon oxide is a common passivation layer, but SiN_x is used very successfully at the front whereas Al_2O_3 is increasingly used at the rear since its negative resident charge provides a field effect passivation for the p-type bulk. The local openings at the rear still have a direct metal/wafer interface that is recombination active (can be improved by a locally formed BSF) but due to the reduced surface, the emitter saturation current density at the rear can be reduced to 35–100 fA/cm^2 [27,28] which makes the front the limiting factor to a maximum V_{OC} of 680 mV [26].

4.3.3 Silicon heterojunction solar cells: concept and status

Since the pioneering work of Taguchi and Tanaka on SHJ solar cells and developed by SANYO, (Japan) in the early 1990s [29], many groups have investigated this promising technology on both *n*-type and *p*-type silicon wafers [30]. In contrast to standard devices that are based on the diffusion of dopant atoms into the crystalline bulk silicon material to form the *p-n* junction, SHJ solar cells are based on the intrinsic and doped a-Si:H layers. Albeit this technology is already being commercialized by Panasonic (Japan), there are still various issues and phenomena that still remain unclear and the full potential of this technology is not completely exploited. Based on the results from Panasonic [31], which show maximum voltages close to the fundamental limit of c-Si (769 mV for a 100 μm wafer [32], or around 749 mV for a 110 μm wafer [33]), the main challenges now are the improvement of the transport properties (including current) while maintaining a high voltage value.

The SHJ conception has the potential for high efficiency at competitive production costs [34]. Recently, Kaneka reported a record efficiency of 26.7% for their standard both-side-contacted SHJ device [9,35], and an interdigitated back-contacted (IBC) device. On the other hand, with a PCE of about 25.6% [31], this IBC device surpassed the record PCE of 25% held until then by the University of New South Wales (UNSW) [36] for about 15 years. This result was obtained on the passivated emitter, rear locally diffused (PERL) solar cells. However, UNSW was the only one hitting a PCE of 25%. Indeed, SunPower commercialized an IBC solar cell and reached 25% on 5-in. wafers as well [37], idem for SHARP (25.1%) achieved however on lab-scale devices [38]. Nevertheless, even if both SHJ and PERL technologies are offering nowadays similar conversion efficiencies, they are showing fundamental differences. In order to maintain a high voltage, both technologies rely on high-quality surface passivation (to minimize the recombination of charge carriers). This, in fact, marks the most crucial difference. In standard cells—we use the PERL cell as reference [36]—the metallic contacts are coated at local openings of the anti-reflective coating (e.g. silicon nitride, SiN_x) and the passivation layer (e.g. silicon dioxide, SiO_2) to form a direct metal–semiconductor contact (see Figure 4.3(a)) (the technology of PERC will be detailed in a separate section). Yet this type of contact provides good transport properties, it is likely the most performance limiting aspect of this technology as it is recombination-active and hence significantly lowers the maximum voltage. Conversely, the benefit of the SHJ concept is the shift of the metallic contacts at a distance from the c-Si interface, hence limiting the recombination effect. This is carried out by inserting a stack of wide-bandgap intrinsic (*i*) and doped (*n* or *p*) a-Si:H layers, as well as a TCO layer—necessary for optical and electrical issues—between the c-Si absorber and the metallic contacts (see Figure 4.3(b)). This device architecture allows reaching very high voltages at open-circuit conditions, even close to the theoretical limit [31,39]. Recently, so-called passivating contacts [40] has emerged as the technology that combines the best of both worlds [37]. However, the gain in surface passivation is accomplished with the sacrifice of the current output [41]. When comparing the configuration of the two technologies as shown in Figure 3.3, this is easily understood. In fact, as the metallic electrodes at the front side directly contact the absorber via localized openings in the stack of highly transparent dielectrics (SiO_2 and $SiNx$)—allowing high passivation between the metal fingers and high light in coupling, the transport is not a limiting factor for the PERC architecture.

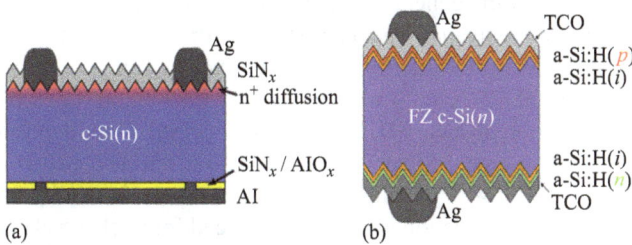

Figure 4.3   Comparison of the structure of (a) n-PERC and (b) SHJ solar cell. Adapted from Ref. [20] with permission from The Royal Society of Chemistry.

Conversely, the SHJ solar cell relies on transport through the intrinsic and doped a-Si:H layers capped with a TCO, which gives rise to parasitic absorption and challenges in carrier transport.

4.3.4   Configuration and fabrication

(a)   **Substrates and surface preparation**

Due to the low-temperature processing of SHJ technology and the fact that no processing-induced improvement in the bulk of the wafer can be expected from impurity gettering [42] or defect hydrogenation [43], the favorite material for SHJ technology is mono- rather than multi-crystalline silicon. High-quality material with millisecond lifetimes throughout the wafers should be used. Moreover, mono-crystalline wafers feature much better-defined surfaces, which may be crucial for conformal film deposition. Habitually, Si (100) substrates are used.

(b)   **a-Si:H film deposition**

Intrinsic amorphous hydrogenated silicon (a-Si:H/i) films are known since a long time for their good passivation property especially for the c-Si [44]. The majority of a-Si:H(i) films are grown by PECVD with silane (SiH_4) —usually diluted in H_2— as a precursor. For device-grade films, typical deposition temperatures and pressures are 200°C and 0.1–1 Torr, respectively. Other techniques have already been reported, including direct-current PECVD [45], hot-wire (also known as catalytic) CVD [46], electron cyclotron resonance CVD [46], and expanding thermal plasmas [47]. Hydrogenated amorphous silicon layers passivate c-Si surfaces mainly by hydrogenation of silicon dangling bonds, reducing thereby the interface defect density [44]. The carrier recombination via defects occurs through the Shockley–Read–Hall mechanism; however, the defect responsible for interface recombination is more likely to be the silicon dangling bond.

(c)   **Doped a-Si:H films**

To fabricate the heterojunction device, doped films (deposited in similar plasma systems) are required to form the emitter and back surface field. For p-type films, either Trimethylboron (TMB) or diborane (B_2H_6) is mixed in the SiH_4 gas flow, while phosphine (PH3) is used for n-type layers. Despite the fact that doped films generate a field effect at the interface with the wafer, their electronic passivation characteristics are often inferior to those occurred by intrinsic films [48]. Minority carrier lifetimes as high as 9 ms (at an excess carrier density of 10^{15} cm^{-3}) were obtained with a 15 nm-thick intrinsic film [49]. As a matter of fact, SHJ devices with doped films deposited directly either on n- or p-type c-Si surfaces (i.e. without intrinsic a-Si:H layer) were limited by their low V_{oc} values [46]. As the proportionality between the doping and defect formation is more pronounced for thick a-Si:H films [50], it is challenging to simultaneously fulfill the surface passivation and doping requirements. For this reason, a few-nanometer-thick intrinsic buffer layer is typically inserted between the c-Si surface and the doped a-Si:H films for device fabrication [51]. Figures 4.4 and 4.5 display a schematic of SHJ as

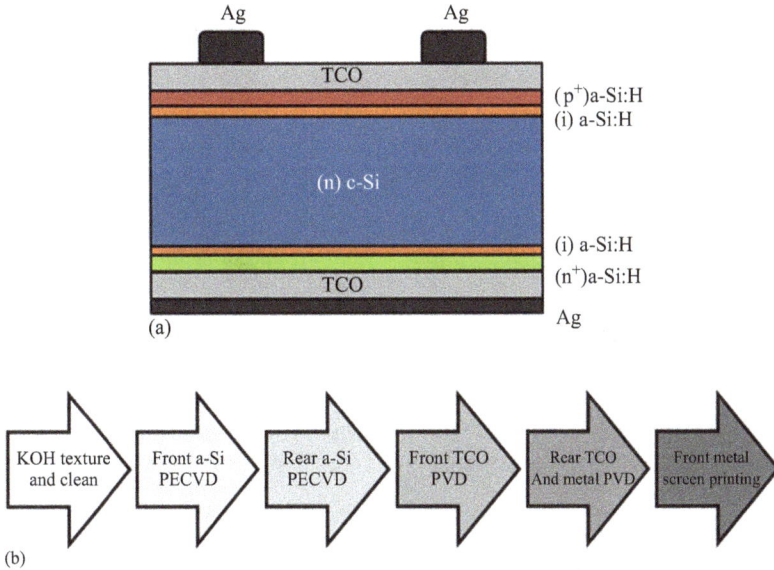

Figure 4.4 (a) Sketch of SHJ as first developed by Sanyo, Japan. The wafer is n-type. (b) Basic process steps for SHJ devices.

developed by Sanyo (Japan) along with the Basic process steps, and the band diagram of amorphous/crystalline SHJ device, respectively.

(d) **Transparent conductive oxide deposition**

As the lateral conductivity of doped a-Si:H layers is low, the front of SHJ devices must be coated with a transparent conductive oxide (TCO) layer to transport charge to the device terminals. The sheet resistance of the TCO must be sufficiently low (typically $R_{sh} < 100$ ohm/□) to avoid deteriorating *FF*. The front TCO also serves as an antireflection coating in SHJ devices and, with a refractive index of about 2 at 600 nm, its thickness is fixed at approximately 75 nm to minimize reflection losses. With thickness pre-determined and mobility limited by material choice, low R_{sh} can only be achieved by increasing the free carrier concentration N. Free carriers, how-ever, absorb parasitically in the IR, so that gains in *FF* are often balanced by losses in J_{sc} [52]. Moreover, further increasing in carrier concentration usually leads to a decrease in the carrier mobility [53]. Optimizing front TCO layers, as well as searching for high-mobility TCO materials, thus represents an important driving factor to further improve device performance.

(e) **Basic considerations**

However, due to the n-type nature of most of TCOs, the n-type and p-type contacts must act either as ohmic and band-to-band tunneling junctions, respectively [54]. The detailed properties of the interfaces, including the band offsets between the a-Si:H layers and the c-Si wafer, are critical for the carrier transport, as they influence band bending in the structure and the carrier

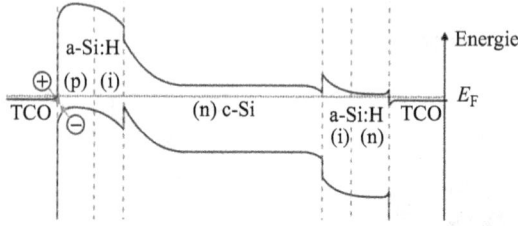

Figure 4.5   Band diagram of amorphous/crystalline silicon heterojunction solar cell

transport properties across the junction. As a matter of fact, for relatively dense a-Si:H films, the a-Si:H/c-Si conduction and valence band offsets are approximately 0.25 and 0.45 eV, respectively [55] and depend as well on the film's deposition conditions. For instance, the valence band offset increases linearly with the hydrogen content of the films, while the conduction band offset remains essentially fixed [55].

4.4   Device operation under various temperatures—experimental results

In this section, we will the review the behavior of different solar cell architectures under high temperature conditions. Figure 4.6 displays the schematic sketches of the different device architectures investigated in this study, namely, p-BSF, p-PERC, n-PERT, and Adv. n-PERT, n-hybrid and n-SHJ. Details about these architectures can be found in Ref. [20]. Two devices based on p-type c-Si are examined, namely BSF and PERC. All other devices are based on n-type c-Si. [20].

4.4.1   Comparison of different technologies

In Figure 4.7, the measured carrier lifetimes as a function of the minority carrier injection with respect to increasing temperatures ranging from 30°C to 120°C are presented for p-PERC, n-PERT and n-SHJ solar cells devices (Figure 4.7(a, b and c), respectively). An increase in the minority carrier lifetime with increasing temperature is observed in all passivated devices. As the temperature dependency of both Auger and radiative recombination in a silicon wafer is negligible in the investigated temperature range [56], changes in the minority carrier lifetime most likely come from the temperature dependency of Shockley–Read–Hall (SRH) recombination statistics in the bulk or at the surfaces.

Schmidt [57] attributed the lifetime increase to the temperature-dependent decrease of the hole capture cross-section of Al-related defects in the silicon bulk material. The temperature-dependent lifetime of silicon heterojunction passivating contacts was investigated by Seif *et al.* [58] who reported an increase in the carrier lifetime with a-Si:H(i/p) and a-Si:H(i/n) passivating layers, but a decrease in carrier lifetime with a-Si:H(i) passivation. However, the origin of the increasing charge carrier lifetime in silicon is not yet fully understood.

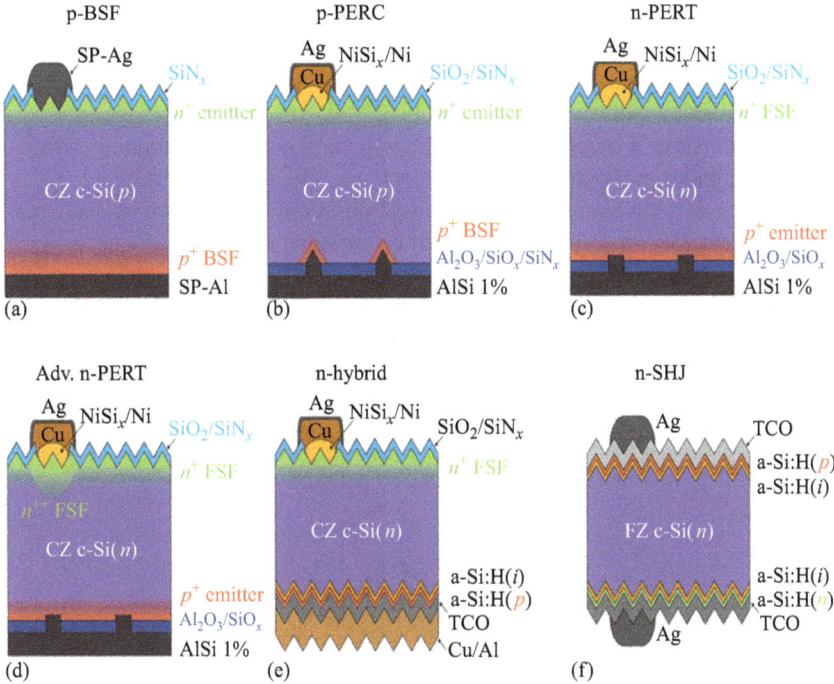

Figure 4.6   Schematic sketches of the different device architectures investigated in this study. Reproduced from Ref. [20] with permission from The Royal Society of Chemistry.

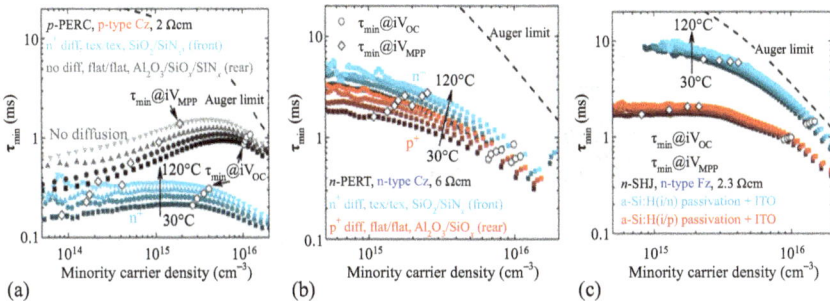

Figure 4.7   Injection-dependent effective minority carrier lifetime, t_{min}, of symmetrically passivated samples as obtained from temperature-dependent minority carrier lifetime measurements on a Sinton WCT-120TS setup. The lifetime curves indicate the passivation schemes in the investigated cells. (a) On p-type c-Si, homojunction passivation, (b) on n-type c-Si (homojunction), and (c) on n-type c-Si (a-Si:H(i/p) and a-Si:H(i/n) passivation). Reproduced from Ref. [20] with permission from The Royal Society of Chemistry.

In Figure 4.8, the temperature-dependent J(V) parameters of all investigated device architectures are shown at standard 1,000 W m^{-2}, with AM 1.5G irradiation. As can be seen, globally all the parameters of the cells linearly follow the temperature increasing. However, for FF and PCE (η%), this is not really the case in the two cell architectures incorporating a silicon heterojunction passivating contact—this effect is more pronounced for the n-SHJ cell. The non-linearity of FF and η% as a function of temperature was previously observed and noticed in such passivated contacts [59] and is usually observed in solar cells devices incorporating thermionic barriers.

Table 4.1 summarizes the relative TCs of all the investigated cells technologies. Relative TCs were normalized to the value at 25°C. The TC$_{Jsc}$ was found to be similar for all investigated cells and has a positive value. Indeed, the J_{sc} increases with respect to the temperature due to the well-known reduced band-gap of silicon at higher temperatures [60] and the associated enhanced absorption of infrared photons [61].

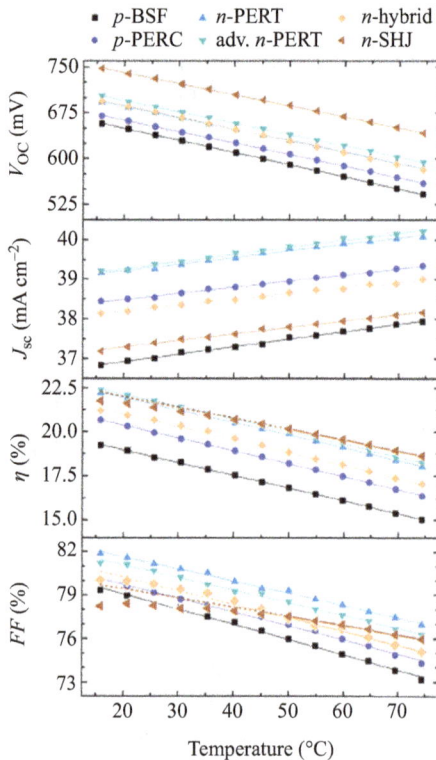

Figure 4.8   Temperature-dependent J (V) parameters of the investigated solar cell architectures. Reproduced from Ref. [20] with permission from The Royal Society of Chemistry.

Table 4.1 Relative TCs at AM 1.5G irradiance of 1,000 W/m² of the devices shown in Figure 4.8, derived from linear fitting between 25°C and 75°C of the temperature-dependent J(V) parameters shown in Figure 4.8. Reproduced from Ref. [20] with permission from The Royal Society of Chemistry

Architecture	TC_{Voc} (%/°C)	TC_{Jsc} (%/°C)	TC_{FF} (%/°C)	TC_{PMPP} (%/°C)	TC_{RMPP} (%/°C)
p-BSF	−0.31	0.05	−0.14	−0.39	−0.39
p-PERC	−0.29	0.04	−0.12	−0.36	−0.37
n-PERT	−0.28	0.04	−0.11	−0.33	−0.34
Adv. n-PERC	−0.27	0.04	−0.11	−0.33	−0.33
n-hybrid	−0.28	0.04	−0.12	−0.35	−0.33
n-SHJ	−0.25	0.04	−0.08	−0.29	−0.30

Table 4.2 Data of commercial modules taken from module data sheets. Reproduced from Ref. [20] with permission from The Royal Society of Chemistry

Manufacturer	Architecture	TC_{Jsc} (%/°C)	TC_{Voc} (%/°C)	TC_{FF} (%/°C)	TC_{PMPP} (%/°C)
SolarWorld	p-BSF	0.04	−0.30	−0.17	−0.43
Trina	p-BSF	0.05	−0.32	−0.14	−0.41
SolarWorld	p-PERC	0.04	−0.30	−0.15	−0.41
Q CELLS	p-PERC	0.04	−0.29	−0.15	−0.40
LG	n-PERT	0.03	−0.28	−0.13	−0.38
Panasonic	n-SHJ	0.03	−0.25	−0.08	−0.30

Tables 4.1 and 4.2 show the relative temperatures coefficients calculated at AM 1.5G irradiance of 1,000 W/m² of the devices shown in Figure 4.6 and commercial ones taken from module data sheets. The technology with the highest TC_{PMPP} was found to be the n-SHJ architecture which is associated to its higher TC_{Voc} and TC_{FF}. For TC_{RMPP}, the value of the n-SHJ cell is also the most favorable. Maintaining a higher characteristic resistance at MPP with increasing temperature helps to avoid the detrimental influence of series resistance.

In sum, and as can be seen, SHJ technology shows the lowest TC among the others (both in Lab and/or commercial modules) which highlight the suitability of this architecture to resist to the degradation induced by temperature increasing.

4.5   CIGS and CdTe thin film solar cells

Since their discovery in the early 1960s, heterojunction thin film solar cells (HTFSC) have often been considered as a promising photovoltaic technology for low cost production of electricity, due to their low material consumption, large flexibility in terms of processes and device configuration, high efficiency and long-term performance stability. Two HTFSC technologies have been so far mature enough to reach the commercialization stage: the Cu (In,Ga)(S,Se)$_2$ (also called CIGS) based solar cells and the CdTe-based solar cells. Although today's world-wide PV market is largely dominated by the crystalline silicon technologies, the HTFSC technologies were representing about 4% of 415 GW installed worldwide in 2017 [62]. The installed annual capacity of CdTe PV modules in 2017 was about 2.2 GW (with First Solar, USA, being the major manufacturer of such modules), while about 1.8 GW of CIGS PV panels were installed that year (mainly manufactured by Solar Frontier, Japan and Hanergy, China/Germany/USA) [62]. Improvement in power output, module production methods and design, and new possibilities for HTFSC device applications are reaching the market every year. Indeed, due to their lightweight and their ease to be processed on different type of substrates, HTFSC can be easily integrated in transportation systems (bicycles, cars, buses, etc.) or building with unusual, curved shape (stadium, etc.) [63,64]. Off-grid products and materials (such as awnings, tents, or jackets) is another possible niche application that was developed for this PV technology [65]. Their competitive cost makes them also suitable for larger application, such as solar farms. As an example, Topaz solar farm and the Desert Sunlight Solar Farm, two of the largest solar farms in the world located in the desert of California, USA, generating annually about 1,300 GWh each, includes respectively about 9 million and 8.8 million CdTe modules from First Solar [66,67]. In Indiana, USA, over 84,000 CIGS modules from Solar Frontier were installed on the Pastime and McDonald solar power projects [68].

   In the present part, the device configuration of both CIGS and CdTe solar cell technologies are introduced, followed by the fabrication aspect of these devices. Finally, the performance of CIGS and CdTe solar cells under standard and high temperature conditions will be compared.

4.5.1   Device configuration

The CIGS and CdTe-based PV technologies adopt the typical structure of p–n junction-based thin film solar cells [69], minimally composed of a back contact, a thin film absorber and a transparent front selective electrode (often called window layer) (see Figure 4.9). In such structure, the absorber layer is a direct bandgap semiconductor (typically p-type) with a high optical absorption coefficient ($\alpha > 10^5$ cm^{-1}), capable of absorbing most of the incident light with a thickness of only a few microns. This absorber converts the incident photons into electron–hole pairs, that are further separated by the p/n junction formed at the absorber/window layers interface. The window layer is typically a highly n-type doped oxide semiconductor

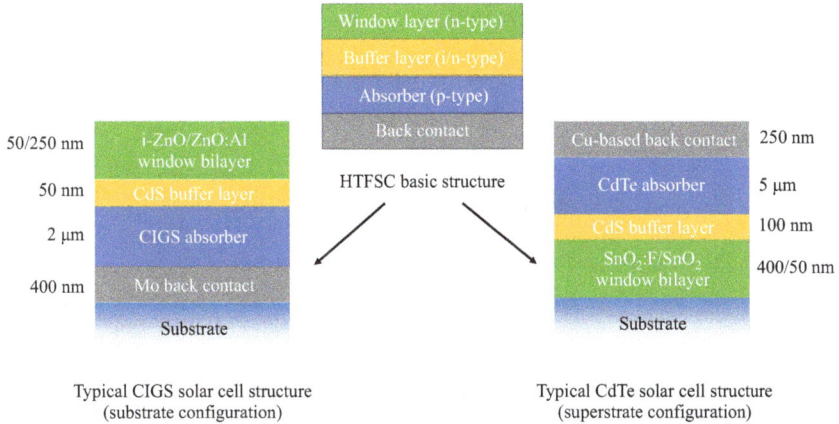

Figure 4.9 contains diagrams labelled:

HTFSC basic structure:
Window layer (n-type)
Buffer layer (i/n-type)
Absorber (p-type)
Back contact

Typical CIGS solar cell structure (substrate configuration):
50/250 nm — i-ZnO/ZnO:Al window bilayer
50 nm — CdS buffer layer
2 μm — CIGS absorber
400 nm — Mo back contact
Substrate

Typical CdTe solar cell structure (superstrate configuration):
Cu-based back contact — 250 nm
CdTe absorber — 5 μm
CdS buffer layer — 100 nm
SnO$_2$:F/SnO$_2$ window bilayer — 400/50 nm
Substrate

Figure 4.9   Typical structures of CIGS and CdTe heterojunction thin film solar cells

thin film, with a band gap wide enough to transmit most of the incident light to the absorber. Using such ideal absorber/window design, the carrier generation is mainly located at or close to the space charge region, and the carrier collection is accomplished by drift, or field-assisted collection. As the recombination rate at the heterojunction due to structural defects can severely affect the efficiency of the device, an additional so-called buffer layer is usually added to the structure to further improve the quality of the absorber/window interface. Although the structure of these solar cells is continuously under improvement, the basic solar cell configurations that have led to the high-efficiency device technology of today are as shown in Figure 4.9.

CIGS solar cells typically adopt a so-called substrate structure (i.e. the substrate used to process the solar cells is the back side of the future panel). For CIGS solar cell, a 400–600 nm Mo thin film is used as back contact, as this material has the advantage to have a high stability at high temperature and a low resistivity [70,71]. The absorber is typically made of a 1–3 μm thick, polycrystalline chalcopyrite CIGS thin film. One of the advantage of such material is the possibility to finely tune the bandgap of the absorber by playing on the [Ga]/([Ga]+[In])(GGI) ratio and [S]/([S]+[Se]) (SSSe) ratio while keeping the chalcopyrite structure. The optimal bandgap for high efficiency CIGS solar cell was found to be about 1.1–1.2 eV, which typically correspond to a GGI ratio of 0.2–0.3 for a pure selenide compound [72,73]. Interestingly, CIGS absorbers can be prepared with a relatively wide tolerance to stoichiometry variation (particularly regarding the [Cu]/([In]+[Ga]) (CGI) ratio, that can easily vary from 0.8 to nearly 1, due to the presence of compensating defects) [74]. Another interesting aspect of polycrystalline CIGS solar cells is that the grain boundaries in the absorber have usually a beneficial role on the electrical properties of the absorber, and their properties can be tuned depending on their composition [75]. It should be noted that the free surfaces of

CIGS films are typically slightly Cu-poor, with a composition close to $CuIn_3Se_5$ phase; this top layer is often called surface defect layer (SDL) [76]. A 50-nm thin CdS layer is then typically used as buffer layer [77]. Although the band gap of CdS (2.4 eV) is not ideal in such structure as blue light photons (<520 nm) can be partially absorbed by the buffer layer before to reach the CIGS absorber, this material has led so far to best performance in terms of efficiency and reproducibility [78]. The window layer is typically composed of a 50–100 nm ZnO (3.3 eV)/ 250–600 nm aluminum doped zinc oxide (AZO) (3.4 eV) bilayer. The typical resistivity values of these layers are 1–100 Ω cm and 10^{-4}–10^{-3} Ω cm, respectively (with an overall transparency above 90%). The resistive ZnO layer helps to prevent the formation of shunt paths within the solar cell structure [79].

CdTe solar cells typically use the superstrate structure (i.e. solar cell configuration where the glass substrate is not only used as supporting structure but also as window for the illumination and as part of the encapsulation). In this structure, the window layer is usually a 200–500 nm AZO, indium tin oxide (ITO), fluorine-doped tin oxide (FTO), or cadmium tin oxide (CTO) transparent conductive layer, with a low resistivity (i.e. sheet resistance below 15 $\Omega/\square$). A 50–100 nm highly resistive transparent (HRT) layer such as ZnO or SnO_2, with an electrical resistivity in the range of 10^2–10^5 Ω cm, can be added on top of the conductive one to prevent any shunting issue through micro-pinholes [80]. The thin n-type CdS layer (50–100 nm) improves the growth and electrical properties between the TCO and the absorber [81]. The CdTe absorber is typically of 2–8 μm. CdTe is a direct bandgap material (1.45 eV), close to the optimal bandgap value for sunlight conversion into electricity using a single junction device. Although the absorber used in such structure is p-type, the conductivity type of CdTe can be changed from p- to n-type by varying the Cd–Te stoichiometry (i.e. n-type for Cd excess, p-type for Te excess) [82]. As for CIGS solar cells, the absorber is the most important part of the solar cell structure as the crystal properties affect both the carrier lifetime and concentration, influencing the performance of the devices. The final part of the device is the back contact, usually made of a bilayer, the first one containing Cu (such as Cu/Au [83], Cu/graphite [84], etc.). The diffusion of Cu through the CdTe layer was found to dope the absorber and improve the overall performance of the device [85].

Due to their similar structure, the efficiency of both CIGS and CdTe solar cells was over the past few years greatly improved using similar modifications in the standard solar cell structures and in the material properties, such as:

1. *Wide bandgap buffer layer*: Although the CdS buffer layer is still used as standard in most research laboratories working on CIGS and CdTe solar cells, the low bandgap of CdS has been considered since a long time as a limiting factor in terms of efficiency; a high number of alternative n-type Cd-free materials with wider bandgap have been tried over the past 20 years for both systems, such as ZnS, Zn(O,S), (Zn,Mg)O, In_2S_3, and ZnSe, that were leading most of the time to lower efficiencies/reproducibility than the CdS reference [78,86]. Recently, remarkable improvement have been made for CIGS solar

cells with efficiencies up to 22.8% obtained using an alternative Zn(S,O,OH)/ (Zn,Mg)O bilayer to replace the traditional CdS/ZnO one [87]. The loss in FF usually observed when replacing CdS by Zn(S,O,OH) is compensated by the increase in J_{sc} due to the reduction of absorption losses in the buffer/window layer. In the case of CdTe solar cells, the thinning of the CdS layer was used to reduce the absorption of blue photons in the buffer, as no alternative buffer layer was found so far to be more efficient than CdS [81].

2. *Graded bandgap absorber*: A graded bandgap composition can be used to improve the electronic properties of the absorber by reducing the recombination losses at the interfaces. For CIGS solar cells, a wide bandgap is located at both side, with a narrow bandgap located at the "notch" point (i.e. V-shape gradient); the wider bandgap located at the front interface would improve the V_{oc} (mainly by improving the band alignment with the buffer layer) while the increased bandgap at the back side improves the carrier collection (J_{sc}) and decrease the recombination at the absorber/back contact interface (i.e. back surface field (BSF) effect); this gradient is usually created by controlling the Ga content trough the CIGS absorber during the process [88]; for CdTe solar cells, a linear gradient can be obtained using Se diffusing from an intermediate CdSe buffer layer at the surface of the absorber, to form the lower bandgap Cd (Te,Se) alloy [89]; this gradient was found to increase the carrier lifetime and thereby the V_{oc} of the devices. A BSF near the absorber/back contact interface can also be created using an intermediate layer of ZnTe, having a higher bandgap than CdTe [90].

3. *Third atoms and post-treatments*: Since a long time, post-treatment have been applied to the thin film chalcogenide absorber to improve their electrical properties. That is typically the case of the $CdCl_2$ treatment applied to CdTe thin film after their deposition, to improve the crystallinity of the absorber by recrystallizing the nanograins (if present) and removing the structural defects in the absorber and at the CdTe/CdS interface [91,92]. More recently, alkali (first Na, then K, Rb, and Cs) post-treatment have been applied to CIGS absorber (though various incorporation strategies), leading to the passivation of defects at the absorber surface and at the grain boundaries [93,94]. By decreasing the concentration of compensating donor defect in the materials, the p-type carrier concentration increases, leading to a lower FF and therefore higher V_{oc} and FF.

4.5.2 Device fabrication

One of the main advantages of HTFSC is the ability of being deposited with the wide diversity of deposition techniques available for thin films [95]. Although in practice a limited number of techniques lead to highly efficient devices, relatively low cost techniques (such as solution processing) are still being explored with the hope to further reduce the production costs of HTF modules [96]. Another advantage of CIGS and CdTe solar cells is that they can be processed in a less controlled environment than typically silicon and perovskite solar cells (that requires a clean room and a glove box, respectively). Thin-film PV technologies have also the advantageous possibility of using monolithic integration for series connection of individual cells within a module. Compared to Si solar cells that must have a front metal grid for solar cell interconnections, the integrated interconnect scheme allows to avoid the use of grids by connecting directly the TCO of one cell with the back contact of the next one.

The processing of CIGS solar cells starts by the selection of the substrates. Soda lime glass substrate is commonly used as it can handle temperature up to 550°C, being therefore compatible with the temperature typically used to prepare CIGS absorber. Flexible substrates (such as aluminum foil, polymers, thin stainless steel sheets, etc.) have been also successfully investigated [97] (PCE up to 20.4% on flexible polymer substrate [98]), although their use reduces the processing temperatures of the solar cells. The deposition of the Mo back contact is commonly done by DC sputtering. The control of the sputtering pressure allows to tune the conductivity the back contact and its permeability to impurities diffusing from the substrates [99]. Extra layers (such as MoNa layer for the supply of Na [100], or impurity-blocking layer, such as SiN and TiN [101]) can also be deposited before, in sandwich or after the back contact using a similar technique. As previously mentioned, the CIGS thin film can be deposited using a wide range of thin film deposition techniques [102,103], either referred as one-step techniques when all the elements are deposited at the same time at high temperature (using e.g. co-evaporation [104] or reactive sputtering [105]) or as two-step processes, where a thin film precursor is first deposited on the back contact (using for instance sequential sputtering [106], solution processing [107], electrodeposition [108], etc.) and then annealed at high temperature under Se and/or S-based atmosphere. The typical annealing temperature used to process CIGS thin films is in the 400–600°C range. The record efficiencies for CIGS solar cells have all been obtained using vacuum-based techniques (using either one or two step processes), since they allow a nice control of the absorber composition as well as of the impurities content compared to solution-based deposition methods. Using such techniques, the formation of compositional gradient can easily be done by tuning the supply of Ga (or S) during the growth of the absorber [88]. Some of the non-vacuum-based techniques have also reached relatively high efficiency (up to 17.3% for the solution-based precursor using hydrazine-based solution [96]). It should be noticed that during the deposition of the absorber, a $MoSe_2$ thin film forms at the Mo surface, with a thickness depending on the processing conditions of both the back contact and the absorber [109]. This $MoSe_2$ layer (1.3 eV, p-type) acts as low-recombinative back surface for the photo-electrons in the $Cu(In,Ga)Se_2$ absorber and at the same time provides a low-resistance contact for the majority carriers [110]. As previously mentioned, alkali element can be used to improve the properties of the absorber. These alkali element naturally diffuses from the glass substrates during the processing of the absorber at high temperature. Their supply can alternatively be done by evaporation of the alkali source (typically NaF, KF, etc.) after the growth of the absorber, during a post-treatment at high temperature, under Se atmosphere to avoid the formation of Se vacancies [98]. Prior to the CdS deposition, a KCN etching wet treatment can be performed on the CIGS absorber to remove any traces of highly conductive CuxSe secondary phases [111]. Safer alternative to the KCN etch have been recently developed [112]. Various deposition techniques can also be applied to deposit the buffer layer. The CdS buffer layer is traditionally deposit by chemical bath deposition (CBD), using a solution containing a Cd salt, thiourea, and ammonia heated at 60–80°C [113]. Alternative buffer layers can also be done using the same process (such as Zn(O,S)) [114]. Alternative deposition techniques that can allow a nice coverage of the absorber surface without damaging it, such as ALD, ILGAR, and sputtering, have been also investigated for the same materials and other alternative materials ((Zn,Mg)O, (Zn,Sn)O, etc.) [78]. Then the deposition of the ZnO/AZO bilayer is commonly done by RF sputtering under Ar plasma [79]. The solar cells stack is often completed by a 50-nmNi/Al grid prepared by evaporation, to easily probe the electrical properties of the solar cells without damaging the device.

The CdTe solar cell fabrication technology seems to be more suitable for large-scale production compared to CIGS, as all the components of the device can be processed using deposition techniques that are fast, reproducible and homogeneous over large areas. In the superstrate configuration, the first layer to be processed on the substrate is the TCO/HRT bilayer, such as FTO/SnO_2 (typically done by CVD) [115], AZO/-i-ZnO (usually done by RF sputtering) [116], etc. As this layer is the first to be deposited in this configuration, the chemical and physical stability of the TCO during the processing of the upper layers is an important parameter that determines the choice of the material and techniques to be used. CdS can be deposited with a wide range of deposition methods [117–119], on a laboratory scale, the highest efficiency has been obtained by CBD but it has been successfully deposited also by close-space sublimation (CSS) technique (CdS:O) or by sputtering (CdS:F). Among the diverse techniques that can be used to growth the CdTe absorber, CSS of a CdTe target under inert atmosphere is the most popular technique since it allows high substrate temperatures and high deposition rate [120]. At the industrial level, this method can be adapted in the form of vapor transport deposition (VTD) [121]. Here the major experimental parameter to control is the substrate temperature that strongly affect the grain growth. Other techniques have also been developed for the growth of CdTe absorber, such as vacuum evaporation [122], atomic layer epitaxy [123], electrodeposition [124], screen-printing [125], and sputtering [126]. An important step in the CdTe-based solar cell fabrication process is the treatment of the CdTe film in presence of Cl [127]. It is generally accepted by the scientific community that the Cl-treatment increases the grain size of CdTe, passivates the grain boundaries, and promotes a mixing between CdS and CdTe at their interface. This treatment can be done by dipping the CdTe film in a saturated $CdCl_2$–methanol solution to deposit a $CdCl_2$ film on top of CdTe, or by evaporation of a thin film of $CdCl_2$ [128]. After the $CdCl_2$ deposition, the sample is annealed in air at about 400°C for 20 min. In order to remove the $CdCl_2$ residuals from the CdTe surface, it is needed to make an etching in a Br–methanol solution, or in a mixture of nitric and phosphoric acids. Alternatively, the $CdCl_2$ treatment can be done using chlorine containing gases [129], or other chlorine carriers such as

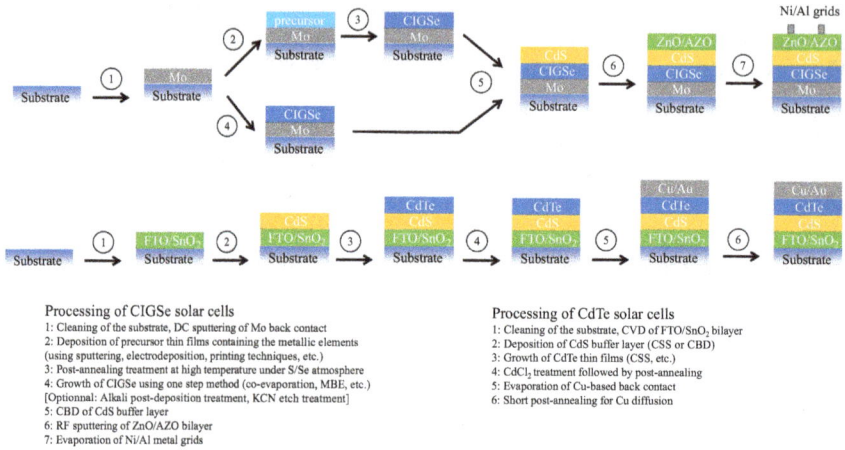

Processing of CIGSe solar cells
1: Cleaning of the substrate, DC sputtering of Mo back contact
2: Deposition of precursor thin films containing the metallic elements
(using sputtering, electrodeposition, printing techniques, etc.)
3: Post-annealing treatment at high temperature under S/Se atmosphere
4: Growth of CIGSe using one step method (co-evaporation, MBE, etc.)
[Optionnal: Alkali post-deposition treatment, KCN etch treatment]
5: CBD of CdS buffer layer
6: RF sputtering of ZnO/AZO bilayer
7: Evaporation of Ni/Al metal grids

Processing of CdTe solar cells
1: Cleaning of the substrate, CVD of FTO/SnO$_2$ bilayer
2: Deposition of CdS buffer layer (CSS or CBD)
3: Growth of CdTe thin films (CSS, etc.)
4: CdCl$_2$ treatment followed by post-annealing
5: Evaporation of Cu-based back contact
6: Short post-annealing for Cu diffusion

Figure 4.10   Typical CIGS and CdTe solar cells processing baselines at the laboratory scale

MgCl$_2$ [130]. To complete the fabrication of the device, the stack is completed with the Cu-based back contact. Evaporated ZnTe:Cu/Au contact (further activated with a rapid thermal processing step to diffuse the Cu atoms into the absorber) is widely used at a laboratory scale, together with MoOx [131,132].

At the module level, further processing steps are needed to complete the device, such as the patterning for the formation of the monolithic interconnections, the optional deposition of large area grids, and the encapsulation of the module. The monolithic interconnection is usually obtained through three different period-ical patterning steps [133]. For example, for CIGS solar cells, these steps are: (1) patterning of the Mo using a IR laser, (2) mechanical or laser scribing patterning after the absorber and buffer layer deposition, and (3) final patterning step after the TCO deposition. The total width of the interconnects and the spacing between the periodical scribes depends on the processing techniques and on the module output requirements. Module encapsulation is important because the stability of the module depends on proper protection against humidity. Typically, the encapsula-tion is done between two glass sheets, using EVA as an encapsulant and desiccant type tape as an edge sealant (to block the moisture ingress). Recently, the use of metal grids deposited on the TCO layer has been investigated, to reduce the para-sitic resistive losses in the TCO at the module level [134].

4.5.3   Device performances

4.5.3.1   Under standard conditions

The evolution of the performances of CIGS and CdTe solar cells is depicted in Figure 4.11 [135–137]. Bonnet and Rabnehorst were the first to report a 6% efficiency CdTe/CdS thin film solar cells in 1972, while Kamerski *et al.* developed the first thin film CIGS solar cell in 1976 that reached 4.5%. Since then, the improvements of the materials and interfaces quality based on empirical findings have been implemented on the solar cell processing baseline, leading to a step-by-step increase of the efficiencies of both technologies over the past 40 years [136,137]. A summary of the best certified device performances measured under the global AM1.5 spectrum (1,000 W/m^2) at 25°C for both CIGS and CdTe-based solar cells and modules can be found in Table 4.3 [138–144]. Certified efficiencies up to 22.9% for CIGS (V_{oc} = 744 mV, J_{sc} = 38.77 mA/cm^2, FF = 79.5 %) and

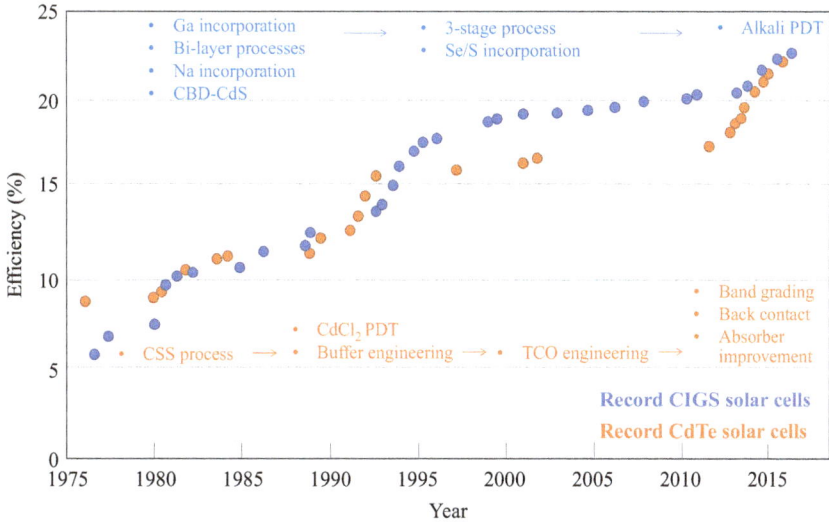

Figure 4.11   *Evolution of the performances of CIGS and CdTe solar cells over the last decades (adapted from [135])*

Table 4.3   *Record power conversion efficiencies obtained for CIGS and CdTe PV technologies at the solar cell and module levels*

Device (area)	V_{oc}	I_{sc} (or J_{sc})	FF	Efficiency	References
CIGS cell (1.041 cm^2)	744 mV	38.77 mA/cm^2	79.50%	22.90%	[138]
CdTe cell (1.0623 cm^2)	875.9 mV	30.25 mA/cm^2	79.40%	21.00%	[139]
CIGS cell (1 cm^2)—not certified	738 mV	39.77 mA/cm^2	79.45%	23.30%	[140]
CdTe cell (0.4798 cm^2)—not certified	887.2 mV	31.69 mA/cm^2	78.50%	22.10%	[141]
CIGS module (841 cm^2)	48.0 V	0.456 A	73.70%	19.20%	[142]
CIGS large module (9703 cm^2)	28.24 V	7.254 A	72.50%	15.70%	[143]
CdTe large module (7038.8 cm^2)	110.6 V	1.533 A	74.20%	18.60%	[144]

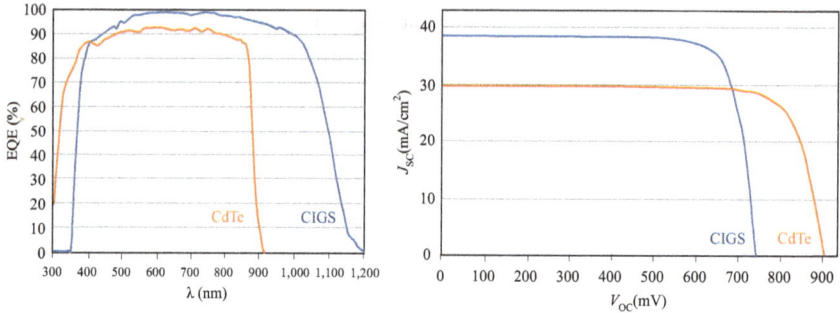

Figure 4.12   EQE and IV curves of the certified record CIGS and CdTe solar cells (adapted from Refs [138,139])

21.0% (V_{oc} = 875.9 mV, J_{sc} = 30.25 mA/cm^2, FF = 79.4%) for CdTe have been obtained on at the solar cell level. Both CIGS and CdTe solar cells are usually characterized by high external quantum efficiency (EQE), leading to relatively high J_{sc} values (corresponding to the integral of the product of the illumination spectrum and the EQE) (see Figure 4.12). The EQE, and therefore the J_{sc}, are mainly controlled by the band gap value of the absorber (that can vary in the case of CIGS depending of the composition of the material), the absorption and band gap of the buffer and window layers, the optical losses (front surface reflection—that can be improved using an evaporated MgF$_2$ anti-reflective layer- and shadowing from collection grids) [133]. Electronic losses (i.e. recombination losses) also affect the J_{sc} and the V_{oc} of the solar cells. In HTFSC as in any inorganic solar cells, the V_{oc} value is proportional to the band gap value (E_g) of the absorber, but is severely affected by the predominant recombination mechanism, the minority carrier lifetime and the doping of the absorber [145]. Factors that influence the FF values of CIGS and CdTe solar cells are the effects from series and shunt resistance, the voltage bias dependence of the current collection, spatially inhomogeneous recombination properties, and unfavorable band offset conditions at the interface [146].

Although these record efficiencies HTFSC are similar to the ones obtained on mc-Si based solar cells [147], the device operation is relatively more complex and can be influenced by a high number of parameters that makes the reproducibility of these results very challenging. As previously discussed, the processing conditions of the solar cells greatly impact the band diagram of HTFSC devices. Figure 4.13 shows the typical possible recombination paths in typical CIGS and CdTe solar cells, i.e. at the interfaces, in the space charge region or in the neutral region [148,149]. The main recombination mechanism is usually the SRH mechanisms, but it can be enhanced by tunneling due to the presence of high electric field. Slight changes in the band diagram (such as the modification of the value of the valence or conduction bands discontinuity at the interface, and the presence of gradient) will directly influence the predominant recombination mechanisms for the photo-generated charge carriers [146]. For this reason, the CIGS and CdTe solar cell performances are greatly influenced by the properties (e.g. chemical composition,

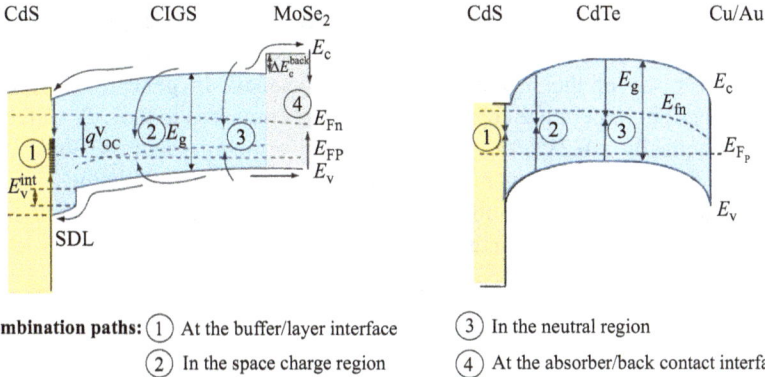

Recombination paths: ① At the buffer/layer interface ③ In the neutral region
② In the space charge region ④ At the absorber/back contact interface

*Figure 4.13 Possible recombination paths in typical CdS/CIGS and CdS/CdTe
solar cells; note that these band diagram have been simplified. With
E_g, the band gap of the absorber, E_v the maximum energy level of the
valence band, E_c the minimum energy level of the conduction band,
E_{fn} and E_{fp} the quasi fermi level for the electrons and the holes,
respectively.*

morphology, thickness, etc.) of each single layer composing the solar cell
structure [150].

One typical feature of CIGS and CdTe solar cells is the cross-over observed
between the dark and light *I–V* curves, usually explained by the presence of photo-
sensitive electrical defects. Another interesting feature of HTFSC is their
metastable electrical performances. The increase in V_{oc} (and sometimes FF and J_{sc})
in CIGSe and CdTe solar cells during illumination at RT is a commonly observed
phenomenon, often called light soaking effect [151]. Although the overall understanding of such metastable behavior is still incomplete, they seem to be linked to
the concentration of point defects in the devices, that can capture/release the photogenerated carriers depending on the analysis conditions and lead to persistent
photoconductivity in the active layers. Consequently, the measured CIGS and CdTe
device performances can vary greatly depending on preconditioning procedures.

4.5.3.2 Under high temperature

Like all other solar cells technologies, the performances of HTFSC are sensitive to
temperature. Increase in temperature typically reduces the bandgap of a semiconductor, thereby affecting most of the solar cell's parameters. The study of the
electrical behavior of solar cells with temperature is important as, in terrestrial
applications, they are exposed to huge temperatures variations from summer to
winter; the operating temperature of PV panels in desert environment can easily go
up to 70°C.

The effect of the temperature on the electrical performances and reliability of
CdTe and CIGS-based solar cells have been investigated both theoretically and
empirically [152–154]. In most cases, the increase of the temperature of the device

induces a severe decrease of the V_{oc}, a slight decrease of the FF values and a short increase of the J_{sc} of HTJSC due to the decrease of the bandgap, leading to a general decrease of the power conversion efficiency. In general, HTFSC have lower values of temperature coefficient for power compared to the c-Si wafer-based module (typically $-0.21\%/°C$ measured on CdTe module, and $-0.36\%/°C$ measured on CIGS module, compared to $-0.48\%/°C$ on c-Si module in the 25°C–60°C range); the reduced V_{oc} loss observed at high temperature in the case of HTFSC seemed correlated with the increased bandgap value at room temperature of the absorber compared to c-Si [155].

Despite the variation of temperature, CIGS and CdTe-based modules show an excellent long-term stability, as their temperature corrected efficiency (i.e. corrected with the temperature coefficient of the module) is very stable [153]. Interestingly, in the case of CIGS solar cells, the temperature dependency of the V_{oc} was found to be dependent on its value at room temperature (a high V_{oc} at 25°C led to a slower loss of V_{oc} when the temperature increases); furthermore, it was observed that the dependence of the J_{sc} to the temperature can be influenced by the processing conditions of the solar cells (that impact the main recombination mechanism in the device), either being positive or negative (see Figure 4.14 as example) [156]. As a general comment, the behavior of the electrical parameters of

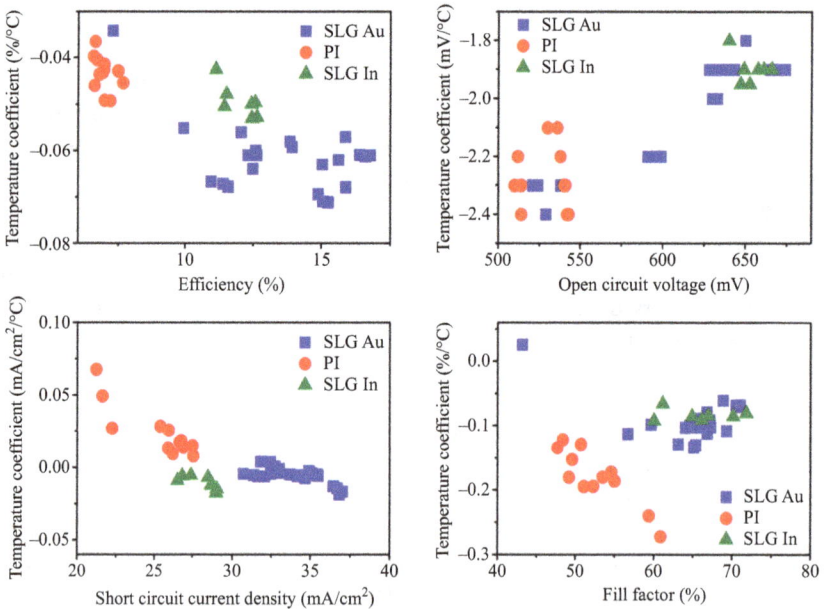

Figure 4.14   *Absolute temperature coefficients of the efficiency, open circuit voltage, short circuit current density, and fill factor as a function of these parameters at room temperature for CIGS solar cells processed using different conditions. Reproduced with permission from Ref. [156].*

CIGS and CdTe solar cells as a function of the temperature is greatly impacted by the processing conditions of the device.

This can be also observed in electrical measurements done at low temperature, usually performed on the device to get some insights on the carrier transport mechanism in the device [133]. For example, current–voltage measurements at low temperature performed on CIGS or CdTe solar cells give access to the V_{oc} trend at low temperature and are usually used to determine the predominant recombination mechanism in the device through the extrapolation of V_{oc} at $T = 0$ K (i.e. an activation energy close to the bandgap value is typically synonym of recombination taking place in the bulk of the absorber, while lower activation energies are expected for recombination mainly happening at the buffer/absorber interface) [157]. Photoluminescence and capacitance–voltage measurement performed at low temperature also give some indication on the nature of the point defect present in the device and on the carrier distribution [158,159].

4.5.4   Device reliability

Compared to Si-based solar cells, the long-term performance of CIGS and CdTe module is relatively hard to assess through accelerated lifetime tests (ALTs) [160]. The literature revealed that a positive or negative outcome of these tests (typically the IEC 61646 standard test) is not necessarily reflecting the aging of the reliability of the modules in real test conditions, mainly due to the limited comparison made between field experience and indoor tests [161–163]. Furthermore, some of the performance changes are reversible and some are not, which makes deployment, testing, and energy-yield prediction more challenging. Nevertheless, certain key factors, such as humidity, have been found to be clearly detrimental for non-encapsulated CIGS and CdTe solar cells [164,165].

Table 4.4 shows the degradation rate of few examples of CdTe and CIGS modules installed in different field locations [166–175]. The tested modules vary from stable (no degradation after several years) to very vulnerable to outdoor

Table 4.4   Examples of degradation rates (assuming a linear decrease of the output) of CdTe and CIGS PV module arrays installed in different field locations

Panel type (manufacturer)	Period	Location	Degradation rate (rel %/year)	References
CIGS (Wurth)	2006–2011	Nicosia, Cyprus	1.9–2.4	[166]
CIGS (Shell Solar)	2006–2011	Colorado, USA	0	[167]
CIGS (ZSW)	2003–2007	Widderstall, Germany	0.2	[168]
CIGS (Siemens Solar)	1988–2006	Colorado, USA	−0.2 to 1.7	[169]
CIGS	2001–2003	South Africa	8.1	[170]
CdTe (First Solar)	2005–2011	Morelos, Mexico	0.76	[171]
Cd Te	1995–2000	Colorado, USA	0.6	[172]
Cd Te	2003–2006	Tucson, USA	0.4–0.8	[173]

exposure. In most cases, the efficiency decrease was explained by the decrease of the FF value, while a minimal deterioration of the V_{oc} and I_{sc} could be observed. Unfortunately, most references only report the changes in electrical parameters without analysis of the modification of the material properties, although the disparity of the lifetime results show that the prediction of the stability cannot be done without a fundamental understanding of the degradation mechanisms in the solar cell structure.

The research done on the failure mechanisms and degradation process occurring in such type of solar cells and modules have revealed the following points [176,177]:

- On the solar cell level, cell degradation mechanisms specific to HTFSC have been observed, such as the increase of recombination at the junction (due to diffusion of impurities and/or electromigration), the shunting (due to the diffusion of metals, impurities, etc.), the increase in series resistance by contact degradation (due to corrosion or diffusion), and the delamination of the back contact (due to lamination stresses).
- On the module level, some specific degradation mechanisms were identified for these technologies, such as the increase of interconnect resistance and the shunting across the isolation scribe (all due to corrosion and/or electromigration), due to the monolithic interconnection architecture of the module.
- Other degradation mechanism, previously observed on Si based modules, could also be identified, such as the encapsulation delamination, the front glass breakage or the loss of high potential isolation.

To better understand the degradation mechanism on the cell level, reliability tests have been performed on the individual component of the solar cell structure, such as ALT including liquid or gaseous phase, like « damp heat test » ($85°C/85\%$ humidity). In the case of CIGS solar cells for instance, the exposure of the Mo back contact to water and oxygen lead to the formation of a resistive Mo oxide and/or hydroxides top layer (possibly containing Na), increasing the resistance of the layer and decreasing its reflectivity [178]. Due to the intrinsic stability of CIGS, the degradation mechanisms identified for the absorber are generally with respect to grain boundaries (GBs) and interfaces with other layers. Alkali element (such as sodium) diffusing from the glass substrate during damp heat tests combines with the presence of water that have been shown to lead to the oxidation of the CIGS absorber [179]. The diffusion of Na rough the GBs and its accumulation in the depletion region during the ALT results in the formation of shunt paths that reduce the shunt resistance of the cells and therefore the V_{oc} of the device [180]. The damp heat test can also lead to the diffusion of the CdS buffer layer into the CIGS and ZnO layers, forming undesirable compounds such as $ZnSO_4$ and similar sulphates [181]. The conductivity and transparency of the TCO layer are also degraded under ALT, mainly due to the diffusion and adsorption of atmospheric species at the grain boundaries [182]. All this degradation processes are greatly influenced by the processing conditions of the individual layers (e.g., using a higher sputter pressure during the deposition of molybdenum led to a denser thin film that will be less

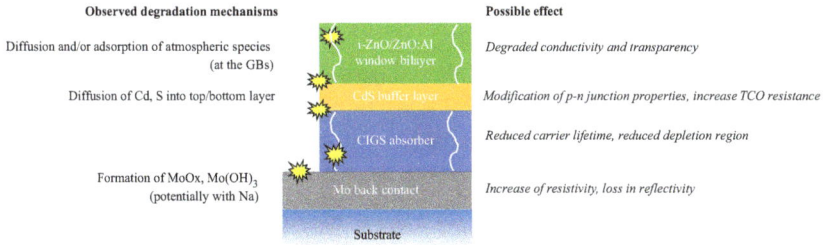

Observed degradation mechanisms

Diffusion and/or adsorption of atmospheric species (at the GBs)

Diffusion of Cd, S into top/bottom layer

Formation of MoOx, Mo(OH)₃ (potentially with Na)

Possible effect

Degraded conductivity and transparency

Modification of p-n junction properties, increase TCO resistance

Reduced carrier lifetime, reduced depletion region

Increase of resistivity, loss in reflectivity

Figure 4.15   Degradation mechanisms identified on CIGS solar cells under ATL tests

subject to corrosion during the damp heat test [183]). Figure 4.15 summarizes the main degradation mechanism observed in CIGS solar cells after ATL tests performed either on single layer or the full structure.

4.6   Organo-metal halide perovskites based solar cells

The star of the second decade of the twenty-first century in solar energy is definitely the hybrid inorganic-organic halide perovskite solar cell. Over the last 10 years, hybrid inorganic–organic perovskite solar cells (HPSCs) emerged as one of the promising solar cell device materials due to suitable electronic and optical properties, high photon absorption, earth abundance, and potential cheap fabrication process. The structure of perovskite crystal has been discussed in the previous chapters. Within a short period of time, the efficiency of HPSCs jumped incredibly from 6.5% to 17.9%. The world record efficiency in 2019 is approaching 24%, which makes it the highest efficiency for the thin film-based solar cells technologies. We would like to mention the effort of pioneer scientists, who first tried to apply HPSCs for organic light emitting diodes and integrated circuits.

In the context of the application of solar cells for the desert environment, it is worth highlighting that the behavior of most solar cells under high temperature is well known, a slight increase in the current and a clear decrease in the open circuit voltage (V_{oc}). This is mainly due to the decrease in the bandgap of the absorber. Recently, it has been shown that for $CsSnI_3$ which has perovskite structure, the band gap increases with increasing the temperature [184]. Therefore, such phenomenon suggests that the open circuit voltage of this solar cell based on the above absorber may increase when increasing the temperature. Hence, the importance of perovskite-based solar cell in it is application to desert climate that could be in the increase of the open circuit voltage expected at high temperature.

Hybrid organic/inorganic perovskite materials have been known for more than a century and have been developed remarkably in the last two decades mainly by Mitzi and co-workers for OLED and other electronic applications [185]. Actually, their applicability for solar cells was anticipated a long time ago but unfortunately not developed since then. However, the generated knowledge is transferable and

being used for perovskite solar cell (PSC). The first application of the hybrid perovskite for solar cells was based on photo electrochemical (PEC) cell. That was
done by Kojima and coworkers, and they reported a 2.2% efficient cell with liquid
electrolyte in 2006 and then 3.9% in 2008 [186]. The prepared devices were suf-
fering from severe instability due to the photodegradation. Many researchers then
used the same absorbing perovskite; but in nano-particle form to replace the dye in
DSSC. Park and colleagues used the same absorbing perovskite material; but
in nano-particle form to replace the dye in DSSC, a 6.5% efficiency was reported in
2011 [187]. This was the turning point that attracted more interest in HPSCs. The
next main step in the development was the use of solid HTM instead of liquid
electrolyte in DSSC. Kim and co-workers used the OLED-developed Spiro-
OMETAD and reported a 9.7% efficient all-solid perovskite solar cell [5]. At the
same time, Snaith and his team used a very thin-layer of perovskites and measured
an efficiency of 10.9% [188]. The second turning point, in the history of perovskite-
based solar cell was done by Snaith *et al.* [189] when they used the perovskite
absorber in thin film architecture. This opened the floor to the OPV community to
use perovskite materials instead of the organic absorber and the field of perovskite
solar cell gained more importance.

4.6.1   Solar cell architectures and designs

The perovskite solar cells have emerged from the field of the dye-sensitized solar
cells, so building the device also matched this consideration. In the beginning, the
absorber was impregnated on the mesoporous ETM (TiO_2) and the HTM was in a
liquid phase. When the liquid HTM was replaced by a solid one (spiro-OmeTAD)
the cells took the form of the solid-state dye sensitized solar cells or of the
extremely-thin-absorber solar cell (ETA). Finally, the solar cell took the thin film
approach and hence used the techniques utilized in the thin film technologies
including the organic solar cells.

Nowadays three main architectures are used to build HPSC. Figure 4.16 shows
the schematic of three different perovskite devices. Two of the three architectures
are like thin film-based solar cells. They consist of three main layers, namely hole
transport layer (HTL), perovskite absorber layer, and electron transport layer (ETL)

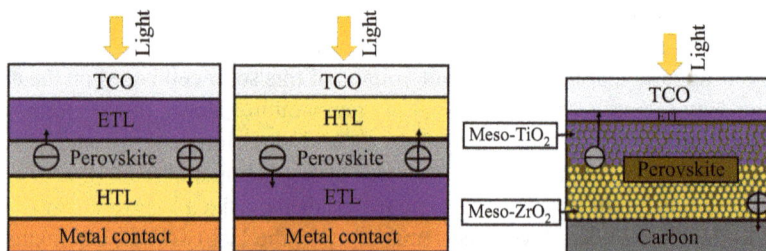

*Figure 4.16   Different configurations of perovskite solar cells. n–i–p (left), p–i–n
(middle), and dye (right).*

in addition to the front and back contacts. The ETL and HTL positions can be swapped in a way to make that n–i–p structure is the ETL that is deposited on the transparent conducting glass (TCO) from where the light will pass through to rich the absorber. In the case of the HTL deposited on the TCO, this architecture is named p–i–n and in the case where the ETL should be deposited on the top of the absorber followed with a back contact. These layers will result in the mechanism of a working solar cell, where light will be absorbed in the absorber layer, then the separation of charges will occur at two different interfaces, which are perovskite/ETL and perovskite/HTL.

The third architecture is the mesoporous one (Figure 4.16(right)) where no HTL is used, and the back contact is assured by a thick carbon film. The preparation of this kind of solar cell is a bit different as all its components are deposited successively and after each deposition the sample must be annealed at high temperature to sinter the film. The introduction of the scaffold mesoporous Zirconium oxide insulating layer is to prevent electrons from reaching the back contact [190,191]. The perovskite film is impregnated through the carbon contact followed by annealing at 100°C to crystalize the perovskite. Efficiencies up to 13% were attained using this structure. It is worth noting here that this kind of solar cells and mini modules are the only one showing long-time stability exceeding one year.

4.6.2 Device fabrication

Not much development has been made for the deposition of the ETM, HTM, and both sides of the electrodes. The ETM, mainly composed of metal oxides, has been deposited by chemical and physical routes. Like the ETM, the HTM deposition is dominated by solution processing routes. Only few groups have used physical methods to deposit HTMs furthermore, new HTMs have been developed as alternatives to spiro-OMeTAD [192]. In this section, we will report about the various deposition process and composition of absorber layers that have been used to grow perovskite films.

(a) Conventional perovskite absorber layer
    The crystal structure of halide perovskites is characterized by the formula ABX_3, where B^{2+} represents a metal cation (typically Pb^{2+} or Sn^{2+}), X^- represents a halide anion (I^-, Br^-, or Cl^-), and A^+ can either be an inorganic or organic cation in fully inorganic and hybrid organic–inorganic halide perovskites, respectively. The [BX6/2] octahedra are the primary structural units of the 3D perovskite inorganic frameworks, connected through corner sharing along all three spatial dimensions. The infinite frameworks are only stable when the size of A cations conforms to the Goldschmidt relation for ionic solids, which restricts their sizes. The violation of this restriction could result in edge- or face-sharing octahedra connectivity or leave some of X^- anions as terminal ligands, leading to various perovskite-like structures with ABX_3 or dimensionally reduced stoichiometries. Although some BX framework types could be obtained in inorganic perovskite-like compounds, organic A cations' substitution provides the greatest diversity. Of particular interest are low-dimensional

perovskite-like structures, such as quasi-2D crystals, which have gained considerable attention. The 2D perovskite-like semiconductors are natural multiple quantum wells (QWs) that display a well-pronounced quantum size effect, increased exciton binding energy and oscillator strength, and are valued for their stability and potential applications in optoelectronics, including photovoltaics and lasers [193,194].

(b) FA-MA perovskite absorber layers

Perovskite solar cells typically use a limited set of chemical components, including methylammonium (MA), formamidinium (FA), and cesium (Cs) as monovalent cations, and Pb(II) and Sn(II) as divalent species, combined with chloride, bromide, or iodide as anions. However, the intrinsic properties of MAPbI$_3$ and FAPbI$_3$ prevent their use as pure materials for industrial applications due to their lack of chemical stability caused by phase behavior. For example, MAPbI$_3$ undergoes a cubic to tetragonal phase transition at about 57°C and can slowly turn white (MAPbI$_3$·H$_2$O) or yellow (MA4PbI6·2H2O) when exposed to humid conditions, while FAPbI$_3$ can transform from its cubic black α-phase to the nonperovskite hexagonal yellow δ-phase in just a few weeks in dry air at room temperature. To create a FA-MA perovskite, a mixture containing 508 mg PbI2, 68 mg PbBr2, 180.5 mg FAI, 20.7 MABr, 720 μL DMF, and 180 μL DMSO are stirred without heating using a magnetic stirrer. The spin coater needs to be programmed to perform a single run at 2,000 rpm and 4,000 rpm for 10 s and 30 s, respectively, with a fixed pipette volume of 40 μL. During the second step, 120 μL of chlorobenzene must be dropped precisely between 10 s and 13 s to ensure film homogeneity. All samples should be annealed at 100°C for 50 min on a hot plate inside a glove box, resulting in a perovskite layer thickness of approximately 300 nm [195].

(c) 2D/3D perovskite absorber layers

To create a 2D perovskite, an organic functional group is added to a 3D perovskite using a crystal structure called Ruddlesden–Popper with the formula (RNH3)2An-1BnX3n+1, where (An−1BnX3n+1)2⁻ represents a conductor layer from the parent 3D perovskite, such as CsPbI3. The conductor layer is separated by an organic spacer cation, such as butylammonium or phenethylammonium. The value of n determines the dimensionality of the structure: $n = \infty$ is a 3D structure, $n = 1$ is a pure 2D structure, and other values of n are quasi-2D structures. The 2D structure is flexible due to the various values of n and organic functional groups, and it has gained much attention in recent years due to its better stability compared to 3D counterparts. However, a 2D/3D hybrid structure can also be formed by adding an organic functional group to three dimensions, which combines the advantages of both dimensions. This research direction is highly attractive for developing high-performance perovskite devices [196].

(d) Carbon-based PSCs

Organic–inorganic halide perovskites are crucial components in modern solar cells. However, the high cost of noble metals and organic hole transport layer (HTL) materials hinder the commercialization of perovskite solar cells. To address this issue, researchers have proposed HTL-free carbon-based perovskite solar cells (CePSCs) to reduce fabrication costs and improve stability. These devices have gained significant attention and have achieved a maximum power conversion efficiency (PCE) of 16.37% with low manufacturing costs and excellent moisture resistance. However, CePSCs have some limitations, such as lower PCE compared to conventional perovskite solar cells. The only interface between the electron-hole pair in CePSCs is the electron transport layer (ETL) and perovskite interface, making the ETL-material essential for enhancing stability and performance by effectively splitting up the electron–hole pair and extracting the photogenerated electrons from the absorbing layer. The ETL also acts as a hole-blocking layer to reduce electron-hole recombination. Consequently, researchers are currently focused on developing optimized ETL materials with advanced features. TiO_2 is one of the extensively used ETL materials in perovskite solar-cells due to its superior opto-electronic properties, high chemical stability, and nontoxicity [197].

(e)   Tin (Sn)-based PSCs

Perovskite solar cells are known to contain lead (Pb), which results in excellent performance. However, the toxicity of lead poses a significant environmental risk and hinders the mass production and commercialization of perovskite solar cells. Therefore, it is crucial to explore Pb-free or Pb-reduced perovskites to make them more affordable. When substituting lead in perovskite unit cells, it is important to consider the stability and ionic radius of the perovskite structure. Divalent metal ions such as calcium (Ca), tin (Sn), barium (Ba), and strontium (Sr) are promising candidates for replacing Pb. Among these, Sn has garnered much attention due to its similar coordination geometry and electron configuration to Pb. Sn-based perovskites with the formula $ASnX_3$ are the best alternative to Pb-based devices. Sn-based perovskite solar cells offer superior carrier mobility, bandgap, low excitation binding energies, short circuit current density, and a theoretical PCE of 33%. However, the efficiency of Sn-based perovskite solar cells is much lower (10%) than that of Pb-based perovskite solar cells. Additionally, the stability of perovskite solar cells is strongly affected by the oxidation of Sn2+ to Sn4+ [198–201].

4.6.3   *Stability and toxicity issues*

The Achill heel of perovskites-based solar cell is their stability. Toxicity is also an important issue as regulation can intervene at any time and band the usage of Pb-based products due to their environment impact. We will discuss in this section these two issues and what are the solutions proposed to these two problems.

4.7   Multijunction solar cells

4.7.1   *Basic principles of multi-junction solar cells*

Solar cells aiming at achieving highest efficiency should use a set of various materials that could absorb the widest possible part of the solar spectrum. Multi-junction solar cells are made of multiple single-junction layers, stacked upon each other, in such a way that each layer—in the sense from the top of the solar cell to its bottom—has a smaller bandgap than the adjacent previous one, to be able to absorb the photons that have energies higher than the bandgap of that layer, but less than that of the next layer, and so on [202]. One of the major factors limiting the multi-junction solar cells concept is the availability of low-defects-density-materials showing optimal bandgaps.

Alloys of groups III and V of the periodic table are excellent candidates for this purpose. In fact, their bandgaps span a wide spectral range, and most of them have direct electronic structure, highly suitable for a high absorption coefficient. In addition, their complex structures can be controllably grown with a high crystalline property by various deposition techniques [203–206].

Figure 4.17 represents the solar energy that can be theoretically employed by single and III–V triple-junction cells. Multi-junction solar cells have been

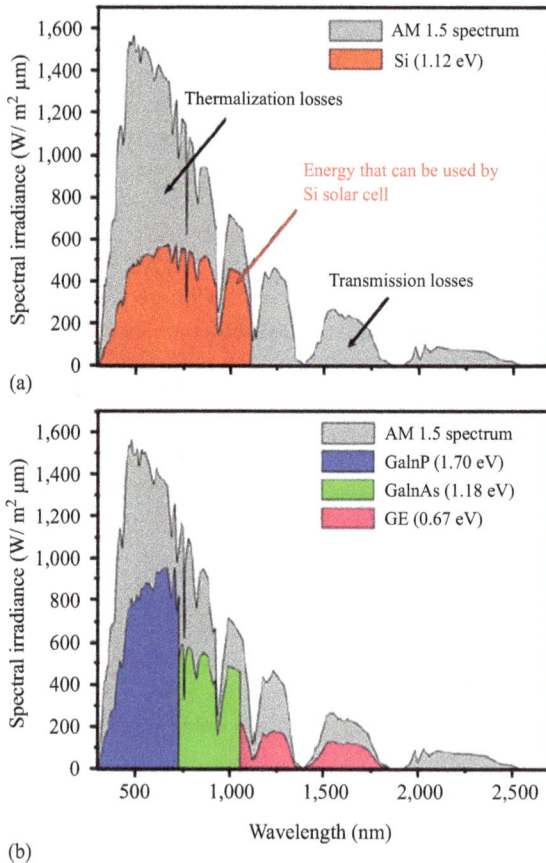

Figure 4.17   *The AM1.5 solar spectrum and the parts of the spectrum that can be used theoretically by (a) Si solar cells, and (b) $Ga_{0.35}In_{0.65}P/$ $Ga_{0.83}In_{0.17}As/Ge$ solar cells, adapted from Ref. [204].*

investigated deeply as earlier as 1960 [207], and the pioneer multi-junction solar cell was successfully demonstrated in early 1980s, where a photoconversion efficiency of about 16% was achieved [208]. In 1994, the US National Renewable Energy Laboratory (NREL) has broken the 30% PCE value and 40.7% efficiency was successfully achieved through a triple-junction cell [209]. Note that theoretically, the maximum PCE limit of multi-junction solar cells is about 86.8% [210].

To optimize the photoconversion efficiency in the solar cells, the PV device should absorb as much as possible of the light spectrum, and the bandgap should be designed accordingly. Moreover, the bandgap-difference between two adjacent layers made with different materials should be engineered to be reduced to the minimum as the excess energy from absorbed light which is converted to waste heat energy is equivalent and proportional to this difference [211].

The photoconversion efficiency as a function of the material bandgap energy is displayed in Figure 4.18. As can be seen, for a conventional solar cell configuration, the GaAs has almost the optimal band gap energy (equivalent to 1.4 eV), which is implicitly limited to PCE below or equal to 25%, at one-sun concentration.

The triple-junction solar cells that are nowadays produced are made mainly of GaInP (1.9 eV), GaAs (1.4 eV), and Ge (0.7 eV), whereas advanced multi-junction solar cells target the use of AlGaInP (2.2 eV), AlGaAs (1.6 eV), GaInP (1.7 eV), GaInAs (1.2 eV), GaInNAs (1.0–1.1 eV) [205]. It is noteworthy to mention that in monolithic multijunction solar cells concept, where various semiconductor layers are grown directly on top of each other using the same substrate, to reach the maximum electrical conductivity and the highest optical transparency, all the deposited layers should obey to the epitaxial property and have similar crystal structure. The mismatch in the crystal lattice constants of different layers may create dislocations in the lattice of the cell layers and significantly decrease the

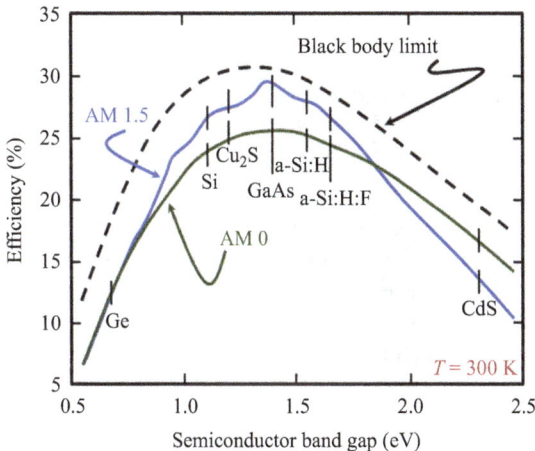

Figure 4.18   Dependency of the conversion efficiency on the semiconductor bandgap. Adapted from Ref. [205]

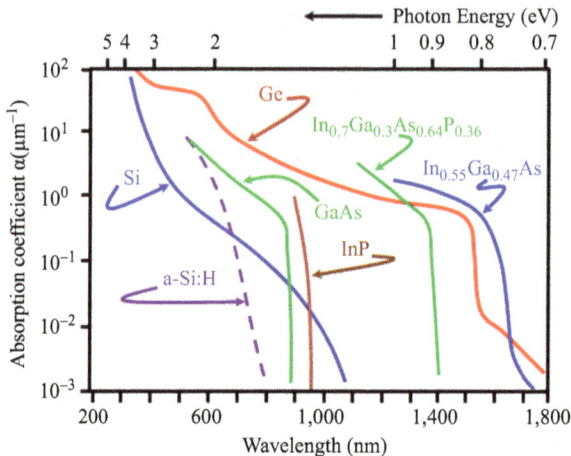

Figure 4.19   Absorption coefficient versus wavelength for various semiconductor materials. Adapted from Ref. [213].

efficiency of the solar cell device. As a matter of fact, a lattice mismatch as small as 0.01% significantly decreases the current produced by the solar cell [202].

Current matching property in these monolithically grown multijunction solar cells is also a highly desirable characteristic [212]. In fact, the output current of the multijunction solar cell is limited to the smallest value of the currents produced by any of the individual junctions. As the current is proportional to the amount of the incident photons with energy exceeding the semiconductor's bandgap and to the absorption constant of the material, and the junction layer must be thin or thick accordingly. As a matter of fact, the design of GaInP/GaAs/Ge solar cell involves a relatively thick Ge layer because of its lower absorptivity, while thickness of other layers varies case by case [202].

Example of the absorption constant for various semiconductors with respect to the photon wavelength is displayed in Figure 4.19.

4.7.2   Fabrication of multijunction solar cells

Molecular organic chemical vapor deposition (MOCVD) or molecular beam epitaxy (MBE) are the main processes employed to fabricate multijunction solar cells. Two main techniques are currently used, namely the mechanical stacking of independently grown layers, or monolithically growth of individual semiconductor layer on top of the other, where the final stack is seen as one single piece. This latter configuration is mainly based on MOCVD or MBE [208,214]. The mechanically stacking technique is a less suitable process due to the lack of accuracy and high cost [202]. MOCVD is often a preferable option to the MBE, because it is easily scalable to large production level, in addition to grow films with a high crystal quality. Electrons transport through different layers in the monolithic multijunction solar cells configuration is ensured through tunnel junctions, which are made of various stacks of highly doped layers that in turns create an effective potential barrier for both electrons and holes. However, high doping is necessary to create a thin depletion region that favors the tunneling across the junction and minimize the optical losses.

For the n-type dopants, the main used elements are Sn, Si, S, Se, Te, C, Ge, while for p-type dopants, Si, Zn, Be, Mg, Cd, C, and Ge are employed [204]. Furthermore, the antireflection coating is habitually made with TiO_2/Al_2O_3, Ta_2O_5/SiO_2 or ZnS/MgF_2, which are all a broadband dual-layer dielectric stack, whose spectral reflectivity characteristics are designed to reduce typically large reflectance (~30%).

One of the most efficient multijunction photovoltaic cells families are made of GaInP, GaAs, and Ge layers, grown on Ge substrate. The schematic of this triple-junction solar cell, containing about 20 layers, is displayed in Figure 4.20. The associated quantum efficiencies of each layer of this cell are shown in Figure 4.21 [202].

Note that the first layer of this cell is composed of GaInP (1.85 eV) that aims at converting short wavelengths such as blue and UV, the second layer is GaAs (1.42 eV) that captures near-infrared light, and the third layer is made of Ge, effectively absorbs the lower photon energies of the IR radiation that are above 0.67 eV [205,206].

Photons of energy below 650 nm wavelength could pass through the GaInP into the GaAs one and hence are not efficiently captured. A big part of the light spectrum is being absorbed by the Ge film, as the difference between the bandgap of the top two layers is 0.4 eV while the difference between the bottom two layers is 0.7 eV [216].

4.7.3 Multijunctions solar cell under high temperature operation

In this section, we point out a scenario of the operation of the multijunction solar cell under desert environment by highlighting the temperature effect on GaInP/GaAs/GaInNAsSb solar cells, experimentally assessed by the group of Arto *et al.* [217] through the measurement of the temperature coefficient parameter.

In this case, the solar cells were fabricated by molecular beam epitaxial (MBE) process (more details can be found in Refs [218,219]). The temperature was then varied from room temperature to 80°C, and the external quantum efficiency (EQE) measurements were carried out with a monochromator system. The *I–V* characteristics were performed via a three-band solar simulator calibrated with single junction GaInP, GaAs, and GaInNAsSb solar cells. The short circuit current densities of single junction cells were calibrated by EQE measurements (AM 1.5), the temperature of the sample mount was set to 25°C. For simulator-based measurements, the cell temperature was varied from 25°C to 150°C [221]. First, the temperature-dependent *I–V* characteristics of a GaInP/GaAs/GaInNAsSb triple-junction cell and GaInP, GaAs, GaInNAsSb single junction cells were determined at one sun conditions. Figure 4.22 shows the structure of a triple-junction photovoltaic cell [216]. Figure 4.21 shows the quantum efficiency of each layer of a GaInP/GaAs/Ge triple-junction solar cell [202].

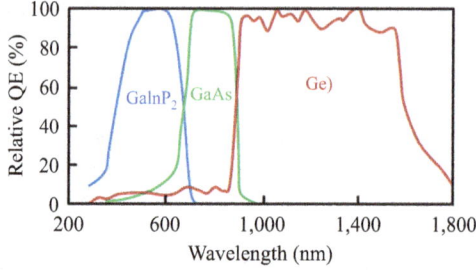

Figure 4.20   Quantum efficiency of each layer of a GaInP/GaAs/Ge triple-junction solar cell. Adapted from Ref. [202].

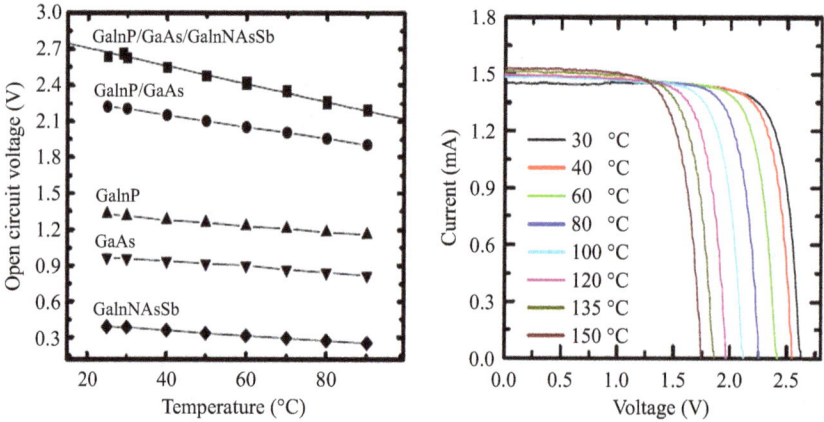

Figure 4.21   (a) Temperature dependence of the open circuit voltage (Voc) for the GaInP/GaAs/GaInNAsSb multijunction cell and single junction cells in the temperature range from 25°C to 90°C. All the cells were measured at AM1.5G except the triple junction cell which was measured at AM1.5D. (b) Measured I–V characteristics of GaInP/ GaAs/GaInNAsSb cell with a three-band solar simulator set to AM1.5D in the temperature range of 30°C to 150°C. Reproduced from Ref. [215].

For all the investigated GaInP/GaAs/GaInNAsSb, GaInP/GaAs, GaInP, GaAs, and GaInNAsSb solar cells configurations, the temperature dependencies of their open circuit voltages (V_{oc}) are displayed in Figure 4.21a. Note that for single junction cell, the V_{oc} was measured by means of a one band Oriel solar simulator at AM1.5G and calibrated to one sun, whereas for the triple-junction cell, the measurements were carried out through 3-band solar simulator calibrated to AM1.5D and are shown in Figure 4.21b.

Furthermore, the EQE measurements obtained at RT and 80°C are displayed in Figure 4.23. The associated J_{sc} values are simply deduced by integrating the values

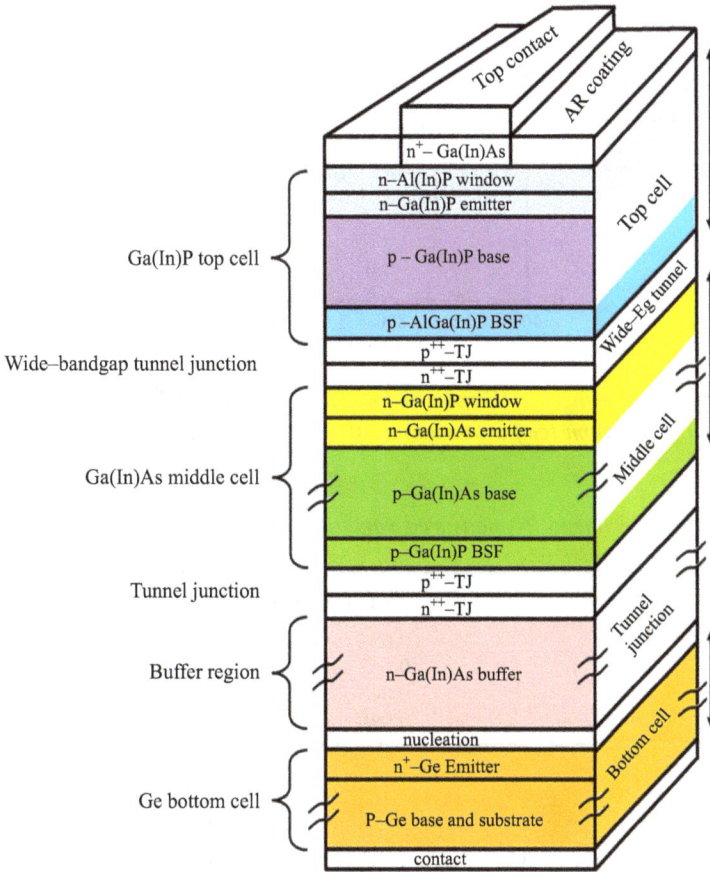

Figure 4.22   Structure of a triple-junction photovoltaic cell. Adapted from Ref. [216].

of the product of the EQE and the spectral irradiance at each wavelength and at each temperature [215]. The calculated J_{sc} values for the GaInP, GaAs, and GaInP/GaAs/GaInNAsSb devices have demonstrated a linear behavior of increased Jsc with respect to the temperature, except for GaInNAsSb sub-cell where the slope was found to be negative (this latter case is associated to the particularities of the measured properties of the AM1.5D spectrum in the 900–1350 nm spectrum, where the operation band of GaInNAsSb sub-cell is set). More details are available in the Ref. [217].

The temperature coefficients (which are calculated by measuring the slopes of the plots) for the J_{sc} and V_{oc} characteristics are shown in Table 4.5. For the open circuit voltage, the measured slope for GaInP/GaAs/GaInNAsSb cell was found to be similar to those reported previously for GaInP/GaAs/Ge cells [222,223], whereas, for the efficiency, the TC was found to be −0.09%/°C.

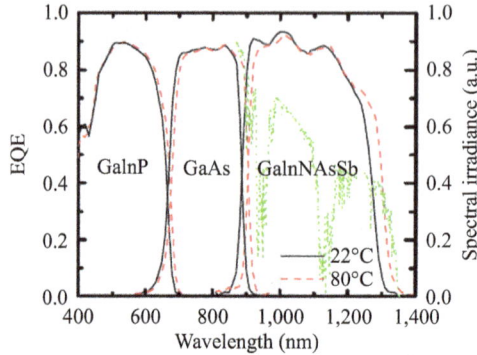

Figure 4.23   Measured EQE spectra for GaInP, GaAs, and GaInNAsSb sub-cells at room temperature and 80°C. Reproduced from Ref. [215].

Table 4.5   Experimentally measured various temperature coefficients

Sub cell	$TC_{J_{sc}}$ (mA/cm^2/°C)	$TC_{V_{oc}}$ (mV/°C)
GaInNAsSb	0.031	−2.0
GaAs	0.073	−2.2
GaInP	0.063	−2.5
GaInP/GaAs/GaInNAsSb	0.040	−7.5

In summary, the triple-junctions solar cell has found to exhibit a photo-conversion efficiency of 31% at AM1.5G. Heating the cell, which is equivalent to its operability under desert environment, has found to reduce the cell efficiency. The impact of the temperature on the multijunction solar cell was expressed through the experimentally measured temperature coefficient for the solar cell efficiency, which was estimated at −0.09%/°C, as measured at AM1.5D excitation, for the triple-junction GaInP/GaAs/GaInNAsSb cell, while the temperature coefficient for the open circuit voltage and the short circuit current were found to be −7.5 mV/°C and 0.04 mA/cm^2/°C, respectively.

References

[1]   S. Kurtz, K. Whitfield, G. TamizhMani, *et al.* Evaluation of high temperature exposure of photovoltaic modules. *Progress in Photovoltaics: Research and Applications*, 19(8), 954–965, 2011.

[2]   D.D. Smith, P.J. Cousins, A. Masad, *et al.* Generation III high efficiency lower cost technology: transition to full scale manufacturing. In *38th IEEE Photovoltaic Specialists Conference*, Austin, TX, 2012, pp. 001594–001597.

[3] T. Mishima, M. Taguchi, H. Sakata, and E. Maruyama. Development status of high-efficiency HIT solar cells. *Solar Energy Materials and Solar Cells*, 95(1), 18–21, 2011.

[4] J.P. Seif, A. Descoeudres, M. Filipic, *et al.* Amorphous silicon oxide window layers for high-efficiency silicon heterojunction solar cells. *Journal of Applied Physics*, 115(2), 0245021–0245028, 2014.

[5] Panasonic, 2015. http://panasonic.net/ecosolutions/solar/hit/.

[6] M.A. Green, K. Emery, and A.W. Blakers. Silicon solar cells with reduced temperature sensitivity. *Electronics Letters*, 18(2), 97–98 (1982).

[7] M.A. Green. Accuracy of analytical expressions for solar cell fill factors. *Solar Cells*, 7, 337–340, 1983.

[8] M. Taguchi, E. Maruyama, and M. Tanaka. Temperature dependence of amorphous/crystalline silicon heterojunction solar cells. *Japanese Journal of Applied Physics*, 47(2), 814–818, 2008.

[9] NREL, 2022. https://www.nrel.gov/pv/.

[10] M.A. Green, Y. Hishikawa, W. Warta, *et al.* Solar cell efficiency tables (version 50). *Progress in Photovoltaics: Research and Applications*, 25(7), 668–676, 2017.

[11] IEA. Trends 2016 in Photovoltaic Applications. Technical Report, IEA, 2016.

[12] K. Yoshikawa, H. Kawasaki, W. Yoshida, *et al.* Silicon heterojunction solar cell with interdigitated back contacts for a photoconversion efficiency over 26%. *Nature Energy*, 2(5), 17032, 2017.

[13] A. Richter, M. Hermle, and S.W. Glunz. Reassessment of the limiting efficiency for crystalline silicon solar cells. *IEEE Journal of Photovoltaics*, 3(4), 1184–1191, 2013.

[14] ISFH Press Release, 2018. https://isfh.de/en/26-1-record-efficiency-for-p-typecrystalline-si-solar-cells/.

[15] A. Richter, J. Benick, F. Feldmann, A. Fell, M. Hermle, and S.W. Glunz. n-Type Si solar cells with passivating electron contact: identifying sources for efficiency limitations by wafer thickness and resistivity variation. *Solar Energy Materials and Solar Cells*, 173, 96–105, 2017.

[16] F. Feldmann, M. Simon, M. Bivour, C. Reichel,M. Hermle, and S.W. Glunz. Efficient carrier-selective p- and n-contacts for Si solar cells. *Solar Energy Materials and Solar Cells*, 131, 100–104, 2014.

[17] S. DeWolf, J. Holovský, S.-J. Moon, *et al.* Organometallic halide perovskites: sharp optical absorption edge and its relation to photovoltaic performance. *The Journal of Physical Chemistry Letters*, 5(6), 1035–1039, 2014.

[18] P. Löper, B. Niesen, S.-J. Moon, *et al.* Organic–inorganic halide perovskites: perspectives for silicon-based tandem solar cells. *IEEE Journal of Photovoltaics*, 4(6), 1545–1551, 2014.

[19] S.W. Glunz, R. Preu, and D. Biro. Crystalline silicon solar cells. In *Comprehensive Renewable Energy*, vol. 1. New York, NY: Elsevier, 2012, 353–387.

[20]   J. Haschke, J.P. Seif, Y. Riesen, *et al.* Energy yield in hot and sunny climates: impact of silicon solar cell architecture and cell interconnection. *Energy & Environmental Science*, 10, 1196–1206, 2017. doi:10.1039/C7EE00286F.

[21]   D.-H. Neuhaus and A. Münzer. Industrial silicon wafer solar cells. *Advances in OptoElectronics*, 2007, 1–15, 2007.

[22]   S. Werner, S. Mourad, W. Hasan, and A. Wolf. Structure and composition of phosphosilicate glass systems formed by POCl3 diffusion. *Energy Procedia*, 124, 455–463, 2017.

[23]   A. Dastgheib-Shirazi, M. Steyer, G. Micard, H. Wagner, P.P. Altermatt and G. Hahn. Relationships between diffusion parameters and phosphorus precipitation during the POCl3 diffusion process. *Energy Procedia*, 38, 254–262, 2013.

[24]   E. Bustarret, M. Bensouda, M.C. Habrard, J.C. Bruyère, S. Poulin, and S.C. Gujrathi. Configurational statistics in a – Si x N y H z alloys: a quantitative bonding analysis. *Physical Review B*, 38(12), 8171–8184, 1988.

[25]   C. Ballif, D.M. Huljić, G. Willeke, and A. Hessler-Wyser. Silver thick-film contacts on highly doped n-type silicon emitters: structural and electronic properties of the interface. *Applied Physics Letters*, 82(12), 1878–1880, 2003.

[26]   A. Cuevas. Physical model of back line-contact front-junction solar cells. *Journal of Applied Physics*, 113(16), 164502, 2013.

[27]   J.P. Seif, G. Krishnamani, B. Demaurex, C. Ballif, and S.D. Wolf. Amorphous/crystalline silicon interface passivation: ambient-temperature dependence and implications for solar cell performance. *IEEE Journal of Photovoltaics*, 5(3), 718–724, 2015.

[28]   M.A. Green, Y. Hishikawa, E.D. Dunlop, D.H. Levi, J. Hohl-Ebinger, and A. W. Ho-Baillie. Solar cell efficiency tables (version 51). *Progress in Photovoltaics: Research and Applications*, 26(1), 3–12, 2018.

[29]   M. Tanaka, M. Taguchi, T. Matsuyama, *et al.* Development of new a-Si/c-Si heterojunction solar cells: ACJ-HIT (artificially constructed junction-heterojunction with intrinsic thin-layer). *Japanese Journal of Applied Physics*, 31(11), 3518–3522, 1992.

[30]   S. DeWolf, A. Descoeudres, Z.C. Holman, and C. Ballif. High-efficiency silicon heterojunction solar cells: a review. *Green*, 2(1), 7–24, 2012.

[31]   K. Masuko, M. Shigematsu, T. Hashiguchi, *et al.* Achievement of more than 25% conversion efficiency with crystalline silicon heterojunction solar cell. *IEEE Journal of Photovoltaics*, 4(6), 1433–1435, 2014.

[32]   T. Tiedje, E. Yablonovitch, G.D. Cody, and B.G. Brooks. Limiting efficiency of silicon solar cells. *IEEE Transactions on Electron Devices*, ED-31 (5), 711–716, 1984.

[33]   A. Richter, M. Hermle, and S.W. Glunz. Reassessment of the limiting efficiency for crystalline silicon solar cells. *IEEE Journal of Photovoltaics*, 3(4), 1184–1191, 2013.

[34]   P. Loper, B. Niesen, M. Soo-Jin, *et al.* Organic-inorganic halide perovskites: perspectives for silicon-based tandemsolar cells. *IEEE Journal of Photovoltaics*, 4(6), 1545–1551, 2014.

[35]   M. Taguchi, A. Yano, S. Tohoda, *et al.* 24.7% Record efficiency HIT solar cell on thin silicon wafer. *IEEE Journal of Photovoltaics*, 4(1), 96–99, 2014.

[36]   M.A. Green. The path to 25% silicon solar cell efficiency: history of silicon cell evolution. *Progress in Photovoltaics: Research and Applications*, 17(3), 183–189, 2009.

[37]   P.J. Cousins, D.D. Smith, L. Hsin-Chiao, *et al.* Generation 3: improved performance at lower cost. In *38th IEEE Photovoltaic Specialists Conference*, 2010, pp. 000275–000278.

[38]   J. Nakamura, N. Asano, T. Hieda, C. Okamoto, H. Katayama, and K. Nakamura. Development of heterojunction back contact Si solar cells. *IEEE Journal of Photovoltaics*, 4(6), 1491–1495, 2014.

[39]   T. Tiedje, E. Yablonovitch, G.D. Cody, and B.G. Brooks. Limiting efficiency of silicon solar cells. *IEEE Transactions on Electron Devices, ED*-31 (5), 711–716, 1984.

[40]   L. Tous, M. Aleman, R. Russell, *et al.* Evaluation of advanced p-PERL and n-PERT large area silicon solar cells with 20.5% energy conversion efficiencies. *Progress in Photovoltaics: Research and Applications*, n–a–n–a, 2014.

[41]   Z.C. Holman, A. Descoeudres, L. Barraud, *et al.* Current losses at the front of silicon heterojunction solar cells. *IEEE Journal of Photovoltaics*, 2(1), 7–15, 2012.

[42]   S.M. Myers, M. Seibt, and W. Schroter. *Journal of Applied Physics*, 88, 3795, 2000.

[43]   F. Duerinckx and J. Szlufcik, *Solar Energy Materials and Solar Cells*, 72, 231, 2002.

[44]   J.I. Pankove and M.L. Tarng. *Applied Physics Letters,* 34, 156, 1979.

[45]   U.K. Das, M.Z. Burrows, M. Lu, S. Bowden, and R.W. Birkmire. *Applied Physics Letters,* 92, 063504, 2008.

[46]   K.V. Maydell, E. Conrad, and M. Schmidt. *Progress in Photovoltaics,* 14, 289, 2006.

[47]   A. Illiberi, K. Sharma, M. Creatore, and M.C.M. van de Sanden. *Physica Status Solidi RRL*, 4, 172, 2010.

[48]   L. Korte and M. Schmidt. *Journal of Non-Crystalline Solids*, 354, 2138, 2008.

[49]   M. Schmidt, A. Schoepke, L. Korte, O. Milch, and W. Fuhs. *Journal of Non-Crystalline Solids*, 338–340, 211, 2004.

[50]   K. Pierz, W. Fuhs and H. Mell. *Philosophical Magazine Part B*, 63, 123, 1991.

[51]   M. Taguchi, M. Tanaka, T. Matsuyama, *et al. Tech. Digest 5th International Photovoltaic Science and Engineering Conference*, Kyoto, Japan, 1990, p. 689.

[52]   Z.C. Holman, A. Descoeudres, L. Barraud, F. Zicarelli, J. Seif, S. De Wolf, *et al. IEEE Journal of Photovoltaics*, 2(1), 7–15, 2012. doi:10.1109/ JPHOTOV.2011.2174967.

[53]   K. Ellmer and R. Mientus. *Thin Solid Films,* 516, 4620, 2008.

[54] A. Kanevce and W.K. Metzger. *Journal of Applied Physics*, 105, 094507, 2009.

[55] T.F. Schulze, L. Korte, F. Ruske, and B. Rech. *Physics Review B*, 83, 165314, 2011.

[56] S. Wang and D. MacDonald. *Journal of Applied Physics*, 112, 113708, 2012.

[57] J. Schmidt. *Applied Physics Letter*, 82, 2178–2180, 2003.

[58] J. P. Seif, G. Krishnamani, B. Demaurex, C. Ballif, and S. De Wolf. *IEEE Journal of Photovoltaics*, 5, 718–724, 2015.

[59] J.P. Seif, D. Menda, A. Descoeudres, L. Barraud, O. Özdemir, C. Ballif, and S. De Wolf, *Journal of Applied Physics*, 120, 054501, 2016.

[60] M.A. Green, *Progress in Photovoltaics*, 11, 333–340, 2003.

[61] M.A. Green. *Solar Energy Materials and Solar Cells*, 92, 1305–1310, 2008.

[62] Photovoltaics Report – Fraunhofer ISE, 2018, https://www.ise.fraunhofer.de/content/dam/ise/de/documents/publications/studies/Photovoltaics-Report.pdf

[63] Press Release, Hanaergy. 10 PV Buses in Norway with Miasolé Flex Modules, Oslo, Norway, 19 July 2018. https://www.hanergy.eu/norway-first-10-pvbuses/

[64] Press Release, Midsummer. Flexible Solar Panels from Midsummer Covers the Logos of the "world's best sports arena", Stockholm, Sweden, December 2, 2015. https://midsummer.se/press/news-blog/

[65] Press Release, Midsummer. Midsummer Solar Cells are Integrated into Tarpon Solar's Canvas – The Solution Takes First Place in the MTI Technology Award, Stockholm, Sweden, February 15, 2018, https://midsummer.se/press/news-blog/

[66] http://www.firstsolar.com/Resources/Projects/Topaz-Solar-Farm

[67] http://www.firstsolar.com/Resources/Projects/Desert-Sunlight-Solar-Farm

[68] Press Release, Solar Frontier. Solar Frontier Announces Total of 20+ Megawatt Module Supply Agreements with Cypress Creek Renewables, Tokyo, December 21, 2015, http://www.solar-frontier.com/eng/news/2015/C051570.html

[69] M.A. Green. *Solar cells: Operating Principles, Technology and System Applications*, Kensington: University of New South Wales, 1992.

[70] J. Scofield. Sputtered molybdenum bilayer back contact for copper indium diselenide-based polycrystalline thin-film solar cells. *Thin Solid Films*, 260 (1), 26–31, 1995.

[71] E. Moons, T. Engelhard, and D. Cahen. Ohmic contacts to p-CuInSe2 crystals. *Journal of Electronic Materials*, 22(3), 1993.

[72] W.E. Devaney, W.S. Chen, J.M. Steward, and R.A. Mickelson. Structure and properties of high efficiency ZnO/CdZnS/CuInGaSe2 solar cells. *IEEE Transactions on Electronics Devices*, ED 37, 428, 1990.

[73] M.A. Contreras, L.M. Mansfield, B. Egaas, *et al.* Wide bandgap Cu(In,Ga) Se2 solar cells with improved energy conversion efficiency. *Progress in Photovoltaics: Research and Applications*, 20, 843–850, 2012. doi:10.1002/pip.2244

[74] T. Haalboom, T. Godecke, F. Ernst, *et al.*, Phase relations and microstructure in bulk materials and thin films of the ternary system Cu-In-Se. In *Institute of Physics Conference Series*, 152E, 1997, p. 249.

[75] C. Persson and A. Zunger. Anomalous grain boundary physics in polycrystalline CuInSe2: the existence of a hole barrier. *Physical Review Letters*, 91, 266401, 2003. doi:10.1103/PhysRevLett.91.266401.

[76] S. Wei, S. Zhang, A. Zunger. *The Journal of Applied Physics*, 87, 1304–1311, 2000.

[77] L. Kazmerski, P. Ireland, F. White, and R. Cooper. *Proceedings of 13th IEEE Photovoltaic Specialist Conference*, 1978, pp. 184–189.

[78] N. Naghavi, D. Abou-Ras, N. Allsop, *et al.* Buffer layers and transparent conducting oxides for chalcopyrite Cu(In,Ga)(S,Se)2 based thin film photovoltaics: present status and current developments. *Progress in Photovoltaics: Research and Applications*, 18, 411–433, 2010.

[79] U. Rau and M. Schmidt. *Thin Solid Films*, 387, 141–146, 2001.

[80] B.E. McCandless and K.D. Dobson. Processing options for CdTe thin film solar cells. *Solar Energy*, 77, 839–856, 2004. http://dx.doi.org/10.1016/j.solener.2004.04.012.

[81] J. Britt and C.S. Ferekides. Thin-film CdS/CdTe solar cell with 15.8% efficiency. *Applied Physics Letters*, 62, 2851–2852, 1993.

[82] F. Kruger and D. de Nobel. *Journal of Electronics,* 1, 190–202, 1955.

[83] C.S. Ferekides, V. Viswanathan, and D.L. Morel. RF sputtered back contacts for CdTe/CdS thin film solar cells. In *Conference Record of Twenty Sixth IEEE Photovoltaic Specialists Conference*, 1997, pp. 423–426. http://dx.doi.org.10.1109/PVSC.1997.654118.

[84] N. Suyama, T. Arita, Y. Nishiyama, N. Ueno, S. Kitamura, and M. Murozono. CdS/CdTe solar cells by the screen-printing sintering technique. In *Proceedings of the 21th IEEE Photovoltaic Specialists Conference*, 1990, pp. 498–503.

[85] J. Perrenoud, L. Kranz, C. Gretener, *et al.* A comprehensive picture of Cu doping in CdTe solar cells. *Journal of Applied Physics*, 114, 174505, 2013. doi:10.1063/1.4828484.

[86] S. Siebentritt. Alternative buffers for chalcopyrite solar cells. *Solar Energy*, 77(6), 767–775, 2004.

[87] Recent Research Progress of High-efficiency CIGS Solar Cell in Solar Frontier | T. Kato | 7thIW-CIGSTech 2016/6/23.

[88] W. Witte, D. Abou-Ras, K. Albe, *et al.* Progress in photovoltaics: research and applications. In *Gallium Gradients in Cu(In,Ga)Se2 Thin-Film Solar Cells*, vol. 23, no. 6. Chichester: Wiley, 2015, pp. 717–733.

[89] J.D. Poplawsky, W. Guo, N. Paudel, *et al.* Structural and compositional dependence of the CdTexSe12x alloy layer photoactivity in CdTe-based solar cells. *Nature Communication,* 7,12537, 2016. http://dx.doi.org/10.1038/ncomms12537

[90] A. Morales-Acevedo. Analytical model for the photocurrent of solar cells based on graded band-gap CdZnTe thin films. *Solar Energy Materials and*

Solar Cells, 95, 2837–2841, 2011. http://dx.doi.org/10.1016/j.solmat. 2011.05.045.

[91]   B.E. McCandless. Thermochemical and kinetic aspects of cadmium telluride solar cell processing. In *Proceedings of the 2001 MRS Spring Meeting*, 2001, pp. H1.6, 1–12.

[92]   B. McCandless, L. Moulton, and R. Birkmire. *Progress in Photovoltaics*, 5, 249–260, 1997.

[93]   Y. Sun, S. Lin, W. Li, *et al.* Review on alkali element doping in Cu(In,Ga)Se2 thin films and solar cells. *Engineering*, 3, 452–459, 2017.

[94]   P. Jackson, R. Wuerz, D. Hariskos, E. Lotter, W. Witte, and M. Powalla. Effects of heavy alkali elements in Cu(In,Ga)Se2 solar cells with efficiencies up to 22.6%. *Physica Status Solidi (RRL) – Rapid Research Letters*, 10, 583–586, 2016. doi:10.1002/pssr.201600199.

[95]   K. Seshan and D. Schepis. *Handbook of Thin-Film Deposition Processes and Techniques*. Amsterdam: Elsevier, 2018.

[96]   T. Zhang, Y. Yang, D. Liu, *et al.* High efficiency solution-processed thin-film Cu(In,Ga)(Se,S)2 solar cells. *Energy & Environmental Science*, 9, 3674, 2016.

[97]   P. Reinhard, A. Chirilă, P. Blösch, *et al.* Review of progress toward 20% efficiency flexible CIGS solar cells and manufacturing issues of solar modules. In *2012 IEEE 38th Photovoltaic Specialists Conference (PVSC) PART 2*, 3–8 June 2012, Austin, TX. doi:10.1109/PVSC-Vol2.2012.6656789.

[98]   A. Chirilă, P. Reinhard, F. Pianezzi, *et al.* Potassium-induced surface modification of Cu(In,Ga)Se2 thin films for high-efficiency solar cells. *Nature Materials*, 12, 1107–1111, 2013.

[99]   P. Bommersbach, L. Arzel, M. Tomassini, *et al.* Influence of Mo back contact porosity on co-evaporated Cu(In,Ga)Se2 thin film properties and related solar cell. *Progress in Photovoltaics: Research and Applications*, 21 (3), 332–343, 2013.

[100]   H.-H. Sung, D.-C. Tsai, Z.-C. Chang, *et al.* An Na source via MoNa intermediate layer for three-stage evaporation of Cu(in, Ga)Se2solar cells. *Materials Science in Semiconductor Processing*, 39, 79–83, 2015.

[101]   B. Shin, Y. Zhu, N.A. Bojarczuk, S. Jay Chey, and S. Guha. Control of an interfacial MoSe2 layer in Cu2ZnSnSe4 thin film solar cells: 8.9% power conversion efficiency with a TiN diffusion barrier. *Applied Physics Letter*, 101, 053903, 2012. https://doi.org/10.1063/1.4740276

[102]   M. Kemell, M. Ritala, and M. Leskel. Thin film deposition methods for CuInSe2 solar cells. *Critical Reviews in Solid State and Materials Sciences*, 30, 1–31, 2005, doi:10.1080/10408430590918341

[103]   S. Niki, M. Contreras, I. Repins, *et al.* CIGS absorbers and processes. *Progress in Photovoltaics: Research and Application*, 18, 453–466, 2010.

[104]   A.M. Gabor, J.R. Tuttle, D.S. Albin, M.A. Contreras, and R. Noufi. *Applied Physics Letters* 65, 198, 1994.

[105]   J. Kim, L. Ho-Sub and N.-M. Park. CIGS thin film solar cell prepared by reactive co-sputtering. In *Proceedings of SPIE – The International Society*

for Optical Engineering, September 2013, p. 8823. doi:10.1117/12.20 21851.

[106] M. Marudachalam, H. Hichri, R. Klenk, R.W. Birkmire, W. N. Shafarman, and J.M. Schultz. *Applied Physics Letters*, 67, 3978, 1995.

[107] B.M. Basol. *Thin Solid Films*, 514, 361–362, 2000.

[108] J. Zank, M. Mehlin, and H.P. Fritz. *Thin Solid Films*, 286, 259, 1996.

[109] T. Wada, N. Kohara, T. Negami, and M. Nishitani, Chemical and structural characterization of Cu(In,Ga)Se2/Mo interface in Cu(In,Ga)Se2 solar cells. *Japanese Journal of Applied Physics*, 35, 1253, 1996.

[110] K.-J. Hsiao, J.-D. Liu, H.-H. Hsieh and T.-S. Jian. Electrical impact of MoSe2 on CIGS thin-film solar cells. *Physical Chemistry Chemical Physics*, 15, 18174–18178, 2013.

[111] Y. Hashimoto, N. Kohara, T. Negami, M. Nishitani, and T. Wada. *Japanese Journal of Applied Physics*, 35, 4760, 1996.

[112] M. Buffière, A.-A. El Mel, N. Lenaers, *et al.* Surface cleaning and passivation using (NH4)2S treatment for Cu(In,Ga)Se2 solar cells: a safe alternative to KCN. *Advanced Energy Materials*, 5, 1401689, 2015.

[113] J. Kessler, K.O. Velthaus, M. Ruckh, *et al.*, Chemical bath deposition of CdS on CuInSe2, etching effects and growth kinetics. In *Proceedings of the 6th International Photovoltaic Solar Energy Conference*, New Delhi, India, 1992, p. 1005.

[114] M Buffière, S. Harel, L. Arzel, C. Deudon, N. Barreau, and J. Kessler. Fast chemical bath deposition of Zn (O, S) buffer layers for Cu (In, Ga) Se2 solar cells. *Thin Solid Films*, 519(21), 7575–7578, 2011.

[115] E. Colegrove, R. Banai, C. Blissett, *et al.* High-efficiency polycrystalline CdS/CdTe solar cells on buffered commercial TCO-coated glass. *Journal of Electronic Materials*, 41(10), 2833–2837, 2012.

[116] Z. Du, X. Liu, Y. Zhang, J. Shuai, and H. Li. *RSC Advances*, 6, 108067, 2016.

[117] D. Abou-Ras, G. Kostorz, A. Romeo, D. Rudmann, and A.N. Tiwari. Structural and chemical investigations of CBD- and PVD-CdS buffer layers and interfaces in Cu(In, Ga)Se2-based thin film solar cells. *Thin Solid Films*, 480–481, 118–123, 2005.

[118] H.R. Moutinho, D. Albin, Y. Yan, *et al.* Deposition and properties of CBD and CSS CdS thin films for solar cell application. *Thin Solid Films*, 436, 175–180, 2003.

[119] N. Romeo, A. Bosio, and V. Canevari. The role of CdS preparation method in the performance of CdTe/CdS thin film solar cell. In *Proceedings of the 3rd World Conference on Photovoltaic Energy Conversion*, 2003, 469–470.

[120] N. Romeo, A. Bosio, R. Tedeschi, A. Romeo, and V. Canevari. A highly efficient and stable CdTe/CdS thin film solar cell. *Solar Energy Materials and Solar Cells*, 58, 209–218, 1999.

[121] R.C. Powell, G.L. Dorer, U. Jayamaha, and J.J. Hanak. Technology support for initiation of high-throughput processing of thin film CdTe PV. *Final Technical Report*, 1998.

[122] A. Salavei, I. Rimmaudo, F. Piccinelli, and A. Romeo. Influence of CdTe thickness on structural and electrical properties of CdTe/CdS solar cells. *Thin Solid Films*, 5, 5–8, 2013.

[123] J. Skarp, E. Anttila, A. Rautiainen, and T. Suntola. ALE-CdS/CdTe-PV-cells. *International Journal of Solar Energy*, 12, 137–142, 1992.

[124] G.C. Morris and S. Das. Some fabrication procedures for electrodeposited CdTe solar cells. *International Journal of Solar Energy*, 12,95–108, 1992.

[125] N. Suyama, T. Arita, and Y. Nishiyama. CdS/CdTe solar cells by the screen-printing sintering technique. In *Photovoltaic Specialists Conference 1990, Conference Record, Twenty First IEEE*, 1990, pp. 498–503.

[126] A.D. Compaan, R.W. Collins, V. Karpov, and D. Giolando. Fabrication and Physics of CdTe Devices by Sputtering, NREL (National Renewable Energy Laboratory) Technical Monitor, Final Report, University of Toledo, Ohio, 2009. https://www.nrel.gov/docs/fy09osti/45398.pdf.

[127] H.R. Moutinho, F.A. Abufoltuh, D.H. Levi, P.C. Dippo, R.G. Dhere, and L. L. Kazmerski. Studies of recrystallization of CdTe thin films after CdCl2 treatment. In NREL/CP-523-22944, 1997.

[128] A. Romeo, M. Terheggen, D. Abou-Ras, *et al.* Development of thin-film Cu(In, Ga)Se2 and CdTe solar cells. *Progress in Photovoltaics: Research and Applications*, 12, 93–111, 2004.

[129] N. Romeo, A. Bosio, A. Romeo, S. Mazzamuto, and V. Canevari. High efficiency CdTe/CdS thin film solar cells prepared by treating CdTe films with a Freon gas in substitution of CdCl2. In *Proceedings of the 1st European Photovoltaic Solar Energy Conference*, 2006, p. 1857.

[130] J.D. Major, R.E. Treharne, L.J. Phillips, and K. Durose. A low-cost non-toxic post-growth activation step for CdTe solar cells. *Nature*, 511, 334–337, 2014.

[131] L. Kranz, C. Gretener, J. Perrenoud, *et al.* Doping of polycrystalline CdTe for high-efficiency solar cells on flexible metal foil. *Nature Communication*, 4, 2013, 2046.

[132] J. Tang, D. Mao, T.R. Ohno, V. Kaydanov, and J.U. Trefny. Properties of ZnTe:Cu thin films and CdS/CdTe/ZnTe solar cells. In *Proceedings of the 29th IEEE Photovoltaic Specialists Conference*, 1997, pp. 439–442.

[133] W.N. Shafarman and L. Stolt. Cu(InGa)Se2 solar cells. In *Handbook of Photovoltaic Science and Engineering*, New York, NY: Wiley, 2003 (*Chapter 13*).

[134] J. van Deelen, Y. Tezsevin, and M. Barink. Multi-material front contact for 19% thin film solar cells. *Materials*, 9, 96, 2016. doi:10.3390/ma9020096.

[135] NREL chart of the Best Research-Cell Efficiencies. 2022. Available at: https://www.nrel.gov/pv/cell-efficiency.html

[136] A. Romeo. CdTe solar cells. In *McEvoy's Handbook of Photovoltaics*, New York, NY: Elsevier, 2018 (Chapter I-3-B). doi:http://dx.doi.org/10.1016/B978-0-12-809921-6.00009-4

[137] M. Powalla, S. Paetel, D. Hariskos, *et al.* Advances in cost-efficient thin-film photovoltaics based on Cu(In,Ga)Se2. *Theresa Magorian Friedlmeier, Engineering*, 3(4), 445–451, 2017.

[138]   J.L. Wu, Y. Hirai, T. Kato, H. Sugimoto, and V. Bermudez. New world record efficiency up to 22.9% for Cu (In,Ga)(Se,S)2 thin-film solar cell. In *7th World Conference on Photovoltaic Energy Conversion* (*WCPEC-7*), June 10–15, 2018, Waikoloa, HI.

[139]   First Solar Press Release. First Solar Builds the Highest Efficiency Thin Film PV Cell on Record, 5 August 2014.

[140]   J.-L. Wu, K. Fai Tai, Y. Iwata, T. Kato, H. Sugimoto, and V. Bermudez. Investigation on alkali-treatment mechanisms for improving energy conversion efficiency of Cu(In,Ga)(Se,S)2 modules. In *PVSEC-27*.

[141]   First Solar Press Release. First Solar Achieves yet Another Cell Conversion Efficiency World Record, 24 February 2016.

[142]   H. Sugimoto. High efficiency and large volume production of CIS-based modules. In *40th IEEE Photovoltaic Specialists Conference*, Denver, June 2014.

[143]   http://www.miasole.com (accessed 22 May 2015).

[144]   First Solar Press Release. First Solar Achieves World Record 18.6% Thin Film Module Conversion Efficiency, 15 June 2015.

[145]   A. Kanevce, M. O. Reese, T. M. Barnes, S. A. Jensen, and W. K. Metzger. The roles of carrier concentration and interface, bulk, and grain-boundary recombination for 25% efficient CdTe solar cells. *Journal of Applied Physics*, 121, 2017, 214506.

[146]   T. Minemoto, T. Matsui, H. Takamura, *et al.* Theorical analysis of effect of conduction band offset of windows/CIS layers on performance of CIS solar cells using device simulation. *Solar Energy Material & Solar Cells*, 67, 83–88, 2001.

[147]   J. Benick, A. Richter, R. Müller, *et al.* High-efficiency n-type HP mc silicon solar cells. *IEEE Journal of Photovoltaics*, 7(5), 1171–1175, 2017.

[148]   I. Rimmaudo. *Study of Structure and Electronic Properties of High Performance CdTe Solar Cells by Electrical Investigation*, Verona: University of Verona, 2013.

[149]   U. Rau and H.-W. Schock. Cu(In,Ga)Se2 thin-film solar cells. In *McEvoy's Handbook of Photovoltaics*, New York, NY: Elsevier, 2018 (Chapter I-3-C). doi:http://dx.doi.org/10.1016/B978-0-12-809921-6.000010-0

[150]   T.D. Leea and A.U. Ebong. A review of thin film solar cell technologies and challenges. *Renewable and Sustainable Energy Review*, 70, 1286–1297, 2017, http://dx.doi.org/10.1016/j.rser.2016.12.028.

[151]   M. Gostein and L. Dunn. Light soaking effects on photovoltaic modules: overview and literature review. In *2011 37th IEEE Photovoltaic Specialists Conference*, 10.1109/PVSC.2011.6186605.

[152]   J. Wysocki and P. Rappaport. Effect of temperature on photovoltaic solar energy conversion. *Journal Applied Physics*, 31, 571–578, 1960.

[153]   H. Mohring and D. Stellbogen. Annual energy harvest of PV systems – advantages and drawbacks of different PV technologies. In *Proceedings of the 23rd EUPVSEC*, 2008, pp. 2781–2785.

[154] P. Singh and N.M. Ravindra. Temperature dependence of solar cell perfor-mance – an analysis. *Solar Energy Materials and Solar Cells*, 101, 36–45, 2012, DOI:10.1016/j.solmat.2012.02.019.

[155] A. Virtuani, D. Pavanello, and G. Friesen. Overview of temperature coef-ficients of different thin film photovoltaic technologies. In *25th European Photovoltaic Solar Energy Conference and Exhibition/5th World Conference on Photovoltaic Energy Conversion*, 6–10 September 2010, Valencia, Spain, pp. 4248–4252.

[156] M. Theelen, A. Liakopoulou, V. Hans, *et al.* Determination of the tem-perature dependency of the electrical parameters of CIGS solar cells. *Journal of Renewable and Sustainable Energy*, 9, 2017, 021205. https://doi.org/10.1063/1.4979963

[157] V. Nadenau, U. Rau, A. Jasenek, and H.W. Schock. *Journal of Applied Physics*, 87, 1, 2000.

[158] S. Shirakata and T. Nakada. Photoluminescence and time-resolved photo-luminescence in Cu(In,Ga)Se2thin films and solar cells. *Physica Status Solidi (C) Current Topics in Solid State Physics*, 6(5), 1059–1062, 2009.

[159] M. Cwil, M. Igalson, P. Zabierowski, and S. Siebentritt. Charge and doping distributions by capacitance profiling in Cu(In,Ga)Se2solar cells. *Journal of Applied Physics*, 103, 063701, 2008.

[160] International Electrotechnical Commission IEC 61646 ed. 2, International Standard, Geneva, Switzerland, 2008.

[161] C. Osterwald and T. McMahon. *Progress in Photovoltaics: Research and Applications*, 17, 11–33, 2009.

[162] D. Jordan and S. Kurtz. *Progress in Photovoltaics: Research and Applications*, 21, 12–29, 2013.

[163] T. Carlsson and A. Brinkman. *Progress in Photovoltaics: Research and Applications*, 14, 213–224, 2006.

[164] F.J. Pern, B. Egaas, B. To, *et al.* A study on the humidity susceptibility of thin-film CIGS absorber. In *2009 34th IEEE Photovoltaic Specialists Conference (PVSC)*, 2009. doi:10.1109/PVSC.2009.5411676.

[165] L.C. Olsen, S. Kundu, and M. Englehard. Effects on moisture on CdTe Cell I-V characteristics. In *2006 IEEE 4th World Conference on Photovoltaic Energy Conference*, 10.1109/WCPEC.2006.279927.

[166] G. Makrides, B. Zinsser, M. Schubert, and G. Georghiou. *Solar Energy*, 103, 28–42, 2014.

[167] D. Jordan and S. Kurtz. *Proceedings of the 37th IEEE PVSC*, 2011, pp. 827–832.

[168] G. Niki, M. Contreras, I. Repins, *et al. Progress in Photovoltaics: Research and Applications*, 18, 453–466, 2010.

[169] D. Tarrant and R. Gay. Process R&D for CIS-based thin-film PV. In NREL/SR-520-38805, 2006.

[170] D. Meyer. NREL Conference Paper NREL/ CP-560-36320, 2004.

[171] E.H. Richards, C. Hanley, R.E. Foster, *et al.* Photovoltaics in Mexico: a model for increasing the use of renewable energy systems, *Advances in*

Solar Energy: An Annual Review of Research and Development, 13, Energy, American Solar Energy Association, Boulder, Colorado, 1999.

[172] B. Marion, J. del Cueto, P. McNutt, and D. Rose. Performance summary for the first solar CdTe 1-kW system. In NREL/CP-520-30942, 2001.

[173] M. Ross, G. Rich, L. Petacci, and J. Klammer. Improvement in reliability and energy yield prediction of thin-film CdS/CdTe PV modules. 1-4244-0016-3/06/$20.00, 2006, IEEE.

[174] A. Ullah, A. Amin, T. Kazmi, M. Saleem, and N. Butt. Investigation of soiling effects, dust chemistry and optimum cleaning schedule for PV modules in Lahore, Pakistan. *Renewable Energy*, 150, 456–468, 2020.

[175] S. Guo. Vertically mounted bifacial photovoltaic modules: a global analysis. *Energy*, 61, 447–454, 2013.

[176] M. Theelen. Degradation of CIGS solar cells. PhD thesis, 2015. http://www.cigsdegradation.com/.

[177] T. McMahon. *Progress in Photovoltaics: Research and Applications*, 12, 235–248, 2004.

[178] F. Pern and R. Noufi. *Proceedings of the DOE SETP Review Meeting*, 4, 17–19, 2007.

[179] D. Braunger, D. Hariskos, and H. Schock. *Proceedings of the 2nd WCPEC*, 1, 511–514, 1998.

[180] M. Theelen, V. Hans, N. Barreau, H. Steijvers, Z. Vroon, and M. Zeman. The impact of alkali elements on the degradation of CIGS solar cells. *Progress in Photovoltaics: Research and Applications*, 23(5), 537–545, 2015.

[181] J. Wennerberg, J. Kessler, M. Bodegard, and L. Stolt. *Proceedings of the 2nd WCPEC*, 1998, pp. 1161–1164.

[182] M. Theelen, S. Dasgupta, Z. Vroon, B. Kniknie, N. Barreau, and J. van Berkum. Influence of the atmospheric species water, oxygen, nitrogen and carbon dioxide on the degradation of aluminum doped zinc oxide layers. *Thin Solid Films*, 565, 149–154, 2014.

[183] M. Theelen, K. Polman, M. Tomassini, N. Barreau, H. Steijvers, and J. van Berkum. Influence of deposition pressure and selenisation on damp heat degradation of the Cu (In, Ga) Se2 back contact molybdenum. *Surface and Coatings Technology*, 252, 157–167, 2014.

[184] C. Yu, Z. Chen, J.J. Wang, *et al.* Temperature dependence of the band gap of perovskite semiconductor compound CsSnI3. *Journal of Applied Physics*, 110, 063526, 2011. https://doi.org/10.1063/1.3638699.

[185] D.B. Mitzi, C.A. Feild, W.T.A. Harrison, and A.M. Guloy. Conducting tin halides with a layered organic-based perovskite structure. *Nature*, 369, 467–469, 1994.

[186] A. Kojima, K. Teshima, Y. Shirai, and T. Miyasaka. Organometal halide perovskites as visible-light sensitizers for photovoltaic cells. *Journal of the American Chemical Society*, 131, 6050–6051, 2009.

[187] H. Kim, C.-R. Lee, J.-H. Im, *et al.* Lead iodide perovskite sensitized all-solid-state submicron thin film mesoscopic solar cell with efficiency exceeding 9%. *Science Report*, 2, 591, 2012. doi:10.1038/srep00591.

[188] M.M. Lee, J. Teuscher, T. Miyasaka, T.N. Murakami, and H.J. Snaith. Efficient hybrid solar cells based on meso-superstructured organometal halide perovskites. *Science*, 338, 643–647, 2012.

[189] M. Liu, M.B. Johnston, and H.J. Snaith. Efficient planar heterojunction perovskite solar cells by vapour deposition. *Nature*, 501, 395–398, 2013.

[190] https://www.nrel.gov/pv/cell-efficiency.html

[191] G. Grancini, C.R. Carmona, I. Zimmermann, *et al.* One-year stable perovskite solar cells by 2D/3D interface engineering. *Nature Communication,* 8, 15684, 2017. doi: 10.1038/ncomms15684.

[192] K. Rakstys, S. Paek, M. Sohail, *et al.* A highly hindered bithiophene-functionalized dispiro-oxepine derivative as an efficient hole transporting material for perovskite solar cells. *Journal of Materials Chemistry A*, 4, 18259–18264, 2016. doi:10.1039/C6TA09028A

[193] N.S. Kumar and K.C.B. Naidu. A review on perovskite solar cells (PSCs), materials and applications. *Journal of Materiomics*, 7(5), 940–956, 2021.

[194] L.A. Muscarella, D. Petrova, R. Jorge Cervasio, *et al.* Air-stable and oriented mixed lead halide perovskite (FA/MA) by the one-step deposition method using zinc iodide and an alkylammonium additive. *ACS Applied Materials & Interfaces*, 11(19), 17555–17562, 2019.

[195] X. Chen, H. Zhou, and H. Wang. 2D/3D halide perovskites for optoelectronic devices. *Frontiers in Chemistry,* 9, 715157, 2021.

[196] L. Zhang, C. Liu, J. Zhang, *et al.* Intensive exposure of functional rings of a polymeric hole-transporting material enables efficient perovskite solar cells. *Advanced Materials*, 30(39), 1804028, 2018.

[197] R. Prasanna, T. Leijtens, S.P. Dunfield, *et al.* Design of low bandgap tin–lead halide perovskite solar cells to achieve thermal, atmospheric and operational stability. *Nature Energy*, 4(11), 939–947, 2019.

[198] F. Matteocci, L. Cinà, E. Lamanna, *et al.* Encapsulation for long-term stability enhancement of perovskite solar cells. *Nano Energy*, 30, 162–172, 2016.

[199] J. Wei, F. Huang, S. Wang, *et al.* Highly stable hybrid perovskite solar cells modified with polyethylenimine via ionic bonding. *ChemNanoMat*, 4(7), 649–655, 2018.

[200] A. Isakova and P.D. Topham. Polymer strategies in perovskite solar cells. *Journal of Polymer Science Part B: Polymer Physics*, 55(7), 549–568, 2017.

[201] Y. Du, X. Wang, D. Lian, *et al.* Dendritic PAMAM polymers for strong perovskite intergranular interaction enhancing power conversion efficiency and stability of perovskite solar cells. *Electrochimica Acta*, 349, 136387, 2020.

[202] B. Burnett. *The Basic Physics and Design of III-V Multijunction Solar Cells*, Golden, CO: National Renewable Energy Laboratory, 2002.

[203] M. Yamaguchi. Present status of R&D super-high-efficiency III-V compound solar cells in Japan. In *Proceedings of the 17th Photovoltaic European Conference*, 2001.

[204] J.M. Román. State-of-the-art of III-V solar cell fabrication technologies, device designs and applications. In *Advanced Photovoltaic Cell Design*, 2004.

[205]   F. Dimroth. Next generation GaInP/GaInAs/Ge multi-junction space solar cells. In *Proceedings of the 17th Photo-voltaic European Conference*, 2001.

[206]   R.R. King. Bandgap engineering in high-efficiency multijunction concentrator cells. In *Proceedings of the International Conference on Solar Concentrators for the Generation of Electricity or Hydrogen*, 2005.

[207]   M. Wolf. Limitations and possibilities for improvement of photovoltaic solar energy converters, *Proceedings of the Institute of Radio Engineers*, 48, 1246–1263, 1960.

[208]   A. Luque and S. Hegedus. *Handbook of Photovoltaic Science and Engineering*, New York, NY: John Wiley & Sons, Ltd, 2003.

[209]   Spectrolab solar cell breaks 40% efficiency barrier, December 7, 2006, http://www.insidegreentech.com/node/454

[210]   F. Dimroth. 3-6 junction photovoltaic cells for space and terrestrial applications. In *Photovoltaic Specialists Conference*, 2005.

[211]   K. Zweibel. *Basic Photovoltaic Principles and Methods*. New York, NY: Van Nostrand Reinhold, 1984.

[212]   G.P. Smestad. *Optoelectronics of Solar Cells*, Bellingham, WA: SPIE Press, 2002.

[213]   S.O. Kasap. *Optoelectronics and Photonics: Principles and Practices*, New York, NY: Prentice Hall, 2001.

[214]   J. Poortmans and V. Arkhipov. *Thin Film Solar Cells: Fabrication, Characterization and Applications*, Hoboken, NJ: Wiley, 2006.

[215]   ASTM G 173–03: *Standard Tables for Reference Solar Spectral Irradiances: Direct Normal and Hemispherical on 37° Tilted Surface*, West Conshohocken, PA: ASTM International, 2003.

[216]   R.A. Sherif, R.A. Sherif, R.R. King, *et al.* The multijunction solar cell: an enabler to lower cost electricity for concentrating photovoltaic systems. In *Proceedings of Solar Power Conference*, 2006.

[217]   A. Aho, R. Isoaho, A. Tukiainen, *et al.* Temperature coefficients for GaInP/GaAs/GaInNAsSb solar cells. *AIP Conference Proceedings* 1679, 050001, 2015.

[218]   A. Aho, V. Polojärvi, V.-M. Korpijärvi, *et al. Solar Energy Materials and Solar Cells*, 124, 150–158, 2014.

[219]   A. Aho, J. Tommila, A. Tukiainen, V. Polojärvi, T. Niemi, and M. Guina. *AIP Conference Proceedings*, 1616, 33–36, 2014.

[220]   A. Aho, V.-M. Korpijärvi, A. Tukiainen, J. Puustinen, and M. Guina. *Journal of Applied Physics*, 116 (21), 2014.

[221]   A. Aho, A. Tukiainen, V. Polojärvi, and M. Guina. *Nanoscale Research Letters*, 9, 1–7 (2014).

[222]   G.S. Kinsey, P. Hebert, K.E. Barbour, D.D. Krut, H.L. Cotal, and R.A. Sherif, *Progress in Photovoltaics: Research and Applications*, 16, 503–508, 2008.

[223]   G. Siefer and A.W. Bett, *Progress in Photovoltaics: Research and Applications*, 22, 515–524, 2014.

Chapter 5

PV module technology and energy yield under desert environment conditions

Ahmer A.B. Baloch[1], Brahim Aïssa[2], Amir A. Abdallah[2] and Nouar Tabet[3]

5.1 Introduction

The standard qualification testing from the International Electrotechnical Commission (IEC), such as, IEC 61215 (terrestrial photovoltaic (PV) modules – design qualification and type approval) provides a baseline for PV module to be installed in the field. However, module failure may occur during short- and long-term operation depending on the combined environmental parameters, e.g., in a desert climate with a typical high solar irradiance, high temperature and high soling rate. Studying the impact of the environmental conditions on the performance and degradation of PV system installed in desert climate is very important for the deployment of PV. Therefore, both indoor and outdoor testing of different PV technologies are essential to obtain PV system performance and financial data. From these data, potential energy generated from PV system lifetime will be estimated and the levelized cost of electricity (LCOE) can be calculated to compare the energy-generated from the different PV technologies. Several studies in the literature have shown the dependence of PV module performance on reliability [1–3]. Degradation rate determine the loss in PV module and/or system performance expressed in percent per year (%/year). Other physical and statistical models have also been used to predict degradation rate for a specific type of failure mode [4,5].

Crystalline silicon technology for instance has shown a degradation rate below 1% per year [6,7]. A degradation rate above 1% per year was also reported in the literature, e.g., Lillo-Sanchez *et al.* [8] showed a degradation rate of 1.4%/year of the power output attributed to a loss in module electrical parameters. The difference in degradation rate published in the literature can be explained by, for example,

[1]Research & Development Centre, Dubai Electricity and Water Authority (DEWA), United Arab Emirates
[2]Qatar Environment and Energy Research Institute (QEERI), Hamad Bin Khalifa University (HBKU), Qatar Foundation, Qatar
[3]Department of Applied Physics and Astronomy, College of Sciences, University of Sharjah, United Arab Emirates

different climatic condition, PV module technology dependence, data quality, orientation mismatch, and sensor issues [2]. Besides, the method used to calculate the degradation rate is of a paramount importance. In a hot desert climate for instance, high operating temperature could result in a reduction of the PV system power output up to 30% [9,10]. Additionally, failure of the PV module materials, such as, solar cell cracking, discoloring of the encapsulant and back sheet material can be used as indicators of PV system degradation. Therefore, determination of PV degradation rate requires also an understanding of PV module failure modes and the impact of the climatic conditions, e.g., combined stress factor applied on the PV system during exposure. We note here, linear degradation rate is caused by, for example, encapsulant discoloration, while non-linear degradation rate, by for example, hotspot [11]. Other defects can be easily identified through the infrared (IR) imaging. In the case of silicon solar cells, a solar module, also called a PV module, consists of a number of solar cells connected in series or parallel depending on the power output required. Several solar modules form a string that are often connected to each other and form an array as illustrated in Figure 5.1.

The module material and manufacturing depends on the type of solar cell technology such as conventional aluminum back surface field (Al-BSF), passivated emitter rear contact (PERC), silicon heterojunction (SHJ) and interdigitated back contact (IBC). These cells, apart from Al-BSF, can be manufactured in both monofacial and bifacial modes. Moreover, the metallization scheme can range from few busbars to smart wires to shingling with full, half and quarter cut cells depending on the size of industrial wafer used. The total voltages depend on the number of solar cells connected in series (typically each solar cell produces 0.6 V), while the current depends on the size of the solar cell (40 mA/cm^2 or 9 A). Unlike the 60 solar cells module, recently, 72 cells module are manufactured with a power output in the range of 500 W. Increasing the number of cells reduce the $/Wp. High efficiency solar cell results in a high efficiency module for the same module material commodity and therefore reduces the area required. As compared with silicon technology, thin film technology is known to be less affected by high

Figure 5.1   Solar module, string, and array at the outdoor test facility (Doha, Qatar)

operating temperature (lower temperature coefficient) and requires more area because of a lower power conversion efficiency.

5.2   PV module materials

The bill of materials for a photovoltaic (PV) module typically includes components such as solar cells, tempered glass, encapsulant materials, backsheet, junction box, frame, and connectors. The glass should be transparent in the wavelength between 350 nm and 1,200 nm. In addition, the glass should, from one side, reject the IR irradiance (wavelength above 1,200 nm) in order to reduce the module temperature, and from the other side, prevent UV irradiance in order to increase the lifetime of the encapsulant material. The glass provides the solar module the rigidity and the resistance to the environmental conditions. This assists in blocking moisture ingress to backsheet and encapsulant materials resulting in lesser degradation. Some thin film manufacturers are using non-tempered low-iron glass, or polymers such as DuPont Tefzel, as the module superstrate material [12]. Recently, different anti-reflective coatings ARC and anti-soiling coatings are used to improve the performance [12].

During solar module manufacturing, an encapsulant material, such as ethyl vinyl acetate (EVA), is fused with the solar cells and the glass at around 150°C (cross-linking temperature) to provide a good adhesion between the solar cell and the glass. The EVA has good dielectric properties and, therefore, together with the glass it works as an electrical insulator and protect the solar cells from the environmental conditions. In addition, the EVA acts as a barrier layers that protect the solar cell from the ingress of water vapor. EVA is known to degrade by UV. Alternatives encapsulant material such as PVB is less permeable and less reactive as compared to EVA, also ionomer, polyolefin, and silicone are used. Ionomer is more expensive, but low water vapor transmission rate (WVTR) at 0.3 g/m^2/day is 100 times lower than EVA and helps protect sensitive electrical connections inside the module [12].

At the rear of the EVA, a white polymer sheet with high toughness and low thermal resistivity is used. This polymer sheet has an additional functionality since it improve light trapping by scattering and reflection of the light that comes between the solar cells. This polymer should be impermeable to water vapor and should resist the high UV irradiance. For a bifacial solar cell technology, this back sheet is replaced by a glass rear glass, i.e., glass–glass solar module or transparent backsheets.

5.3   PV module design

During the manufacturing of a conventional glass-back sheet PV module, the solar cells are connected in series using a solder-coated copper ribbon in a tabbed stringer process, which is used to connect the bottom of the first solar cell contacts to the top of the next solar cell contacts. The next step is the process of soldering the interconnection of a string of solar cells using a metal leads or a tabs. This followed by assembling the string on top of the EVA and the glass substrate. The solar cells string is now connected to form a complete solar module. This followed by adding the front

EVA and laminate the composite at a controlled heat and pressure. The last step is mounting the aluminum frames to the module and attaching the junction box.

PV module technology trend is considering new material and module design to reduced cost further, while improving module performance and reliability. In this section, different module designs will be presented and discussed in terms of module reliability. Literature has shown high failure percentages that are coming from the corrosion and interconnection failures [13]. For instance, in a back-contact PV module such as metal wrap through (MWT), where a MWT solar cells are directly in contact with an electric circuit printed on a copper foil through an electrically conductive adhesive, i.e., replacing the tabber string process step. Despite the low cell-to-module (CTM) losses, the MWT costs approximately five times the conventional module manufacturing step [12].

Different approaches could be used to reduce the thermal stress induced in the interconnection due to different operating temperatures causing as a sequence a drop in the module power output. For instance, by reducing the interconnector's coefficient of thermal expansion (CTE) to match the silicon wafer material [14], using electrically conductive adhesive (ECA) or smart wire could resulted in different stress in the solar cells.

Silicon bifacial module configuration with glass/glass or glass/transparent back sheet with or without aluminum frames have shown several advantages as compared with conventional standard mono-facial module configuration. In a frameless glass–glass module, solar cells are encapsulated between two glass substrates. This reduces from one hand the module material commodity cost and from the other hand it proved to reduce the UV effect on PV module material. In addition, since the bifacial solar cell does not require a full aluminum back contacts, the absorption of the IR irradiance is reduced, as a result, the bifacial module temperature decreases and the energy yield increases [15]. The bifacial module is known to generate electricity from both sides using bifacial solar cells. Depending on the ground albedo (reflected solar irradiance), bifacial technology could generate 10–30% more energy yield when compared with the conventional mono-facial PV module with the same watt peak [16–18]. In desert climate, Qatar for example, the diffused solar irradiance could reach up to 30% from the total solar irradiance and therefore, bifacial technology could benefit from the utilization of the diffused solar irradiance.

Weather a glass/glass configuration or a glass/transparent back sheet is used depends on the module material commodity cost, material water permeability. Water ingress into the solar cells may resulted in delamination and degradation [19]. In a glass/glass configuration, the edge should be sealed, e.g., silicone. Glass/glass configuration has several advantages as compared with the glass/back sheet module configuration. The former improves the mechanical rigidity (reduce cell cracks), resistance to high temperature and humidity ingress to the module material and allows the usage of frameless module design. The glass/glass are exploited when using an encapsulant material that does not contain additives that may cause solar cell degradation, e.g., peroxides or acetic acids as in the case of EVA [19]. Another advantage of the glass/glass bifacial technology is apparent with vertical mounting configurations; this feature has opened the PV market for new applications such as in building integrated PV (BIPV), e.g., in building facades and as a noise barrier on highways [19]. For applications with high voltages, modules without Al frame are also known to reduce the potential-induced degradation (PID) failure mode [19].

Other technological advancements such as halfcells, shingling and glass/glass design are also found to be significant for desert installation. In *Shingled* module design, a trade-off between contact shadowing and electrical losses is obtained. Recently, smart wire technology developed by Meyer Burger [20] has shown high energy yield thanks to the low CMT electrical losses, reliable, and low total cost of ownership. Its also important to understand the *effect of module design on cell temperature*. The measured module temperature of bifacial module is found to be higher than the conventional standard module by 5–9°C [20]. Glass/glass for monofacial is hotter than glass/back sheet.

In the next section, the application of different back sheet materials, colored encapsulant materials and the design of the distance between solar cells are assessed to maximize the optical gain from the module rear side [21].

5.4 Cell-to-module (CTM) performance

PV modules are made up of solar cells connected in series and encapsulated between layers. These layers are typically composed of glass, encapsulation, cell and back sheet material with different optical properties. On the surface of solar cell, interconnections such as fingers and bus-bars are required to collect generated current and complete the circuit. This encapsulated assembly increases the module's power and voltage above that of a single solar cell and protects solar cells from the environment. The main goal of module technology is to provide safe, efficient and reliable panels extending PV lifetime to around 20–25 years. However, by the process of modularization the operating conditions for the solar cells change as a result of optical losses due to parasitic absorption and reflection in module sub-layers. Furthermore, the solar cell interconnections cause additional power reduction due to resistance and joule heating which reduces its electrical performance. As a result, in the modularization process the power of the module is usually lesser than the summation of the power of all the solar cells connected to manufacture the module. This difference in total cell power and module power is called a cell to module (CTM) gap [22]. A key measure of CTM gap and CTM ratio (fraction of module power to cell power) is employed for assessing the effectiveness of the modularization process as highlighted by (5.1) and (5.2). A high CTM ratio will therefore ensure the lower production cost by enhancing the efficiency at the system level. Therefore, the ultimate aim is to design a module with a CTM ratio of 100% such that the module that are installed in field performs at their best.

$$CTM\ gap = Cell\ Efficiency - Module\ Efficiency \tag{5.1}$$

$$CTM\ ratio = \frac{Module\ Power}{\sum_{n} Cell\ Power} \tag{5.2}$$

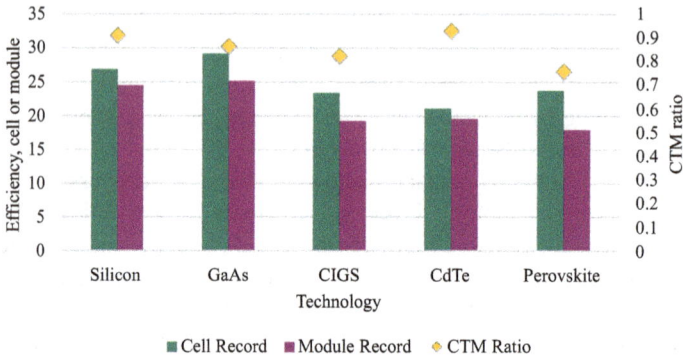

Figure 5.2   Record cell and module efficiencies with a CTM gap [2]

The importance of higher CTM ratio can be explained by Figure 5.2 where a general trend of efficiency gap from CMT i.e. CTM gap is evident. Figure 5.2 shows the state-of-the-art record efficiencies along with their CTM ratio for different solar cell technologies namely Silicon, CIGS, CdTe, GaAs, and perovskite [23]. Currently, the absolute efficiency gap i.e. CTM gap is in the range of 2.3–9.3%. Even for matured technology such as mono-silicon, the CTM gap is 2.3% and a CTM ratio of 91% showing 9% CTM losses is observed [23]. For other technologies, CTM ratio is in much lower range showing the potential to enhance the output at the module level.

To understand and minimize various mechanisms responsible for the loss/gain by upscaling from cell to module, it is vital to understand how each module component alters the performance. Figure 5.3 shows the efficiency drop mechanisms and the key losses/gains incorporated during the fabrication of a module. Integrating solar cells into a PV module results in gains and losses that change the module's power and efficiency compared to the original solar cells. The output of the module is governed by the module design and material selection. An optimum module design will require minimum CTM gap which will result in lower cost per watt. Therefore, to develop such a module, it is important to understand the role of individual component in the module. It should be noted that these losses are primarily evaluated at standard testing conditions (STC) i.e. a cell temperature of 25°C, a wind velocity of 1 m/s, and an incident radiation of AM1.5 1,000 W/m^2. The nominal operating efficiency of modules and CTM losses are assessed under these STC conditions. Although the focus of module technology is on STC efficiency, it is also vital to understand the dependence of energy yield on environmental parameters differing from STC such as insolation, cell temperature, and spectral variability. The operating conditions of the modules under desert environment such as high temperature environment with more incident radiation dose will cause the module losses to be higher due to higher performance degradation.

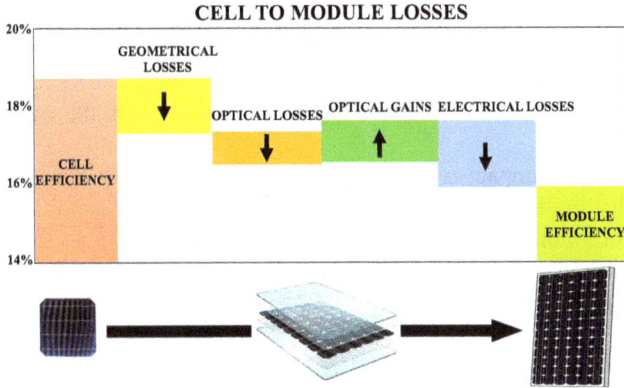

CELL TO MODULE LOSSES

Figure 5.3   Mechanisms responsible for the loss/gain by upscaling from cell to module under STC

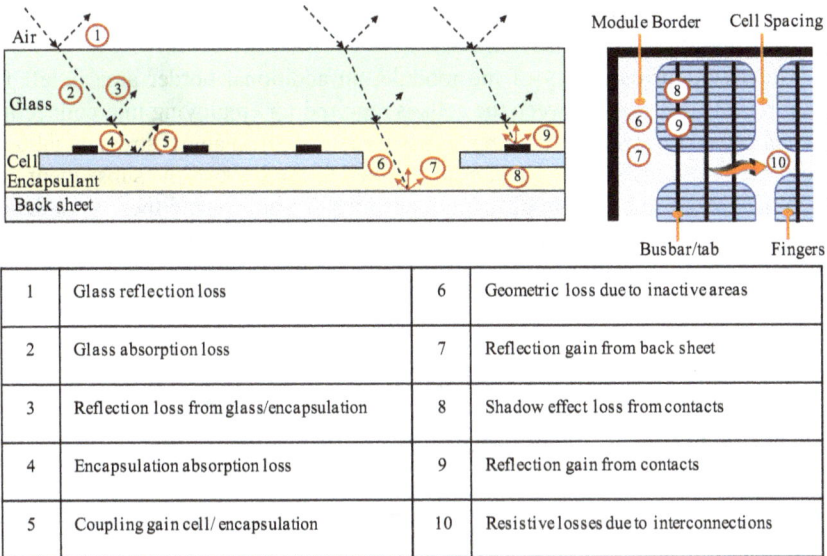

1	Glass reflection loss	6	Geometric loss due to inactive areas
2	Glass absorption loss	7	Reflection gain from back sheet
3	Reflection loss from glass/encapsulation	8	Shadow effect loss from contacts
4	Encapsulation absorption loss	9	Reflection gain from contacts
5	Coupling gain cell/ encapsulation	10	Resistive losses due to interconnections

Figure 5.4   Detailed loss analysis involved for evaluating CTM

In principle, the conversion losses and gains in the CTM development can be characterized based on geometry, optics, and electric effects. Figure 5.4 shows the detailed schematic of the gain and loss mechanism responsible for CTM conversion. They are influenced by:

- *Geometrical*: The losses associated with the geometry of the cell and module are related to the active and inactive areas responsible for photoconversion.

- *Optical*: This effect can result in both losses and gains as a result of material selection based on the target function. They include parasitic absorption and reflection of sub-layers. However, there are not only losses: additional light reflection on the solar cell or reduced light reflection from the encapsulated cell results in gains.
- *Electrical*: The impact is primarily due to resistive losses due to interconnections. Shading effect and reflection from tabs are also important for current collection. In the following sub-sections, the contribution from each effect are discussed in detail. The underlying mechanism are highlighted with potential improvement areas.

5.4.1   Geometry-related losses

Geometrical losses from cell to module results from the presence of inactive areas in the module. A cell when incorporated in the module is spaced apart for optimization of circuit and energy output which reduces the effective area for energy conversion from photons. Solar cells in PV modules are often employed in pseudo-square, square, or round wafers depending on the manufacturing process and the type of material as shown in Figure 5.5.

Moreover, for durability of the module, an additional border area is left for higher rigidity. Spacing between the cells is required for employing interconnection between cells. However, the spacing area is limited by mechanical stress on tabbing wires as they usually pass from front of one cell to the back of the subsequent cell which may result in breakage of interconnections. To understand the loss in detail, it is important to relate it to power conversion efficiency (η) of cells and modules. Efficiency of the module is dependent on the overall module area. It is defined as the fraction of power output to radiation input and incident area as shown by (5.3)

$$\eta = \frac{Power\ out}{Incident\ radiation \times Total\ area} \tag{5.3}$$

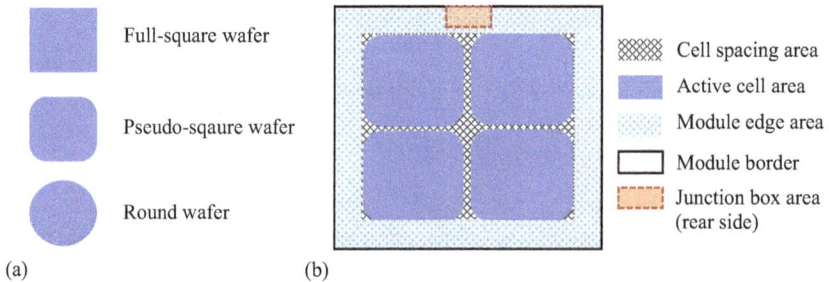

(a)   (b)

Figure 5.5   (a) Pseudo-square, full square, or round wafers used for making solar cells and (b) assembly of a module showing the effect of wafer type on active and inactive areas

where

- *Total area = Active areas + Inactive areas*
- *Active area = Cell area*
- *Inactive areas = Module edge area + Cell spacing area*

The losses associated with the geometry of the cell and module are related to the inactive areas responsible for photoconversion. It includes:

- Module edge inactive area
- Cell spacing inactive area
- Junction box (for bifacial)

Traditionally for mono-facial modules, these inactive areas include the module edge area, cell spacing and frame area. For bifacial, junction box also acts as an inactive area as no light is converted from that region. Practically, cell spacing ranges from 2 to 3 mm which is essential for ribbons quality and prevents breakage it typically enhances the inactive area by 2–3% depending on the type of wafer employed [24]. Shingled modularization is also another avenue that increases the active area of the cell by connecting the top of one cell to the bottom of second. This technology has been found to increase the usable area of module by 90% when compared to conventional tabbing and stringing method.

Figure 5.6 shows the effect of cell spacing on the aperture module efficiency for full square and pseudo-square wafers. Pseudo-square or round wafer yields higher inactive area losses whereas full square employs lesser due to higher coverage of cells. As the cell spacing increases, inactive areas increase which results in

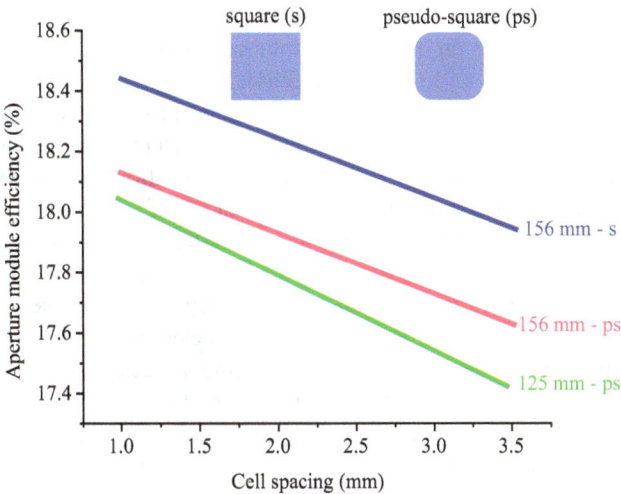

Figure 5.6   Effect of cell spacing on the aperture module efficiency for full square and pseudo-square wafers [25]

a drop of efficiency. By employing back contact cells, the cell spacing can be reduced to 1 mm [24]. This is due to the ease of using ribbons not limited by stresses especially in MWT, emitter wrap through (EWT), and interdigitated back contact (IBC) metallization design. These inactive areas should be reduced for increasing the CTM ratio which will eventually reduce the material usage, production costs, and levelized cost of electricity (LCOE).

Wide range of module border areas are available commercially depending on the size of module, and number of cells [25]. The inactive areas from the cell spacing the border is around 27 mm towards three edges and 35 mm at the junction box edge which results in an efficiency drop of approximately 5% [24]. To reduce inactive area from module edges, larger modules, novel edge-sealing methods and frameless panels should be employed. Larger modules have less percentage of inactive areas and results in lower percentage loss. This should assist designers in reducing the material usage and balance of system costs. Although, the cell spacing, and module border area should be minimized for increasing CTM ratio. However, there are also limitations from PV module safety standards [26] for the minimum allowable dimension of a module based on the electrical and mechanical parameters. The increased inactive area offers two compensations in particular; first is delay in moisture ingression and second is reflection from spacing that may strike back of the cell resulting in indirect coupling gain. An example for such optimization is utilization of white back sheet for increasing the current gain from inactive areas by additional illumination from back sheet.

5.4.2   *Electrical losses*

Electrical losses called as resistive losses result from cell and string connections required in the module design. These resistive losses (I^2R) occurring at the module are dependent on the current passing through the tabs as shown by (5.4). The resistance in turn depends on the design of the connections and their resistivity by the relation (5.5). Therefore, to enhance electrical performance, there are two areas that can be focused. First is the reduction in the current as it increases losses quadratically. This can be achieved by advanced concepts such as half cut cells. Another way is to reduce the resistance of the connections by manipulating the geometrical parameters such as length (L) and area (A) of the tabbing wires. This in principal can be decreased by employing smaller tabs with wide cross-sections. Although employing thicker tabs results in lower resistive losses, this would increase the optical losses due to shadowing thereby increasing the inactive area. Therefore, metallization design must consider both electrical and optical effects to obtain the optimum solution for these conflicting processes. Progress in novel designs of modules and metallization schemes provides a way to reduce CTM gap and achieve optimized performance by reducing resistive losses by not compromising the optical gains. Generally, the main electrical losses include:

- String interconnection
- Cell interconnection
- Mismatch losses

$$P_{loss} = I^2R \tag{5.4}$$

$$R = \rho\frac{L}{A} \tag{5.5}$$

Total ohmic losses result from various interconnections in the module and leakage currents at different points. The power losses from different resistive components include contact resistance between emitter and tabs, base and tabs, interconnection between adjacent cells, string connections and cable losses as shown in Figure 5.7 and (5.6). The series resistance primarily affects the fill factor (FF) of the solar cell thereby reducing the efficiency of the module. This results in the shifting of operating points of modules as the maximum power points moves due to additional resistance thereby reducing FF. Moreover, as the resistive losses are related with current response, the CTM losses for high efficiency cells become extremely important since they operate at higher current values. Bifacial PV module due to increased current gain in real-time operating conditions is therefore prone to higher electrical loss reducing the CTM ratio further. For increasing the CTM performance from electrical point, optimized contact design is used such as back-contacts cells.

$$P_{loss,module} = P_{loss,base} + P_{loss,emitter} + P_{loss,gap} + P_{loss,sc} + P_{loss,cable} \tag{5.6}$$

To understand the electrical losses associated with CTM, it is important to consider the testing of cells and modules. At the cell level, flash lamps are used to analyze the power which uses point contacts at bus bars and controlled regions [27]. The current in this case does not have to travel large distances through different loss modes of interconnections and the rear side is usually contact through full metal-lization contact thereby reducing ohmic losses. On the contrary, at the module level the current needs to travel from fingers to bus bars to interconnections and tabs

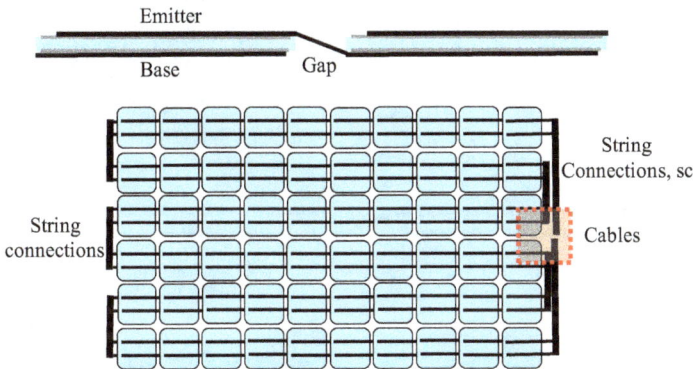

Figure 5.7   *Power losses from different resistive components include contact resistance between emitter and tabs, base and tabs, interconnection between adjacent cells, string connections and cable losses*

resulting in higher ohmic losses than cells even at the flash testing system. This reduces the CTM ratio due to additional resistive components.

To enhance the CTM performance, novel module structures require a reconsideration of the tabs/wires design and geometry. The main objective is to reduce the resistance from the components without compromising the active area of the solar cell as this would increase optical losses.

Main routes through which electrical performance can be enhanced and ohmic losses can be reduced in the modularization process are:

- Smaller tabs/wires, to reduce the contact resistance. In the case of smart wire connection.
- Broader tab width, however, this would increase optical/shading effect.
- Thicker tabs, however, this would increase cell cracking. They could work with conductive epoxy instead of soldering.
- Conductive back sheet approach removes the necessity for soldering of interconnection.
- Back-contact cell decreases resistance at interconnections through collection of current at positions distributed on cell.
- Half-cut cell reduces the current in half thereby resulting in 75% less I^2R losses.
- Shingled module interconnection, as shown in Figure 5.8, found to increase 1.86% abs. efficiency.

Mismatch losses occur in module due to the difference in current in series connected solar cells. The cell with the minimum current limits the performance of the entire string connected in series which in turn reduces the maximum power output. This induces another CTM loss which is more focused on operating conditions than standard testing conditions. Non-homogenous illumination is the main cause of current-mismatching losses in real world operating conditions. They can result from shadowing due to foreign particles, dusts, debris, leaf and reduced height for the case of bifacial modules. For optimal performance under field, this loss is minimized at the modularization level. PV module producers resolve this issue by incorporating bypass diodes to reduce the impact of variable current due to non-uniform radiation as shown in Figure 5.9.

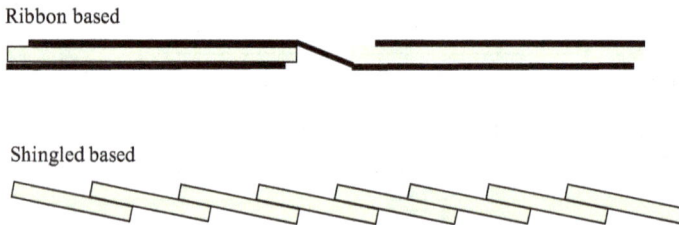

Ribbon based

Shingled based

Figure 5.8   Ribbon-based interconnection and shingled-based interconnection technology

To further understand, let us consider an example for current mismatching in a 60-cell module connected with and without bypass diodes. By employing three bypass diodes connected in parallel, we have 20-cell string that operates at the same maximum power based on the radiation level or any damage as shown in Figure 5.10. To quantify, we assume 250 W_p PV panel with the top string (cell operating at standard testing condition of 1,000 W/m^2 whereas the second and third

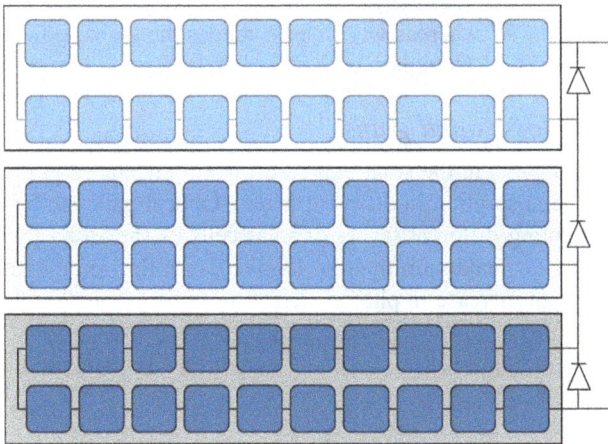

Figure 5.9   Current mismatching due to non-uniform radiation in a 60-cell module connected with three bypass diodes

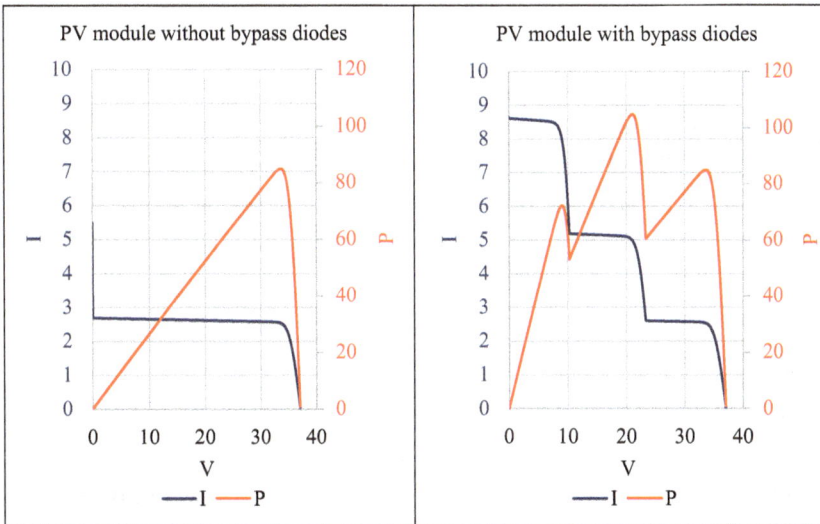

Figure 5.10   IV and the PV curve of the module (left) without bypass diodes and (right) with bypass diodes

string operates at 600 and 300 W/m^2 radiation, respectively). In this case, the IV and the PV curve of the module will be without and with bypass diodes as shown in Figure 5.10. If the PV module is operating without bypass diodes (Figure 5.10—left), the current in the series-connected module will be limited by the lowest illumination of 300 W/m^2. However, by using bypass diodes, we observe that each string operates at its own maximum power point as highlighted by three maximum power peaks in Figure 5.10—right corresponding to each effective radiation. When these modules are connected, the global maximum power for PV without bypass diodes would be 83.5 W whereas with bypass diodes the global power max would be 104 W which is 24.55 % higher.

5.4.3   Optical losses and gains

Optical losses and gains in the solar cell result from reflectance at different interfaces and parasitic absorption in different layers. Optical gain can be achieved by appropriately coupling different layers of solar cells as shown by (5.7) and (5.8). Using tabs/fingers for maximum internal Lambertian reflection may also increase CTM ratio due to an increase in photogeneration of charge carriers:

$$R_{\theta=0°} = \frac{(n_2 - n_1)^2}{(n_2 + n_1)^2} \tag{5.7}$$

$$n_{ARC} = \sqrt{n_{ARC-1} \times n_{ARC+1}} \tag{5.8}$$

This effect can result in both losses and gains as a result of material selection based on the target function. They include parasitic absorption and reflection of sub-layers.

- Glass reflection
- Glass absorption
- Reflection from glass/encapsulant interface
- Absorption in encapsulant material
- Shading from contacts and tabs.

However, there are not only losses: additional light reflection on the solar cell or reduced light reflection from the encapsulated cell results in gains. Main gains due to reflection and coupling includes

- Coupling anti-reflection (AR) coating
- Coupling fingers
- Coupling interconnection
- Coupling back sheet

At first, it generates optical losses by absorbing and reflecting the irradiated sun light in or at the covering layer (e.g. glass and EVA). However, there are also optical gains, which arise by embedding the cell in a material with an intermediate refractive index, which lies between the one of air and the cell surface. Between the module layers, there are several optical effects interacting with each other. Starting

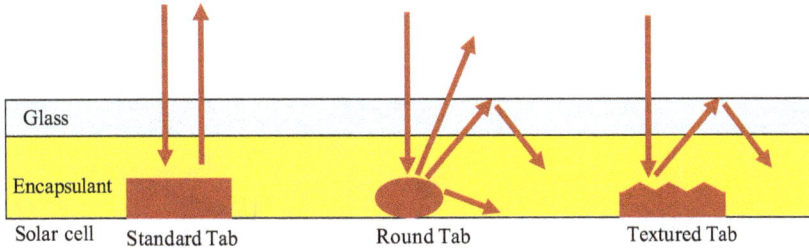

Figure 5.11 Different tab design for current collection: standard rectangular, round tabs, and textured tabs

with the glass, optical losses result from the effective reflectivity of the air–glass interface. It depends on the material refractive index, the surface structures and possibly the antireflective coating. The bulk absorptions of the glass and EVA material generate further optical losses. The optical coupling gains can be separated in direct and in indirect gain mechanisms. The direct gain arises from using refractive indices of adjacent such that there is no reflection. Indirect coupling results from using innovative tab design for total internal reflection as highlighted in Figure 5.11 with different tab designs.

5.5 Module energy yield and reliability under desert environment

5.5.1 Module installation parameters for mono-facial and bifacial modules

The operating conditions of PV modules play a huge role at the cell and module level efficiency [22]. The energy gain from a bifacial module over a mono-facial module is due to the additional albedo collection on the rear side of the module. This energy gain depends on a number of factors, including installation parameters, characteristics of the incident irradiance, module rear-side current response, etc. [23–25]. The parameters which can be controlled during cell (or module) fabrication is the module rear-side current response [23,24] called bifaciality and the behavior of the device to temperature called temperature coefficient. The remaining factors depend on the location and the installation conditions. Due to the various factors affecting the energy yield of a bifacial PV module, the end-use gain from this module type compared to the standard mono-facial modules is not easily quantified. In the literature, a gain in the range of 10–50% is reported [26], depending on the location of installation and the installation parameters. Higher gains are possible if the installation conditions as shown in Figure 5.12 are optimized for a particular location. This is one of the major challenges with bifacial modules. Defining standard guidelines for installation can help the end-use benefits.

Performance parameters for installation

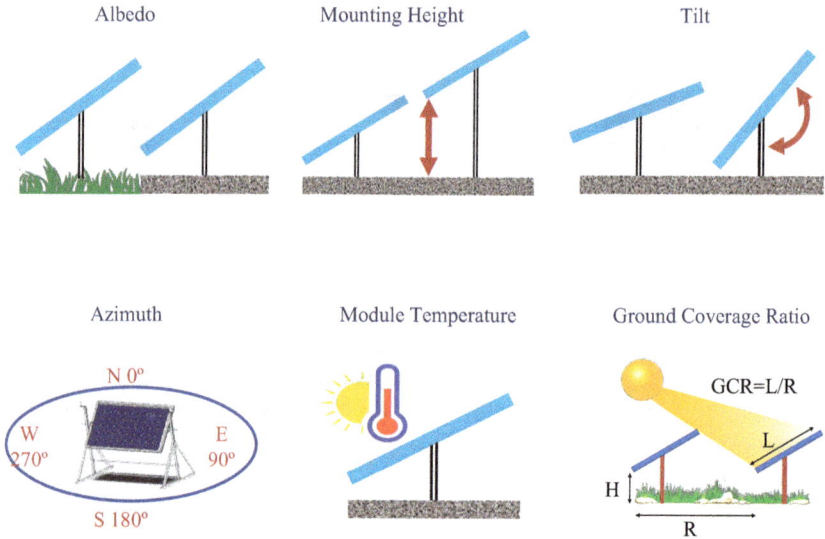

Figure 5.12   *Performance parameters for mono-facial and bifacial PV module installation*

At the cell level, energy yield is affected by:

- Bifaciality
- Temperature coefficient
- Power conversion efficiency (PCE)

At the module level, energy yield is affected by:

- Module design
- Nominal operating cell temperature
- Back sheets i.e. glass–glass or glass–tedlar
- Encapsulation
- Junction design
- Type of frame
- Number of diodes

At the field, following installation parameter affects the performance:

- POA irradiance
- Albedo
- Height
- Tilt
- Azimuth

- Module temperature
- Ground coverage ratio

For modules mounted at the conventional tilt angle (latitude), the installation parameters affecting the energy yield are module height, tilt angle, ground albedo, and the irradiance characteristics (cloudy or sunny day, diffuse content) [27].

Optimizing the installation parameters results in an increase in energy yield that can eventually decrease the LCOE of power plants [28]. Following subheadings shows the impact of individual performance parameter on the energy yield of mono-facial and bifacial modules installed at hot and sunny desert environment of Qatar.

5.5.2   *Effect of mounting height*

One of the most important installation parameters for bifacial PV is the mounting height of the module from the ground as it determines the amount of irradiation on the rear side. The module height is defined as the height between the lower edge of the module and the ground. It affects the output power of mono-facial and bifacial PV modules, but the impact is more significant in the latter.

Figure 5.13 shows the impact of mounting height on the mono-facial and bifacial energy yield operating under desert environment. Mono-facial module's energy yield is not sensitive to height as they are optimized for front side radiation which necessarily does not change with height. However, as we increase the mounting height of bifacial PV modules, the power gain increases due to additional rear side irradiation. However, it reaches the maximum then starts to drop. The reason is the conflicting nature of incoming rear radiation from ground albedo (that decreases with height) and additional diffuse radiation (that increase with height). The impact of the self-shadow on the power gain is highly related with the nature of the irradiance (direct/diffuse). In the case of high diffuse radiation, the contribution from self-shadow would be much lower and higher energy yield can be expected.

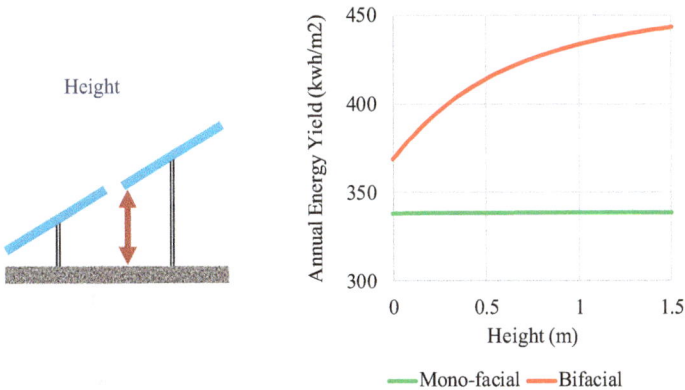

Figure 5.13   Effect of mounting height on mono-facial and bifacial modules annual energy yield

5.5.3   Effect of ground albedo

With a proportional relation, materials with high albedo coefficient showed better performance by reflecting more light on the rear side of the panel. The albedo coefficient is defined as the amount of reflected light on the surface of a given material with respect to the incident light. The most important parameter for performance enhancement of bifacial solar cells is its albedo as shown in Figure 5.14. It is the solar radiation component reflected from the ground and scattered by diffused radiation. Higher albedo implies that more radiation is collected at the rear face of the solar cell resulting in an increase of generation current and overall bifaciality ratio. Bifacial solar cells are albedo collecting devices and are capable of decreasing the LCOE of power plants. Utilizing albedo radiation is preferential for bifacial installations as both beam and diffuse radiation from the atmosphere and earth's surface can be collected. There is a linear dependence (with a positive slope) between the PV power gain and albedo coefficient due to direct relation of radiation with short circuit current.

5.5.4   Effect of tilt angle

The tilt angle can be defined as the angle that the PV module makes compared to horizontal ground surface. For mono-facial modules, a tilt angle equal to the latitude angle will maximize the power gain on an annual basis [29]. On the other hand, the tilt angle of bifacial PV modules must be optimized as that the amount of irradiance coming from the rear and the front sides gives an important power gain. Due to the incoming radiation, the tilt angle affects both of mono-facial and bifacial PV modules. Higher tilt angles are desired for bifacial modules as shown in Figure 5.15 to collect additional rear side radiation from surrounding. Generally, for bifacial, when the tilt angle is low, less rear side irradiation is received. This is due to high albedo contribution from the self-shadow region than the surroundings. This lowers the power gain. Optimal tilt shifts to a higher angle than latitude to harness more albedo radiation. For Qatar, this translates to an angle of 34° for bifacial and 24° for mono-facial.

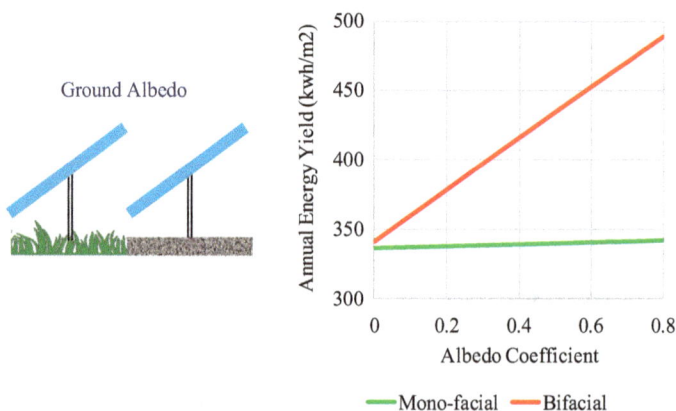

Figure 5.14   Effect of ground albedo on mono-facial and bifacial modules annual energy yield

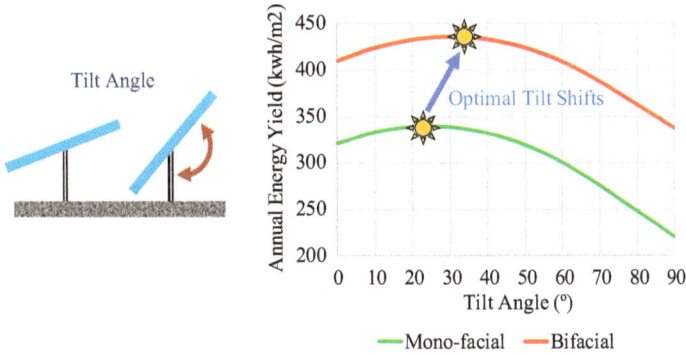

Figure 5.15 Effect of tilt angle on mono-facial and bifacial modules

5.5.5 Effect of module temperature

The difference in module temperature and ambient temperature ($T_m - T_{amb}$) against solar irradiance from the front side can be used to compare module temperature for the given climate conditions [30]. As per IEC [31] and ASTM [32] standards, the nominal module operating temperature (NMOT) is defined as the temperature of module under standard operating conditions such as ambient temperature of 20°C, wind velocity with a magnitude of 1 m/s, and a maximum POA irradiance of 800 W/m². Since it is difficult to meet such standard conditions, NMOT is used to screen and compare field data with solar irradiance above 400 W/m², wind velocity in the range of 0.25–1.75 m/s, and ambient temperature of 5–35°C. For such purpose, glass–glass mono-facial and bifacial modules were tested in the field. The gathered data was then plotted with front side solar irradiation as shown in Figure 5.16: orange dots represent mono-facial module temperature while green dots represent bifacial module temperature. Results showed that both modules were found to be working with similar operating conditions and the global average module temperature difference ($\Delta T = T_{bi} - T_{mono}$) between mono-facial and bifacial cells was observed as 0.31°C for same device configuration. However, in-depth analysis showed that ΔT changes with solar irradiation. For lower solar irradiance, mono-facial PV performed thermodynamically better than bifacial ones ($T_{bi} > T_{mono}$) and for higher solar irradiances bifacial PV module tempearture is lower than monofacial PVs ($T_{mono} > T_{bi}$). This is because bifacial gain is higher from the rear side due to the increased temporal distribution of albedo radiation at lower solar irradiance, which results in augmented heat generation [33].

5.5.6 Effect of ground coverage ratio

The ground coverage ratio (GCR) is defined as the ratio of the module area to the land area; this ratio represents active PV area, which is used for solar energy conversion. GCR calculation can be simplified for PV panels with the east west orientation scheme as the fraction of module length to row-to-row pitch/spacing as

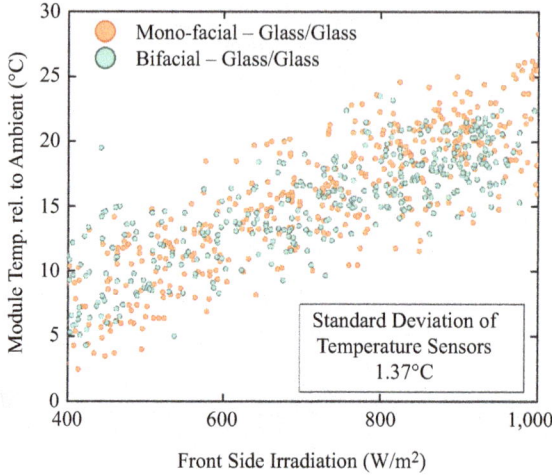

Figure 5.16   Module temperature of glass–glass mono-facial and bifacial PV modules

shown by (5.9). For both mono-facial and bifacial PV modules, GCR is critical factor for optimizing PV field in order to increase the plane of array irradiance and rear side influence (G_{rear}/G_{front}). In addition, it reflects land costs (CAPEX) that contributes to levelized cost for real-life PV systems. The shading characteristics associated with GCR depend on module tilt and height. These factors change the shading between rows, hence affecting the bifacial gain:

$$GCR = \frac{Module\ Length}{Row\text{-}to\text{-}Row\ Spacing}. \tag{5.9}$$

Figure 5.17 shows the effect of tilt angle/GCR (left panel) and elevation/GCR (right panel) on the bifacial gain (BG%) respectively. It is evident that lower GCR, higher tilt angle and increased mounting height would improve bifacial gain, which means higher energy yield with the same number of panels. Although higher energy yield is attractive but lower GCR requires more land area and need higher CAPEX. Thus, for PV systems, installation in regions where land costs are low such as Gulf countries [34], a low value of GCR could be implemented for optimum performance. However, for regions where land is expensive, higher value of GCR along with increased height would result in improved energy yield per unit area.

5.5.7   Effect of azimuth orientation

The azimuth angle expresses sunlight direction, for north, it is 0° and for south, it is 180° as shown in Figure 5.18, left panel. The energy yield of both mono-facial and bifacial modules is affected by azimuth orientation and can significantly alter the power profile. Generally, north of equator locations tend to face the module south i.e. azimuth angle of 180° and vice versa. Many studies are present in the literature

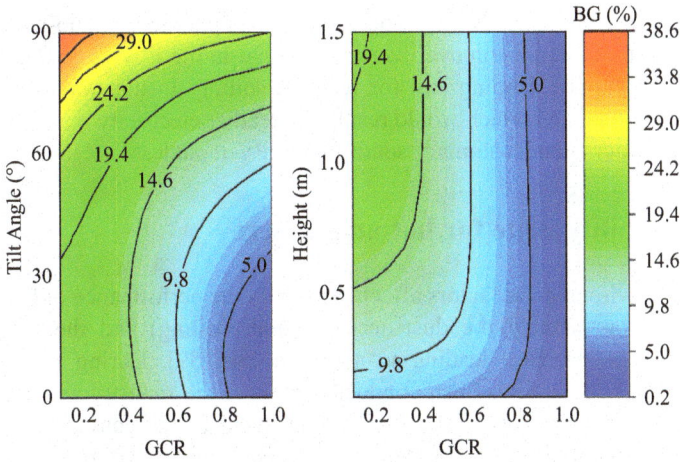

Figure 5.17   (Left) Ground coverage ratio (GCR) and its impact on bifacial gain (BG) with a tilt angle and (right) mounting height

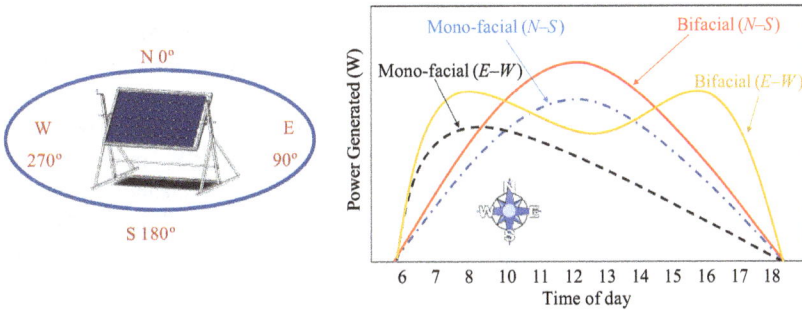

Figure 5.18   (Left) Schematic showing the azimuth angle and (right) typical power profiles for E–W and N–S mounted bifacial and mono-facial modules

that are focused on developing optimal tilt and azimuth orientation for PV modules and solar collectors' applications in different locations [35–37]. Figure 5.18 right panel shows the typical power generation profile during daylight hours for different mono-facial and bifacial arrangements. The PV modules in the south–north (S–N) azimuth orientation produce traditional yield profile for the maximum output at noon; energy generation from bifacial module is more than the mono-facial ones. Nevertheless, the shape of the profile for both types are similar. For the east–west (E–W) azimuth orientation, the energy generation rises as the day starts and reaches at peak before noon then it tends to decrease for the rest of the day in mono-facial configuration as the direct sunlight reaching to PV module decreases. However, for the bifacial systems, the power generation starts again to rise after noon, as it receives additional sunlight from the rear side unlike mono-facial. Therefore, the energy output for bifacial module would be much higher than mono-facial ones in the E–W configuration. Moreover, vertical bifacial in the E–W layout can be used for peak load profile shaping and for reduced soiling. In this way, energy yield profile can be flattened which would results in steadier electricity generation during the daylight hours than customary south facing PV modules.

5.6   Reliability issue for hot arid desert

Two major environmental factors affect drastically the performance of PV panels in the Gulf region, namely the dust accumulation (soiling) and the temperature-induced voltage decrease leading to a power loss [59]. During the last years, intensive R&D activity has been conducted by both the industry and academia to address these major issues. Temperature coefficient of the conversion efficiency represents an important figure of merit for the energy yield of a given photovoltaic technology, especially if planned to operate in Gulf region.

5.6.1   *Infrared thermography of PV modules to identify defects*

The outdoor test facility is a 35,000 m^2 field station for testing solar-energy technologies in Doha, Qatar. It operates more than 80 PV test systems, most of which were installed in 2012, and a meteo station (Figure 5.19).

  In this study, test data from 26 PV systems for the whole of 2017 was analyzed. Each system comprised module strings of approximately 1.5 kW$_p$, mounted at fixed tilt 22° south, connected to identical grid-tied inverters, cleaned once every 2 months. The number systems of each PV technology type are shown in Table 5.1. Due to confidentiality agreements with PV manufacturers that provided the modules, they are not identified by name. All modules are full-size commercial products (not prototypes or mini-modules). All of the c-Si modules were BSF technology except for one mono-Si module which was IBC.

Figure 5.19   PV test systems at the outdoor test facility in Doha, Qatar

Table 5.1   Module technologies of PV test systems used in the study

PV technology	Number of test systems
Mono-Si	2
Poly-Si	10
Bifacial	1
HIT	2
CIGS	8
CdTe	1
μSi-aSi	2
Total	26

String measurements used in this study were DC power (P) and module temperature (T_{mod}), being the average of the rear-side temperatures of three modules in each string. The only meteo measurement used was plane-of-array (22° tilt) irradiation (G_{POA}). All measurements were recorded once per minute. The study intentionally did not investigate effects of environmental parameters (e.g. wind speed, relative humidity) on T_{mod}, because its goal was to examine how different modules technologies performed under similar operating conditions, regardless of how those conditions were reached.

The power capacity (W_p) of each system at Standard Test Conditions (STC) was obtained from module flash-test reports provided by the manufacturer (i.e. before shipping and installation of the modules). The exception was the bifacial system, for which manufacturer flash data was not available, and whose W_p value was taken as the nameplate rating of front-side power.

Specific power (SP) is the DC power output of the PV string normalized by its W_p, and is here reported in units of [W/W$_p$]. Specific yield (SY) is the daily DC energy output (E) per W$_p$, units [Wh/day/W$_p$]. Performance ratio (PR) is SP normalized by the relative irradiation, a dimensionless fraction. Mathematically:

$$SP = P/Wp \tag{5.10}$$

$$SY = E/Wp \tag{5.11}$$

$$PR = SPG_{POA}/G_{STC} \tag{5.12}$$

where G_{STC} is the STC irradiation of 1,000 W/m^2.

Test systems occasionally exhibited faults during the year, resulting in near-zero PR. Each system's data was filtered out of analysis when its PR was less than 0.1. For properly-functioning PV systems, this was approximately 5–10% of daytime records; one CIGS system was intermittently offline which resulted in 24% rejected records.

As noted, a basic expectation is that PV modules that are less temperature sensitive will produce more energy in hot desert conditions. The average SY for each of the 26 test systems was plotted against its power temperature coefficient (PTC) (Figure 5.20). Contrary to expectation, there was essentially no correlation between modules' energy yield and their PTC. This observation also holds when considering modules of the same type—indeed multi-Si modules with higher-magnitude PTC actually produced more energy than those of lower-magnitude PTC. The high SY of the bifacial system is largely because its W_p was based only on its front-side power.

Because the results were against expectation, the analysis was repeated for 2013 field data, when most modules were less than one year old. The 2013 data is not as reliable as the 2017 data due to higher incidence of technical faults and modules coming online at different times, and is not presented here. However, it reproduced the overall finding that systems' SY were not correlated with their PTC. Also, the fact that a large number of test systems (26) was used in the study increases confidence in the results. The clear conclusion from the data is that PV systems' comparative energy yield in the field is not well predicted by their nameplate power temperature coefficients and that other factors are more important. The field measurements from this study also allowed energy "losses" attributable to low-light and W_p discrepancy to be quantified.

With the decrease in the cost of solar PV module globally by about 55% since 2013, the uptake of solar PV has increased dramatically. As these modules are exposed to ambient conditions in the field, they can develop defects or faults. These defects can affect the output power of the PV module and overall system output. To mitigate this, early and easy detection of defects is considered critical for operation and maintenance. Table 5.2 consolidates typical defects found in PV modules with their associated power losses. These include: soiling losses, micro-cracks, inactive strings, shading, snail trail. Some defects can be easily identified through the infrared (IR) imaging and the presence of hot-spots on the PV module. We highlight

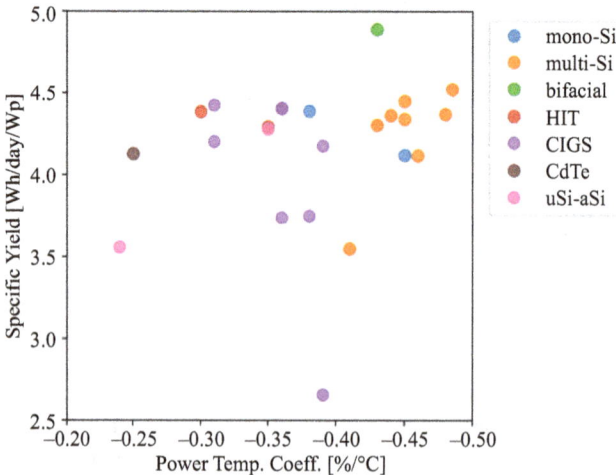

Figure 5.20   Average daily SY versus nameplate PTC of 26 PV test systems, for one year (2017). Adapted from our Ref. [12].

Table 5.2   Various defects types with their associated power losses

Defect type	Power reduction due to defect
Soiling/dust	10% [59]/4.5% [60]
Micro-cracks	2.5–10% [61]
	0.82–3.21% (poly-crystalline)
	0.55–0.9% (mono-crystalline) [62]
Inactive substrings	>2% [63]
Shading	30% [64]
Snail-trail	20% [65]

here the best operational and environmental conditions for conducting IR imaging of PV module to detect defects. With a minor temperature difference of 1.3–1.4°C compared to the adjacent healthy cells likely indicate the presence of internal defects such as shunt. These hot-spots only appeared at lower irradiance conditions irrespective of the cloud condition. The IR imaging on partially and cloudy days showed that the presence of intermittent clouds, high ambient temperature, and low wind speed helps the detection of these internal defects in the PV module [38].

This urges the necessity of having an inspection technology that could periodically monitor the health status of the PV modules without interrupting the normal operation. Also, under the circumstances when the performance of the PV plant reduces significantly to identify the cause and undertake measures accordingly to resolve the issue immediately.

According to previous research [39,40], infrared (IR) imaging could be the most competitive inspection technology which provides thermal or IR image of the PV module. The existing defects in the PV module appear as hot-spots, i.e., bright region compared to adjacent cells, in the IR images which quickens the process of detecting and locating defect. The major advantages of IR imaging-based inspection technology are (i) uninterrupted inspection of the PV module while operating under normal operation condition, (ii) requires less inspection time and reduces the involved costs, (iii) rapid detection of existing and premature defects, and (iv) flexibility of conducting rapid inspection of the system outweighs the other inspection technologies such as visual or field inspection, real-time monitoring of current–voltage (*I–V*) characteristics of the system, and electroluminescence (EL) imaging [41].

Automatic detection of defective PV modules from IR images using *image processing* tools is the future for rapid health assessment of PV modules of the PV plant to undertake required preventive or fault maintenance decisions [42–46]. The machine learning or image identification algorithms rely on good quality images to interpret. Therefore, it is important to generate high quality IR images that could be effectively analyzed for successful detection of defects through the detection of genuine hot-spots. The use of IR imaging as a tool to identify defects in PV modules is well established, however, the appearance of hotspots or other thermal signals has not been well investigated under non-ideal environmental conditions with consideration to the appearance of the hotspots.

Earlier studies [41,47–49] highlighted several concerns related to the operational conditions of IR imaging of the PV module which have direct impact on the creation of reliable IR images such as:

- Interference of sky reflectance.
- Addition of reflected radiation from surroundings with the reflected radiation from the PV module surface.
- Temperature measurement accuracy of hot-spots and good adjacent cells.
- Reliability of the appearance of hot-spots resulted because of internal and external defects.
- Position of the sun, solar irradiance, and ambient conditions has direct impact on the construction of reliable IR images.

Moretón *et al.* [48] stated the selection criterion for considering PV modules as defective (more than 10°C) or non-defective (less than 10°C) based on the normalized temperature difference (Δt) of the identified hot-spots. However, from the presented data, it was found that 25 and 31 hot-spots showed Δt of less than 10°C and more than 20°C, respectively, of which 4 and 10 hot-spots, respectively, caused output voltage loss of more than 10%. This signifies that a hot-spot with small Δt could be responsible for high power loss, and therefore, it is important to identify all defects in a PV module.

Solar irradiance of more than 600 W/m^2 [39,43–46] or 700 W/m^2, low wind speed of less than 4 m/s [44] or 6 m/s [43] or 1 m/s [49], clear sky condition, stable ambient temperature [39], and camera viewing angle of less than 60° [45,48] have been stated as the best operating conditions for IR imaging of the PV module. However, studies are missing to understand the impact on the appearance of internal and external defects of PV modules as hot-spots in the IR images captured under both different environmental and observational conditions.

Therefore, it is important to understand the conditions of capturing the IR image to detect the hot-spots. This study [50] aimed to answer the question, "Which environmental or observational conditions produce high quality IR images that can be used for identifying defects through image identification techniques." The following questions will be answered to address this aim:

1. How the detectability of hot-spots changes in the IR images of the PV module captured under different environmental conditions or observational conditions of IR imaging?
2. Are the detected hot-spots in the IR images of the PV module due to external or internal defects?
3. What is the impact of changing observational conditions of IR imaging on the temperature measurement of PV cells that appeared as hot-spots in the IR images?

Therefore, the main objective of this study is to investigate and identify the best environmental and operational conditions for IR imaging of the PV module to generate quality IR images and detect hot-spots. This study also investigates the impact of environmental and operational conditions of IR imaging on the detection of minor and major hot-spots in the IR images of the PV module resulted because of the presence of internal and external defects, respectively.

The stated temperature or temperature difference (Δt) between cells described in this chapter is determined based on the extracted temperatures of the respective cells from the IR image(s) are titled as "determined temperature." The measured temperatures of the respective cells determined by placing thermocouples at the back of the PV module are titled as "thermocouple temperature."

For consistency, all IR images of the PV module in this study were measured over a range of 35–55°. Only a few exceptions were required, mainly for the IR images captured at early or late solar hours of the day where a temperature scale of 30–45°C/55°C or 25–40°C/45°C or 35–45°C were applied to avoid losing the visual presence of hot-spots (Figure 5.21).

The identified environmental considerations are listed below:

(i) The appearance of minor hot-spots in the IR images because of internal defect such as shunted cell defect stimulates during IR imaging of the PV module under solar irradiance of close to or less than 600 W/m^2.

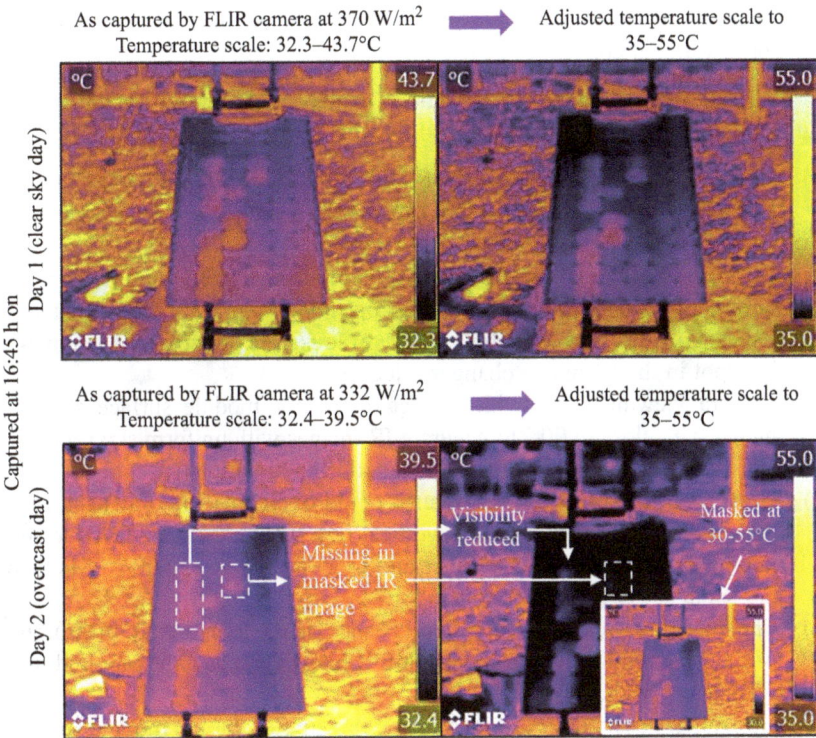

Figure 5.21 *An example to highlight the importance of temperature masking of IR images with an appropriate temperature scale. Adapted from Ref. [50].*

(ii)   Ambient condition such as high ambient temperature and low wind speed are highly suitable conditions for IR imaging of the PV module to detect minor hot-spots in the IR images even if captured under clear sky and high solar irradiance conditions. Also, helps to generate IR images with low sky reflectance.

(iii)   The presence of intermittent cloud in the sky view of the PV module surface was found to be a favorable condition for IR imaging of the PV module to detect minor hot-spots, however, IR imaging should be avoided after sky view of the PV module experiences extended cloud interference.

(iv)   Irrespective of the environmental conditions, maximum number to minor hot-spots with higher temperature difference is only visible in the IR images captured at late solar hours of the day.

(v)   Sky reflectance impact in the IR images can be avoided by IR imaging of the PV module on a clear sky day during and after the solar noon, this observation is not suitable for cloudy or partially cloudy conditions.

(vi)   IR images with about similar and comparatively low sky reflectance can be attained by performing IR imaging from a horizontal observation angle located in the north–west quarter while the sun is in the north–east quarter, vice versa.

(vii)   Importantly, minor hot-spots that resulted because of internal defects are highly sensitive to the environmental conditions while independent to the observational conditions of IR imaging.

The identified observational conditions which are highly suitable for IR imaging of the PV module are listed below:

(i)   Horizontal observation angles of $-30°$ to $+30°$ are highly suitable for IR imaging of the PV module and enhancing the accuracy of detecting, locating, and temperature measurement of the hot-spots.

(ii)   IR imaging of the PV module should be performed from multiple horizontal observation angles to ensure that the appeared hot-spots in the IR images are the genuine hot-spots, it will be a false or lookalike hot-spot if location of the hot-spot in the IR image changes, vice versa.

(iii)   Camera viewing angle with respect to the PV module surface should be higher, i.e., close to $90°$, to generate IR images with uniform sky reflectance and avoid the dominance of external reflected radiations.

(iv)   Higher vertical observation distance could overcome the drawback of low camera viewing angle, a higher vertical observation distance helps to maintain higher camera viewing angle and to avoid the appearance of lookalike hot-spots in the IR images.

(v)   The impact of vertical observation distance on the temperature measurement of the PV module surface is very low while IR images are captured at late solar hours of the day.

(vi)   Horizontal observation distance for IR imaging of the PV module should be selected in such a way that a full view of the PV module surface appears in the IR images. For accurate temperature measurement, short observation distance is better.

In summary, temperature difference analysis could be used as an effective tool for detecting and classifying hot-spots. For a clear sky condition, it is found that the measured temperature of individual cells should be compared with the temperature of the cells located at the top left edge to determine the temperature difference which will ensure the detection of minor hot-spots which might show a very small temperature difference. In conclusion, visual inspection of the IR images of the PV module is very important before processing with the automated image processing tool to ensure the hot-spot detection accuracy. The observations found in this study require further investigations in other regions to understand the applicability and differences.

5.7 Performance and reliability of crystalline-silicon photovoltaics in desert climate

In our study of [12], we summarized the performance of multi-crystalline, mono-crystalline, and silicon heterojunction PV arrays operating in desert climate. Indoor characterization of all PV modules from seven arrays was performed to determine the degradation rate. Visual and electroluminescence (EL) inspections were performed to gain more insight on the module failure. Encapsulant yellowing, back sheet cracking, and cell cracking were the most prominent failure modes observed after 5-years operation in the desert climate. At the OTF, the average daily insolation at standard tilt of $22°$ is 6.5 kWh/m^2/day and the annual average is 2,043 kWh/m^2/day. The average ambient air temperature is $35°C$, while module temperature may exceeds $70°C$ during summer months (in open rack mounting). The average relative humidity varies between 30% and 60%.

Our main findings show the following.

5.7.1 PV performance monitoring and degradation rate

The RdTools, well-established method in the field created by the national renewable energy laboratory (NREL), have been implemented for the PV system degradation analysis using the actual field data for the PV arrays under study. The degradation rate calculation uses Year-On-Year (YOY) method where a distribution of degradation rate is calculated for each PV system [51]. The YOY compares the normalized daily yield separated for each year and, therefore, robust for the seasonal variations. For more details about the RdTool, the reader is referred to Refs. [2,52]). Generally, due to seasonal variations, at least 3–5 years is needed to obtain an accurate degradation rate calculation [53]. In our study, the PV system analysis was done based on the data from 2014-01-01 to 2018-12-31. The input parameters being used were location latitude and longitude, time zone, weather data, plane of array (POA) irradiance data, PV module technology temperature coefficient rated power, PV array installation tilt angle, PV array installed capacity, and AC generated power. The installed capacity is given in Table 5.3.

Table 5.3   Electrical parameters and temperature coefficient (TC) of the various
           PV modules in OTF

PV module ID	PV module electrical parameters						Number of module [–]	Installed capacity [Wp]
	P_{max} [W]	I_{sc} [A]	V_{oc} [V]	TC P_{max} γ[%/°C]	TC V_{oc} β[%/°C]	TC I_{sc} α[%/°C]		
Multi-crystalline silicon (Multi_A)	295	8.75	44	−0.43	−0.32	0.059	6	1,800
Multi-crystalline silicon (Multi_B)	247	8.56	37	−0.45	−0.32	0.076	8	1,963
Multi-crystalline silicon (Multi_D)	280	8.33	44	−0.44	−0.33	0.055	7	2,016
Multi-crystalline silicon (Multi_E)	235	8.35	37	−0.48	−0.34	0.034	8	1,906
Multi-crystalline silicon (Multi_F)	235	8.54	37	−0.45	−0.37	0.060	8	1,883
Mono-crystalline silicon (Mono_G)	260	8.91	38	−0.45	−0.33	0.040	8	2,095
Silicon hetero-junction (SHJ_H)	235	5.84	51	−0.30	−0.24	0.029	8	1,954

The calculated data is further processed through the techniques of normal-ization, filtering, aggregation, and then the degradation analysis. During the degradation analysis, YOY method is utilized in order to calculate the rate of degradation based on two points, which are appearing at the same time in succes-sive years. The AC power and degradation rate are shown in Figure 5.22. The performance of the PV system can be observed from the central tendency of the generated histograms. The results showed that the PV system degradation rates of (Multi_D), (Multi_E), (Multi_F), (Mono_G), and (SHJ_H) arrays were −2.65%/year, −1.91%/year, −0.14%/year, −1.59%/year, and −0.62%/year, respectively. The measured degradation rate of Multi_F, Mono_G, and SHJ_H agrees with the values reported in the literature [54].

The average P_{max} from the measured IV curve at STC for all modules were compared with the initial values provided by the manufacturer. The annual PV module degradation rate was calculated and presented in Figure 5.23 for each silicon technology. Degradation rate presented in a negative value to indicate the decrease in the PV module IV parameters with time. The lowest annual degra-dation rate was measured for the multi-crystalline silicon module (Multi_A), (Multi_B), and (Multi_E). Multi-crystalline silicon (Multi_D) and mono-crystalline silicon (Mono_G) showed the highest PV module power degradation of −3.0%/year and 6.3%/year, respectively. This drop is attributed to a decrease in the measured FF (not shown in this plot) of (Mono_G) and (Multi_D) which were 62% and 60%, respectively, and mainly due to high series resistance of 0.6 O caused by a degradation of the interconnects.

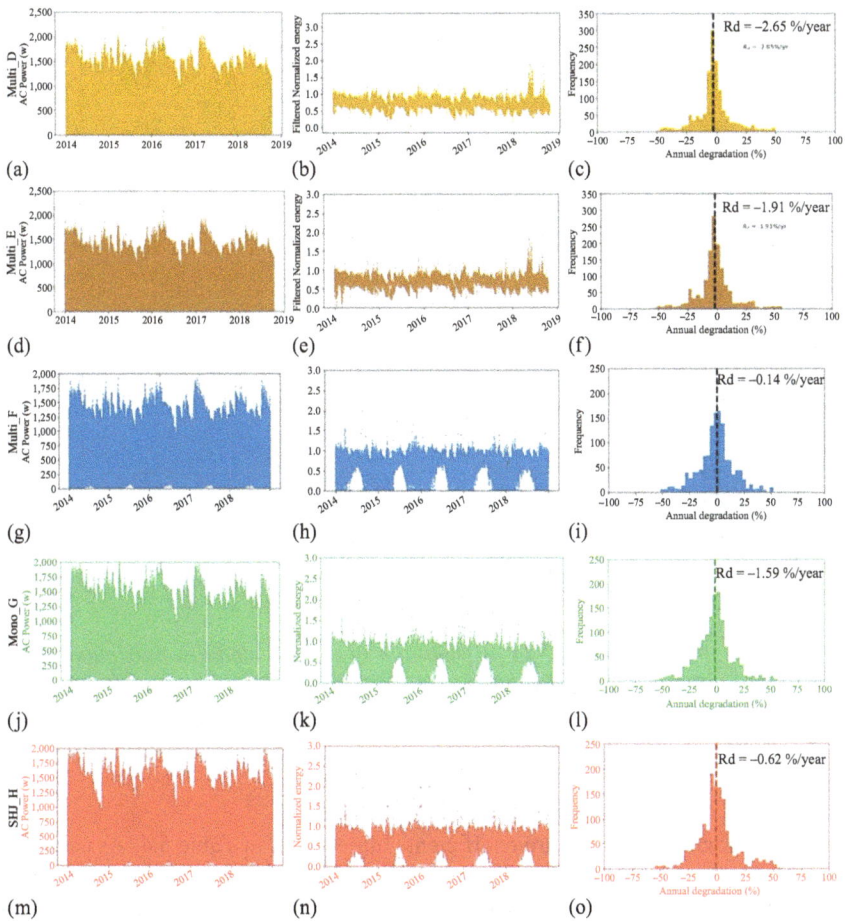

Figure 5.22   For the PV silicon technology under study, the AC power, normalized energy and the calculated degradation rate for: Multi_D, Multi_E, Multi_F, Mono_G and SHJ_H, respectively. Adapted from our Ref. [12].

5.7.2   Temperature coefficients (TC) measurement

In a desert climate, the operating temperature of PV modules may approach 70°C during summer months. Measuring the electrical parameters of PV modules under various operating temperature is important to determine the expected energy yield generated from the PV modules and to assess the performance of the elevated operating temperature. Temperature performance at 15°C, 25°C, 50°C, and 70°C and a constant irradiance of 1,000 W/m^2 was measured. The drop of PV module voltage and power with increasing the module temperature is clear.

For all silicon PV technology under study, the PV module electrical parameters (P_{max}, I_{sc}, V_{oc}, and FF) dependency on temperature are plotted in Figure 5.24 with

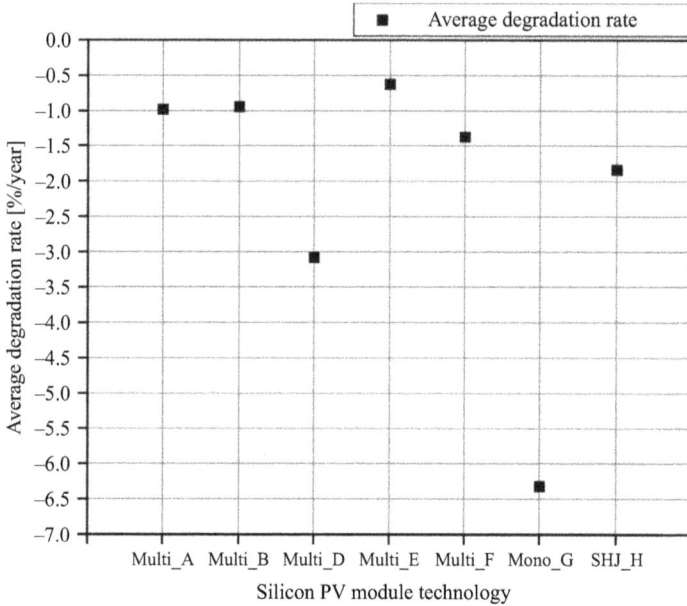

Figure 5.23   Annual PV module degradation of the PV modules maximum power (P_{max}) of different silicon PV technologies installed in desert climate between 2014 and 2018. Each point is an average of PV modules measured (see number of module in each array, Table 5.3).

their corresponding temperature coefficients highlighted in Table 5.4. As expected, the module performances P_{max}, V_{oc}, and FF degrade with increasing temperature. The decrease in V_{oc} with temperature is associated with the increase in saturation current density [55]. A slight increase in the I_{sc} is observed in Figure 5.24(b). At high operating cell temperature, the bandgap of the solar cell material decreases, explaining the slight increase in the I_{sc} [55].

5.7.3   Visual inspection

The visual inspection revealed yellowing of the encapsulant and cracked back sheet of some of the PV modules under study (Figure 5.25). Out of the seven PV module manufacturers, only (Multi_F) and (Multi_D) arrays showed yellowing of the encapsulant. The first indication of yellowing was visible during the visual inspection conducted on February 7, 2017. Figure 5.25(a) shows yellowing of encapsulant material over large area of a number of PV modules. Yellowing of the encapsulant material is a known failure mode observed in various PV sites with high UV irradiation at lower wavelength and high operating temperature. The main cause of this type of discoloration of the encapsulant is the degradation of the ethylene-vinyl acetate (EVA) upon UV radiation in combination with high temperature and water ingress [56,57]. This type of failure mode may result in loss of

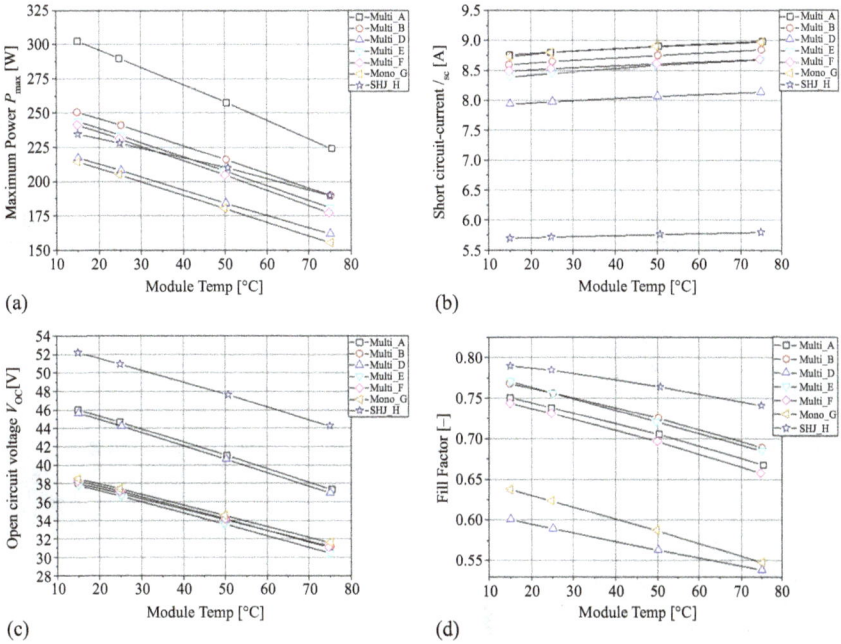

(a)

(b)

(c)

(d)

Figure 5.24 *Electrical parameters temperature dependence of different silicon PV module technologies. The temperature coefficient of (a) P_{max}, (b) I_{sc}, (c) V_{oc}, and (d) FF were determined from the linear fitting of the normalized measurement points. Adapted from our Ref. [12].*

(a)

(b)

Figure 5.25 *(a) Visual inspection photographs showing yellowing observed over large area at the front multi-crystalline silicon array (module Multi_F). (b) Visual inspection photographs PV module with polyamide-based back sheet cracking (indicated by arrows) after exposure for 5-years in desert climate (Multi_B). Not all PV module in this array showed back sheet cracking. Adapted from our Ref. [12].*

Table 5.4   Measured temperature coefficient of the silicon PV module technologies

Empty cell	Multi_A	Multi_B	Multi_D	Multi_E	Multi_F	Mono_G	SHJ_H
P_{max} (γ) [%/°C]	−0.44	−0.42	−0.45	−0.45	−0.46	−0.48	−0.31
V_{oc} (β) [%/°C]	−0.32	−0.31	−0.33	−0.33	−0.32	−0.31	−0.25
I_{sc} (α) [%/°C]	0.042	0.048	0.043	0.057	0.038	0.044	0.033

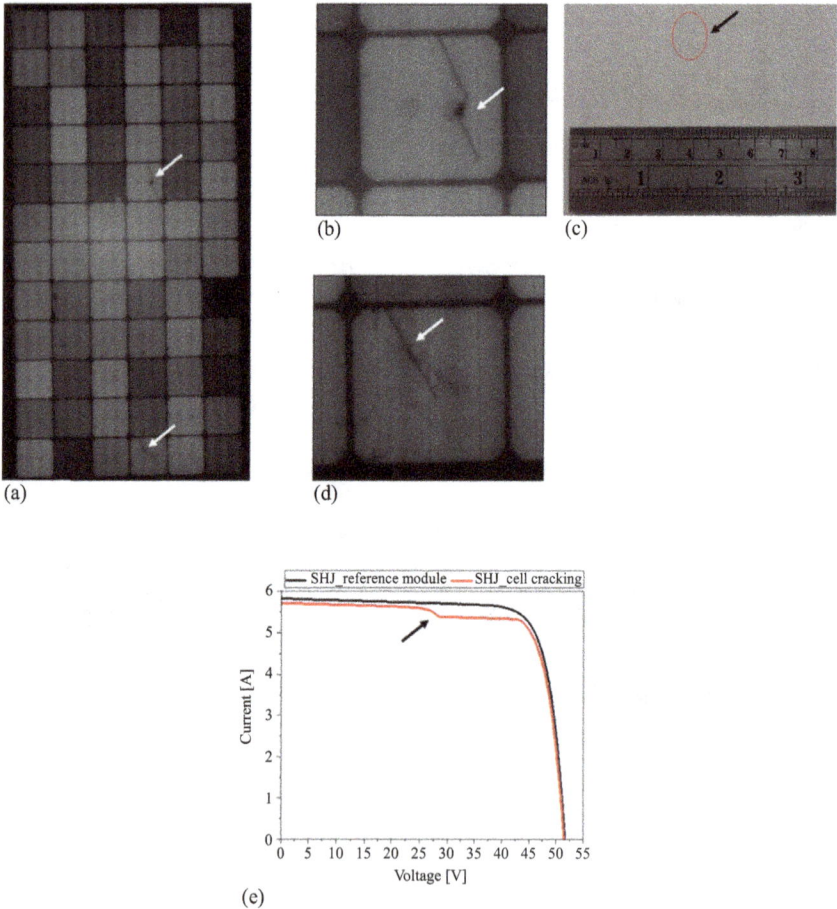

Figure 5.26   (a) Indoor EL image of a silicon heterojunction PV module (SHJ_H) after 5-years exposure in the field. (b) and (d) The magnified image of the cracked cell shown in (a). (c) The rear side of the PV module where an indent with a sharp object (indicated by a red circle). (e) Comparing the IV curve of the reference module and the module with cracked cells. Adapted from our Ref. [12].

light transmittance to the solar cells and/or delamination of the encapsulant material from the solar cell and the glass, which ultimately leads to the reduction in I_{sc} of the module and decreases its power output. However, for (Multi_F) and (Multi_D), no change of the measured I_{sc} was observed after 5-years exposure in the field.

From the seven technologies, only a number of modules from the multi-crystalline silicon array have shown back sheet cracking, for example (Multi_B) did not show back sheet cracking. Figure 5.8(b) shows the back sheet cracking observed in the field. The first back sheet cracking was spotted during the visual inspection conducted on December 15th, 2018. Not all PV module in this array showed back sheet cracking suggesting each module may use different back sheet Bill of Material (BOM). No significant drop in the power output was observed for the (Multi_B). However, a combination of moisture penetration, back sheet racking may results in electrical safety issue of the PV array and as a result loss of the insulation properties of the PV module.

5.7.4   Electroluminescence imaging

The electroluminescence (EL) imaging in combination with IV measurement was used to understand module degradation. Figure 5.26 shows the EL images of the mono-crystalline silicon (Mono_G) and multi-crystalline silicon (Mulit_D) with the measured degradation rate of -4.3%/year and -3.5%/year, respectively. The EL image in Figure 5.26 is showing the shunting indicated by dark patterns, i.e., reduced electroluminescence effect. This is evidence with the IV curve measurement at STC, both (Mono_G) and (Mulit_D) modules show a low shunt resistance of 60 and 81 O, respectively, and a high series resistance of 0.69 and 0.65 O, respectively. The latter is mainly due to the degradation of the solar cells interconnection. Similar results as presented in Figure 5.26 was reported recently by Gaulding *et al.* [54] where a difference in metallization BOM led to a high-series resistance and power degradation.

Cell cracking was observed in the EL image (Figure 5.26(a)). In Figure 5.26 (a), the crack initiated from the mid of the cell and run across the cell to the interconnecting busbar. Investigating the module rear side (see Figure 5.26(c)), an indent with a sharp object was visible and is believed to cause the cell cracking observed by the EL image in Figure 5.26(b). As a result of the cell cracking, a drop of the module current was observed on the IV curve as shown in Figure 5.26(e) (see the break point in the upper part of the *IV* curve). Cell mismatch was observed when taking EL images at a lower I_{sc} and indicated by the difference in the cell luminescence, i.e., dark cells means less luminescence where no or less current is generated.

The *IV* curve of the cracked cells shown in Figure 5.26(e) is compared with the reference PV module without cell cracking. A breakpoint was observed on the *IV* curve of the defective module. Further, the electrical parameters of the reference module are $P_{max} = 237$ W, $V_{oc} = 51.6$, and $I_{sc} = 5.8$ A. The electrical parameters of the cracked PV module are $P_{max} = 230$ W, $V_{oc} = 51.4$, and $I_{sc} = 5.7$ A. The drop in the P_{max} is approximately 4%. Not only cell cracks are responsible of this drop, we believe that several parameters could contributed to this drop in power output after 5-years of operation in the field. We found an increase of the series resistance Rs by 20% when comparing the Rs of the reference module Rs = 0.5 Ohm with Rs = 0.6 Ohm for the module with the cell cracking. Since the FF of both modules remain the same 78%, there is no significant change in the module current. The shunt resistance of the reference module is Rsh = 253 Ohm, while for the cracked module showed a decrease in the module shunt resistance Rsh = 242 Ohm, which suggest that the cell cracking and shunt contribute to the drop in the P_{max}. It should be noted here that the above module failure observations could be used as an indication of possible causes of module degradation.

5.8   Conclusions

PV system performance and reliability in desert climate depend on many parameters, among them, the effect of the high-operating temperature and the high-UV irradiance. In this chapter, we compared the performance and reliability of different silicon PV module technologies after 5-years of operation in desert climate.

From the indoor IV measurement of the all PV modules returned from the field, the PV module power loss is quantified. Out of seven different silicon manufacturers, multi-crystalline silicon (Multi_A), (Multi_B), and (Multi_E) showed the lowest annual degradation rate below 1%/year. Multi-crystalline silicon (Multi_D) and mono-crystalline silicon (Mono_G) showed the highest PV module power degradation of −3.0%/year and 6.3%/year, respectively. The degradation of the FF contributed significantly to the drop in module power degradation. Silicon heterojunction showed the lowest temperature coefficient. From the outdoor performance, PV arrays degradation rate for (Multi_D), (Multi_E), (Multi_F), (Mono_G), and (SHJ_H) of −2.65%/year, −1.91%/year, −0.14%/year, −1.59%/year, and −0.62%/year, respectively, was calculated using the RdTool. These results were found to agree with recent publication for the average degradation rate in hot climate. It is important to note here that, having a large sample size for indoor measurements is required to be able to perform a comprehensive statistical analysis to enable accurate determination of module degradation rate and possible main cause of module failure and to compare with outdoor degradation rate.

Back sheet cracking and encapsulant yellowing were the most common failure mode observed in the field during the first 5-years of operation. These findings agreed with the published PV system failure, which generally occurred during the 1–6 year of operation and at the end of the PV system life. The high UV irradiance and high operating temperature are suggested to be responsible for the occurrence of these failures. Electroluminescence imaging of silicon PV modules returned from the field revealed the presence of solar cell cracking for some of the modules. Power loss is correlated to a combination of low shunt resistance, high series resistance, solar cell cracking, and cell mismatch.

References

[1] Jordan, D.J, Kurtz, S.R., VanSant, K., and Newmiller, J., 2016. Compendium of photovoltaic degradation rates. *Prog. Photovoltaics Res. Appl.* 24, 978–989. https://doi.org/ 10.1002/pip.

[2] Jordan, D.C., Wohlgemuth, J.H., and Kurtz, S.R., 2012. Technology and climate trends in PV module degradation. In *27th European Photovoltaic Solar Energy. Conference and Exhibition.* https://doi.org/10.4229/27thEUP VSEC2012-4DO.5.1.

[3] Maghami, M.R., Hizam, H., Gomes, C., Radzi, M.A., Rezadad, M.I., and Hajighorbani, S., 2016. Power loss due to soiling on solar panel: a review. *Renew. Sustain. Energy Rev.* 59, 1307–1316.

[4] Kaaya, I., Ascencio-Vasquez, J., Weiss, K.A., and Topic, M., 2021. Assessment of uncertainties and variations in PV modules degradation rates and lifetime predictions using physical models. *Solar Energy* 218, 354–367.

[5] Phinikarides, A., Kindyni, N., Makrides, G., and Georghiou, G.E., 2014. Review of photovoltaic degradation rate methodologies. *Renew. Sustain. Energy Rev.* 40, 143–152. https://doi.org/10.1016/j.rser.2014.07.155.

[6] Jordan, D.C., Smith, R.M., Osterwald, C.R., Gelak, E., and Kurtz, S.R., 2010. Outdoor PV degradation comparison. In *Conference Record of the IEEE Photovoltaic Specialists Conference.* https://doi.org/10.1109/PVSC.2010.5616925.

[7] Kurtz, S.R. and Jordan, D.C., 2013. Photovoltaic degradation rates—an analytical review. *Prog. Photovoltaics Res. Appl.* 21, 12–29. https://doi.org/ 10.1002/pip.

[8] Lillo-Sanchez, L., López-Lara, G., Vera-Medina, J., Perez-Aparicio, E., and Lillo-Bravo, I., 2021. Degradation analysis of photovoltaic modules after operating for 22 years. A case study with comparisons. *Solar Energy* 222, 84–94. https://doi.org/10.1016/j.solener.2021.04.026.

[9] Al-Otaibi, A., Al-Qattan, A., Fairouz, F., and Al-Mulla, A., 2015. Performance evaluation of photovoltaic systems on Kuwaiti schools' rooftop. *Energy Convers. Manage.* 95, 110–119. https://doi.org/10.1016/j. enconman.2015.02.039.

[10] Al Harbi, Y., Eugenio, N.N., and Al Zahrani, S., 1998. Photovoltaic-thermal solar energy experiment in Saudi Arabia. *Renew. Energy* 15, 483–486. https://doi.org/10.1016/ s0960-1481(98)00209-2.

[11] Jordan, D.C., Silverman, T.J., Sekulic, B., and Kurtz, S.R., 2017. PV degradation curves: non-linearities and failure modes. *Prog. Photovoltaics Res. Appl.* 25, 583–591. https:// doi.org/10.1002/pip.2835.

[12] Abdallah, A.A., Ali, K., and Kivambe, M., 2023. Performance and reliability of crystalline-silicon photovoltaics in desert climate. *Solar Energy* 249, 268–277.

[13] Wohlgemuth, J.H., Cunningham, D.W., Monus, P., Miller, J., and Nguyen, A., 2006. Long term reliability of photovoltaic modules. In *Conference Record 2006 IEEE 4th World Conference on Photovoltaics Energy Conversion,* pp. 2050–2053.

[14] Zemen, Y., Prewitz, T., Geipel, T., Pingel, S., and Berghold, J., 2010. The impact of yield strength of the interconnector on the internal stress of solar cell within a module. In *25th European Photovoltaic Solar Energy Conference and Exhibition/5th World Conference on Photovoltaic Energy Conversion*, 6–10 September, Valencia, Spain.

[15] Sugibuchi, K, Ishikswa, N., and Obara, S., 2013. Bifacial-PV power output gain in the field test using "EarthON" high bifaciality solar cells. In *Proceedings of the 28th European Photovoltaic Solar Energy Conference*, Paris, France, pp. 4312–4317.

[16] Comparotto, C., Noebels, M., Popescu, L., *et al.*, 2014. Bifacial n-type solar modules: indoor and outdoor evaluation. In *Proceedings of the 29th EU PVSEC*, Amsterdam, The Netherlands.

[17] R. Guerrero-Lemus, R., Vega, R., Kim, T., Kimm, A., and Shephard L.E., 2016. Bifacial solar photovoltaics – A technology review. *Renewable and Sustainable Energy Reviews* 60, 1533–1549.

[18] Abdallah A., Abotaleb A., Buffière M., Stephan G., Richter, S., and Hagendorf C., 2018. Delamination of CIGS thin film photovoltaic module in desert climate. *Environment. 35th European Photovoltaic Solar Energy Conference and Exhibition*.

[19] Kopecek, R., Veschetti, Y., Gerritsen, E., *et al.*, 2014. Bifaciality: one small step for technology, one giant leap for kWh cost reduction. *Photovoltaics Int.* 26, 32–45.

[20] Söderström, T., Papet, P., and Ufheil, J., 2014. Smart wire connection technology. In *28th European Photovoltaic Solar Energy Conference and Exhibition*.

[21] Halm, A., Aulehla, S., Schneider, A., *et al.*, 2014. Encapsulation losses fro ribbon contacted n-type IBC solar cells. In *29th European Photovoltaic Solar Energy Conference and Exhibition*.

[22] Castillo-Aguilella, J.E. and Hauser, P.S., 2016. Multi-variable bifacial photovoltaic module test results and best-fit annual bifacial energy yield model. *IEEE Access* 4, 498–506. doi:10.1109/ACCESS.2016.2518399.

[23] Singh, J.P., Guo, S., Peters, I.M., Aberle, A.G., and Walsh, T.M., 2015. Comparison of glass/glass and glass/backsheet PV modules using bifacial silicon solar cells. *IEEE J Photovoltaics* 5, 783–791.

[24] Singh, J.P., Aberle, A.G., and Walsh, T.M., 2014. Electrical characterization method for bifacial photovoltaic modules. *Sol Energy Mater Sol Cells* 127, 136–142. doi:10.1016/j.solmat.2014.04.017.

[25] Yusufoglu, U.A., Pletzer, T.M., Koduvelikulathu, L.J., Comparotto, C., Kopecek, R., and Kurz, H., 2015. Analysis of the annual performance of bifacial modules and optimization methods. *IEEE J Photovoltaics* 5, 320–328. doi:10.1109/JPHOTOV.2014.2364406.

[26] Guerrero-Lemus, R., Vega, R., Kim, T., Kimm, A., and Shephard, L.E., 2016. Bifacial solar photovoltaics – a technology review. *Renew Sustain Energy Rev.* 60, 1533–1549. doi:10.1016/j.rser.2016.03.041.

[27] Obara, S., Konno, D., Utsugi, Y., and Morel, J., 2014. Analysis of output power and capacity reduction in electrical storage facilities by peak shift control of PV system with bifacial modules. *Appl Energy* 128, 35–48. doi:10.1016/j.apenergy.2014.04.053.

[28] Ooshaksaraei, P., Sopian, K., Zulkifli, R., Alghoul, M.A., and Zaidi, S.H., 2013. Characterization of a bifacial photovoltaic panel integrated with external diffuse and semimirror type reflectors. *Int J Photoenergy* 2013, Article ID 465837. doi:10.1155/2013/465837.

[29] Faiman, D., Berman, D., Bukobza, D., *et al.*, 2003. A field method for determining the efficiency of each face of a bi-facial photovoltaic module. In *3rd World Conference on Photovoltaic Energy Conversion, Proceedings* 2003, 2, 1988–1991.

[30] Neises, T.W., Klein, S.A., and Reindl, D.T., 2012 Development of a thermal model for photovoltaic modules and analysis of NOCT cuidelines. *J Sol Energy Eng* 134, 011009. doi:10.1115/1.4005340.

[31] IEC61215. Crystalline Silicon Terrestrial Photovoltaic (PV) Modules–Design Qualification and Type Approval, 2005.

[32] ASTM-E1036-08. Standard Test Methods for Electrical Performance of Nonconcentrator Terrestrial Photovoltaic Modules and Arrays Using Reference Cells, Annual Book of ASTM Standards. 2008.

[33] Lamers, M.W.P.E., Özkalay, E., Gali, R.S.R., *et al.* 2018 Temperature effects of bifacial modules: hotter or cooler? *Sol Energy Mater Sol Cells* 185, 192–197. doi:10.1016/j.solmat.2018.05.033.

[34] Apostoleris, H., Sgouridis, S., Stefancich, M., and Chiesa, M., 2018. Evaluating the factors that led to low-priced solar electricity projects in the Middle East. *Nat Energy* 3(12), 1109–1114. doi:10.1038/s41560-018-0256-3.

[35] Yan, R., Saha, T.K., Meredith, P., and Goodwin, S., 2013. Analysis of year-long performance of differently tilted photovoltaic systems in Brisbane, Australia. *Energy Convers. Manag.* 74, 102–108. doi:10.1016/j.enconman.2013.05.007.

[36] Jafarkazemi, F and Saadabadi, S.A., 2013. Optimum tilt angle and orientation of solar surfaces in Abu Dhabi, UAE. *Renew. Energy* 56, 44–49. doi:10.1016/j.renene.2012.10.036.

[37] Bakirci, K., 2012. General models for optimum tilt angles of solar panels: Turkey case study. *Renew. Sustain. Energy Rev.*16, 6149–6159. doi:10.1016/j.rser.2012.07.009.

[38] Allet, N., Baumgartner, F., Sutterlueti, J., Schrier, L., Pezzotti, M., and Haller, J., 2011. Evaluation of PV system performance of five different PV module technologies. In *26th European Photovoltaic Solar Energy Conference*, 5–9 September 2011.

[39] Jahn, U., 2018. Review on Infrared and Electroluminescence Imaging for PV Field Applications, Report IEA-PVPS T13-10:2018.

[40] Kandeal, A.W., Elkadeem, M.R., Kumar Thakur, A., *et al.*, 2021. Infrared thermography-based condition monitoring of solar photovoltaic systems: a mini review of recent advances. *Sol. Energy* 223, 33–43.

[41] Rahaman, S.A., Urmee, T., and Parlevliet, D.A., 2020. PV system defects identification using Remotely Piloted Aircraft (RPA) based infrared (IR) imaging: a review. *Sol. Energy* 206 (June), 579–595. https://doi.org/10.1016/j.solener.2020.06.014.

[42] Kim, D., Youn, J., and Kim, C., 2016. Automatic photovoltaic panel area extraction from UAV thermal infrared images. *J. Korean Soc. Surv. Geod. Photogramm. Cartogr.* 34(6), 559–568. https://doi.org/10.7848/ksgpc.2016.34.6.559.

[43] Zefri, Y., ElKettani, A., Sebari, I., and Lamallam, S.A., 2018. Thermal infrared and visual inspection of photovoltaic installations by UAV photogrammetry—application case: morocco. *Drones* 2(4), 41. https://doi.org/10.3390/drones2040041.

[44] Kirsten Vidal de Oliveira, A., Aghaei, M., and Rüther, R., 2020. Aerial infrared thermography for low-cost and fast fault detection in utility-scale PV power plants. *Sol. Energy* 211, 712–724.

[45] Salazar, A.M., 2016. Hotspots detection in photovoltaic modules using infrared thermography. In *MATEC Web of Conferences*, vol. 10015.

[46] Jaffery, Z.A., Dubey, A.K., Irshad, Haque, A., 2017. Scheme for predictive fault diagnosis in photo-voltaic modules using thermal imaging. *Infrared Phys. Technol.* 83, 182–187. https://doi.org/10.1016/j.infrared.2017.04.015.

[47] Alvarez-Tey, G., Jiménez-Castañeda, R., and Carpio, J., 2017. Analysis of the configuration and the location of thermographic equipment for the inspection in photovoltaic systems. *Infrared Phys. Technol.* 87, 40–46. https://doi.org/10.1016/j. infrared.2017.09.022.

[48] Moretón, R., Lorenzo, E., and Narvarte, L., 2015. Experimental observations on hot-spots and derived acceptance/rejection criteria. *Solar Energy* 118, 28–40. doi: 10.1016/j. solener.2015.05.009.

[49] Moreton, R., Lorenzo, E., Leloux, J., and Carrillo, J.M., 2014. Dealing in practice with hot-spots. In *Proceedings of the 29th European Photovoltaic Specialists Conference*, Amsterdam, pp. 2722–2727.

[50] Rahaman, S.A., David, T.U., and Parlevliet, A. 2022. *Solar Energy*, 245, 231–253.

[51] Hasselbrink, E., Anderson, M., Defreitas, Z., *et al.*, 2013. Validation of the PVLife model using 3 million module-years of live site data. In *Conference Record of the IEEE Photovoltaic Specialists Conference*, pp. 7–12. https://doi.org/10.1109/PVSC.2013.6744087.

[52] Deline, C., Muller, M., Deceglie, M., *et al.*, 2020. PV Fleet Performance Data Initiative: March 2020 Methodology Report.

[53] Osterwald, C.R., Adelstein, J., Del Cueto, J.A., Kroposki, B., Trudell, D., and Moriarty, T., 2006. Comparison of degradation rates of individual modules held at maximum power. In *2006 IEEE 4th World Conference on Photovoltaic Energy Conversion, WCPEC-4*, pp. 2085–2088. https://doi.org/10.1109/WCPEC.2006.279914.

[54] Korgaonkar, R. and Shiradkar, N., 2021. Origin and mitigation of extreme performance loss rate values calculated using RdTools. In: *2021 IEEE 48th*

Photovoltaic Specialists Conference (PVSC), pp. 2534–2538. https://doi.org/ 10.1109/ pvsc43889.2021.9518592.

[55] Green, M.A., 2003. General temperature dependence of solar cell performance and implications for device modelling. *Prog. Photovoltaics Res. Appl.* 11, 333–340. https://doi.org/10.1002/pip.496.

[56] Oreski, G. and Wallner, G.M., 2009. Evaluation of the aging behavior of ethylene copolymer films for solar applications under accelerated weathering conditions. *Sol. Energy* 83, 1040–1047. https://doi.org/10.1016/j. solener.2009.01.009.

[57] Wenger, H.J., Schaefer, J., Rosenthal, A., Hammond, B., and Schlueter, L., 1991. Decline of the Carrisa Plains PV power plant: the impact of concentrating sunlight on flat plates. In *The Conference Record of the Twenty-Second IEEE Photovoltaic Specialists Conference*, pp. 586–592. https://doi. org/10.1109/pvsc.1991.169280.

[58] Gaulding, E.A., Mangum, J.S., Johnston, S., *et al.*, 2022. Differences in printed contacts lead to susceptibility of silicon cells to series resistance degradation. *IEEE J. Photovoltaics* 12, 690–695. https://doi.org/10.1109/ JPHOTOV.2022.3150727.

[59] Zeedan, A., Barakeh, A., Al-fakhroo, K., Touati, F., and Gonzales, A.S.P., 2021. Quantification of PV power and economic losses due to soiling in Qatar. *Sustainability* 13(6), 3364.

[60] Tanesab, J., Parlevliet, D., Whale, J., and Urmee, T., 2018. Energy and economic losses caused by dust on residential photovoltaic (PV) systems deployed in different climate areas. *Renew. Energy* 120, 401–412.

[61] Kontges, M., Kunze, I., Kajari-schr, S., Breitenmoser, X., and Bjørneklett, B., 2011. The risk of power loss in crystalline silicon based photovoltaic modules due to micro-cracks. *Sol. Energy Mater. Sol. Cells* 95, 1131–1137.

[62] Bdour, M., Dalala, Z., Al-Addous, M., Radaideh, A., and Al-Sadi, A., 2020. A comprehensive evaluation on types of microcracks and possible effects on power degradation in photovoltaic solar panels. *Sustain* 12(16), 6416.

[63] Dalsass, M., Scheuerpflug, H., Fecher, F. W., Buerhop-Lutz, C., Camus, C., and Brabec, C. J., 2016. Correlation between the generated string powers of a photovoltaic power plant and module defects detected by aerial thermography. In: *2016 IEEE 43rd Photovoltaic Specialists Conference (PVSC)*, pp. 1–6. doi:10.1109/ PVSC.2017.8366737.

[64] Dolara, A., Lazaroiu, G.C., Leva, S., and Manzolini, G., 2013. Experimental investigation of partial shading scenarios on PV (photovoltaic) modules. *Energy* 55, 466–475. https://doi.org/10.1016/j.energy.2013.04.009.

[65] Quater, P.B., Grimaccia, F., Leva, S., Mussetta, M., and Aghaei, M., 2014. Light Unmanned Aerial Vehicles (UAVs) for cooperative inspection of PV plants. *IEEE J. Photovolt.* 4(4), 1107–1113. https://doi.org/10.1109/ JPHOTOV.2014.2323714.

Chapter 6

Bifacial solar technology and module installation

Ahmer A.B. Baloch[1], Brahim Aïssa[2] and Nouar Tabet[3]

6.1 Introduction

As the solar cells approach theoretical limits, the room for improvement is reducing due to diminishing returns on conversion efficiency. To push the state-of-the-art solar cells for their peak performance and overcome limiting factors, innovative approaches are required at each stage of the photovoltaic (PV) chain to decrease the efficiency gap and the levelized cost of electricity. Bifacial PV is one such technology that can harness incoming solar radiation from both front and rear sides to produce more energy yield than its counterpart traditional monofacial PV [1–4]. It has the potential to minimize the negative soiling effect and enhance energy generation under hot desert environment [5–9]. Vertical bifacial module facing east–west is one such configuration that can produce a broader power profile (i.e., relatively high power in the morning and afternoon), which may result in less peak shaving and soiling. Addition of bifacial systems into the existing electrical network can provide advantages including improved reliability, higher energy yield, and power consistency. The market share of bifacial modules is expected to reach 40% by 2028 [10]. This is due to the current interest of the international renewable agencies, industrial workshops, and bifacial PV pilot plant setups and standardization [10–12]. With this growing attention, there are few areas that need to be explored to prove its reliability in the field and minimize investment risk for large-scale deployment in hot and sunny climates.

The performance of solar cells is primarily determined by the properties of the absorbing material. However, the cell architecture also affects the cell characteristics. Silicon solar cells make 95% of the cell market today [10]. The silicon industry has achieved great progress in minimizing the energy losses in the device and maximizing the extraction of the photocurrent. The basic structure of the conventional cell is the back-surface field (BSF) cell. However, the passivated emitter rear contact (PERC) structure as shown in Figure 6.1(a) currently dominating along with other bifacial structures such as TOPCON (also known as

[1]Research & Development Centre, Dubai Electricity and Water Authority (DEWA), United Arab Emirates
[2]Qatar Environment and Energy Research Institute (QEERI), Hamad Bin Khalifa University (HBKU), Qatar Foundation, Qatar
[3]Department of Applied Physics and Astronomy, College of Sciences, University of Sharjah, United Arab Emirates

MONOFACIAL PERC SOLAR CELL   BIFACIAL PERC+ SOLAR CELL

(a)   (b)

MONOFACIAL PV MODULE*   BIFACIAL PV MODULE

(c)   (d)

*Glass-Tedlar-can also be in glass-glass.

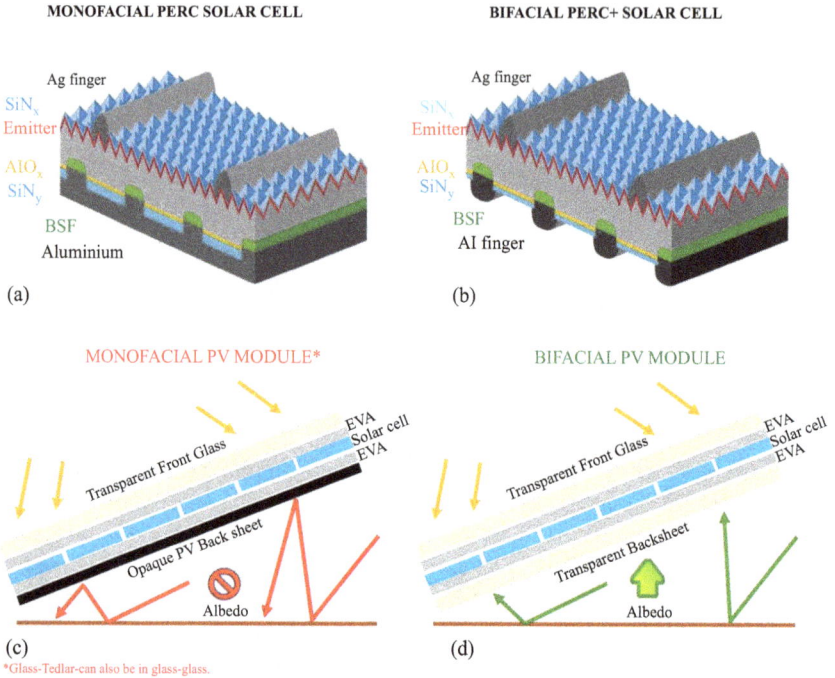

Figure 6.1   Difference between monofacial and bifacial PV at cell and module levels. (a) Passivated emitter rear cell (PERC) monofacial with full aluminum coverage from rear side. (b) Bifacial PERC+ solar cell with aluminum fingers at rear side allowing radiation to pass. Ref. [12]. (c) Schematic for monofacial PV module with opaque back sheet not allowing albedo reflections. (d) Bifacial with a transparent back sheet for collecting albedo radiation.

passivated contact) solar cell and Silicon Heterojunction solar cell. Typically, in PERC structure, A thin layer of AlO_x is added at the back surface to reduce the carrier recombination (passivation layer). PERC can be made bifacial, called PERC+, by replacing the aluminum back contact by a limited number of Al fingers as highlighted in Figure 6.1(b). This difference in bifacial design is at the cell level. Similar concept is applied for different device architecture to make them bifacial. However, to make true bifacial modules, transparent back sheets are also required at the rear while fabricating the module. Figure 6.1(c) and (d) shows the module consisting of a monofacial and bifacial solar cell placed in a sandwich between two layers of encapsulant and back sheet. Ethyl vinyl acetate (EVA) is usually used as encapsulating material. In the monofacial module, a glass sheet is added on the front side of the module while an opaque Tedlar (polymer) sheet is attached at the back as shown in Figure 6.1(c). In the case of the bifacial module highlighted in Figure 6.1(d), transparent sheets (glass or transparent polymer) are added on the front and back surfaces of the module.

The operating conditions of bifacial PV modules affect the cell and module efficiencies [13,14] significantly. Unlike monofacial PV, the performance of bifacial PV is strongly dependent on system-level installation specifications such as albedo radiation, height, tilt, system size, ground coverage ratio, and azimuthal orientation [3,14–16]. The energy gain from a bifacial module over a monofacial module is fundamentally due to the additional rear-side gain of the module. The only parameter which can be controlled during cell fabrication is the module rear-side current response [15,16], which determines the bifaciality of the solar cell.

Bifaciality is defined as the ratio of rear power to front power at standard testing conditions (STC) of plane of array (POA) insolation of 1,000 W/m^2 and cell temperature of 25°C under AM 1.5 spectrum. It is the main parameter that characterizes the front and rear sides of solar cells and should be 100% for a perfect bifacial cell. The remaining factors depend on the location and the installation conditions. Due to the various factors affecting the energy yield of a bifacial PV module, the end-use gain from this type of module compared to the standard monofacial modules is not easily quantified. Bifacial power gains in the range of 10–30% have been reported depending on the location and the installation parameters [11]. Higher gains are possible if the installation conditions are optimized for a particular site [13,14]. However, this optimization problem remains one of the significant challenges with the installation of bifacial modules. A critical application of bifacial PV modules is a vertical installation. The performance of vertically installed PV modules mainly depends on their bifaciality, latitude, diffuse fraction, and albedo [16–18]. The most vital parameter for performance enhancement of bifacial modules apart from the site location is ground albedo. Higher reflective surfaces imply more radiation collection at the rear face of the module increasing the overall bifacial gain.

Bifacial modules have a massive potential for hot and sunny deserts. It includes tapping the bifacial gain from high-albedo ground material and installation of vertical modules, which may yield higher energy due to reduced soiling. Besides, bifacial gain does not significantly increase the temperature of the solar cell. In fact, at higher irradiance and gains, bifacial PV is supposed to operate at a lower temperature than its complement monofacial PV [19]. Vertical bifacial modules can also be used for demand-side management and peak shifting of load curve due to its broad energy yield profile [9,20]. This can be accomplished by optimizing the array configuration and aligning the panel in an azimuth direction of east–west, respectively. All these benefits make bifacial solar modules the promising contender for future installations in extreme climate. Despite the advantages, the performance of bifacial modules is highly sensitive to the installation conditions such as albedo, height, tilt, azimuth angle, and frame type. Therefore, real-time performance monitoring and complete parametric analysis of bifacial modules are timely needed to support the adoption of this technology specifically in the desert regions.

Considering the pressing need, this chapter will present a comprehensive bifacial cell to module (CTM) to field study for the hot and sunny desert environment of Qatar and the MENA region.

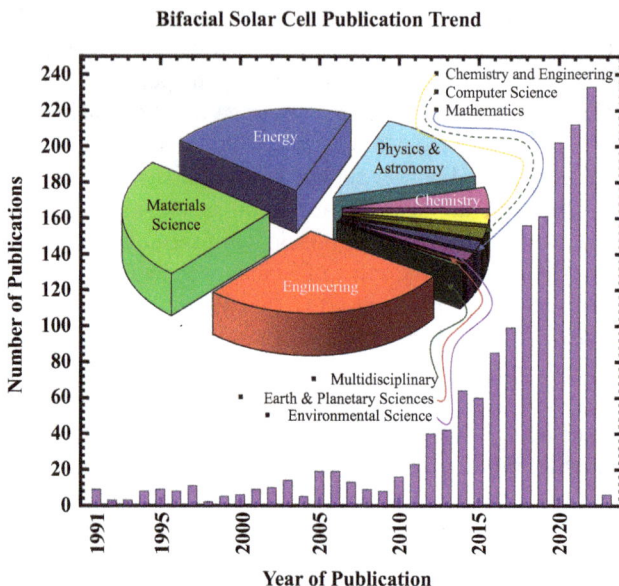

Figure 6.2   *Number of refereed publications on bifacial solar cell publications per year, along with their associated distribution of the published papers per discipline. Data were collected from 1991 to December 2022, inclusively from Scopus®.*

6.2   Technological progress

Bifacial PV has different types of solar cells depending on the type of material and optical configuration of the device. The point of common interest in all these technologies is that they allow light to enter from the rear side of the solar cell. Bifacial solar cells have been in the research field since 1960 when the first substrate was manufactured using single and multicrystal silicon [21]. Figure 6.2 shows the technological progress of bifacial solar cells in research and academia. The dominant role in the semiconductor industry has been from silicon wafers technology followed by thin films. For bifacial solar cells, glass and transparent conductive oxides have been used to enhance the bifaciality of PV [6,7]. Thin films based on chalcogenides such as copper indium gallium selenide (CIGS)/CdTe have been given relatively less attention due to their toxic nature and end-of-life-cycle issues.

Bifacial solar cell publication trend

The most common type of architecture is of PERC solar cell for bifacial panels. This is due to existing production lines for monofacial whereby retrofitting bifacial solar cells can be easily manufactured with an efficiency of up to 22% [22]. Similarly, passivated emitter and rear totally diffused (PERT) technology also has the advantage of using available manufacturing setup for producing bulk bifacial solar cells. A device structure for PERC and PERT consists of an emitter, absorber, antireflection coatings, and transparent contacts. Metallic contacts then collect the photocurrent. Heterojunction intrinsic thin-layer (HIT) solar cells employ thin films of crystalline silicon under low temperature and yield higher power conversion efficiency (PCE) of up to 26.7% as achieved by Kaneka [23]. This is due to the high open circuit voltage achieved from the passivation layer. Currently, the industry shows interest in HIT technology due to its high efficiency and lower temperature coefficient (TC), which are desirable for hot climates. Optical and radiation arrangement provides two types of device structure. Bifacial module with glass/glass and glass/back sheet layers is possible for operation. Glass/glass yields the maximum bifaciality gain from the PV as the rear light is utilized in increasing power. On the contrary, glass/back sheet structure is also available in the market due to its inherent resemblance with mono-facial [16]. From the thermal point of view, glass/back sheet performs better than glass/glass modules. This is due to the higher specific heat capacity of glass, which increases the module temperature.

Recently, a new class of light absorber material called hybrid perovskites solar cell (PSC) has emerged in the PV research activities as an active competitor. As a result, its PCE has shown an unprecedented benchmark increasing from 3.8% in 2009 [24] to 25.8% in just fifteen years [25]. The critical feature, which has enabled such PV family to reach high performance within a short period of time, is the conjunction of the best of both conventional inorganic and organic solar cells by utilizing the efficient operation principles from the first one while using relatively simple fabrication techniques from the second one, thereby providing a possible pathway toward economical solar energy alternatives [26–28]. The high efficiency achieved by PSC has been attributed to its convenient optoelectronic properties such as suitable band gap [29–34], high absorption coefficient over solar spectrum [28,35], low exciton binding energy [36,37], and efficient charge transport characteristics [37–41]. Interestingly, an extensive range of device designs and fabrication methods has resulted in efficient cells. Besides illustrating how rich the field is, this indicates more importantly that there is still a large room for the optimization of the device. Moreover, this has prompted the focus of researchers to work toward tandem solar cells with silicon solar cells and PSCs to reach higher efficiencies.

The design principles and operating conditions of bifacial PV play a huge role in the cell- and module-level efficiencies [2]. The energy gain depends on some factors at different scales in the PV supply chain, that is, from CTM to field. The parameters which can be controlled during cell (or module) fabrication is the module rear-side current response [15,16] called "bifaciality" and the behavior of the device to the temperature called "temperature coefficient." At the module level, the thermo-mechanical properties of different stack layers impact the performance [3,15,16]. At the field level, location and the installation conditions alter the output. Due to the various factors affecting the energy yield of a bifacial PV module, the end-use gain from this module type compared to the standard monofacial modules is not easily quantified. This is one of the significant challenges with bifacial modules. Defining standard guidelines can help the end-use benefits from these devices make the bifacial PV technology more popular. The following sections provide a literature review of key areas relevant to this thesis.

6.3   Bifacial performance parameters

Bifacial PV is a promising solar energy technology that can harvest light from both front and rear sides to produce more energy yield than monofacial PV modules [1–4]. It has been one of the most active themes of research in recent years in the field of solar energy. The inherent capability to harness albedo radiation, improve energy yield [42,43], and minimize negative soiling effect under desert environment shows a potential to decrease the levelized cost of energy [5–7]. Vertical bifacial module facing east–west is another prospect that can generate broader energy yield profile (i.e., relatively high power in the morning and afternoon), which may result in less peak shaving and soiling [6–13]. Addition of these systems into the existing electrical network can provide advantages including improved reliability, higher energy yield, and power consistency. The market share of bifacial modules is expected to reach 40% by 2028 [10]. This is due to the current interest of the international renewable agencies, industrial workshops, and bifacial PV pilot plant setups and standardization [10–12]. Recently, below 3 cent $/kWh has become a norm in the Middle Eastern region with the lowest ever bid record of 1.79 cent $/kWh for bifacial plant in Saudi Arabia [44,45]. However, as the bifacial technology has been less tested in local desert conditions, the particular bid was not chosen due to technological risk and bankability issues. With this growing attention, there are a few areas that need to be explored to prove its reliability in the field and minimize investment risk for its large-scale deployment, especially in hot and sunny climates.

The main parameters to enhance efficiency and energy yield are mentioned below:

At the cell level, energy yield is affected by

- Bifaciality
- Temperature coefficient
- Power conversion efficiency

At the module level, energy yield is affected by

- Module design
- Nominal operating cell temperature
- Back sheets, that is, glass/glass or glass/Tedlar
- Encapsulation
- Junction design
- Type of frame
- Number of diodes

At the field, the following installation parameters affect the performance:

- Climate—incident solar radiation, ambient temperature, wind speed, spectrum
- Ground albedo

- Module temperature
- Height
- Tilt
- Azimuth
- Ground coverage ratio
- System size

For bifacial modules mounted at the conventional tilt angle (latitude), the installation parameters affecting the energy yield are module height, tilt angle, ground albedo, and the irradiance characteristics (cloudy or sunny day, diffuse content) [9]. Figure 6.3 shows the most important parameter for bifacial technology, that is, spectral albedo radiation spectrum for different materials. Albedo is the solar radiation component reflected from the ground and scattered by diffused radiation. Higher albedo implies more radiation is collected at the rear face of the solar cell increasing current generation and overall bifaciality ratio. However, the spectral response of the ground albedo should also match the spectral response of the solar cells for harnessing optimal radiation. A critical application of bifacial PV modules is a vertical installation. The literature shows that the performance of vertically installed PV modules mainly depends on the bifaciality, latitude, diffuse fraction, and albedo [16–18].

Bifacial solar cells are albedo collecting devices and are capable of decreasing the levelized cost of electricity (LCOE) of power plants [47]. Utilizing albedo radiation is preferential for bifacial installations as both beam and diffuse radiation

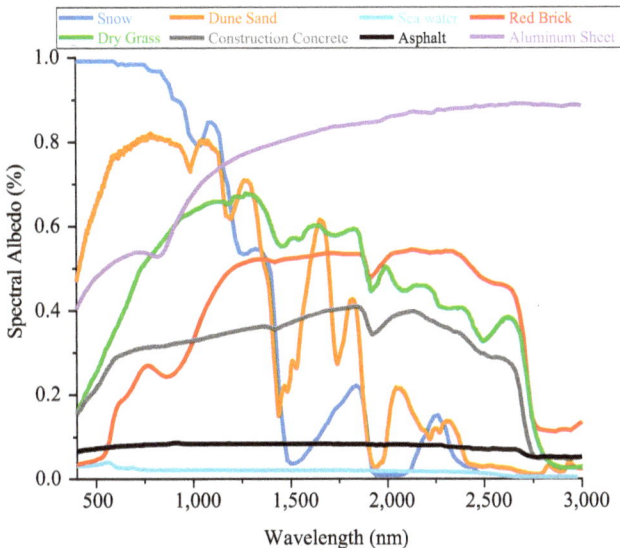

Figure 6.3   Spectral albedo radiation spectrum for various materials. Spectral albedo data collected from Ref. [46].

from the atmosphere and earth's surface can be collected. Based on the albedo values, snow due to high reflectivity and inclination toward high band gap was found to increase the performance of amorphous solar cells. A white membrane as a base material for ground reflectance revealed an enhancement of 54% for a single panel with a bifacial gain ratio of 0.89 [11]. Frank *et al.* [48] carried out an extensive study on the effect of albedo and concluded that standard bifacial PV performs 20% better than monofacial PV. Using innovative and optimized designs, a 30% power enhancement has also been reported by employing special module arrangement.

6.4   Potential applications

Bifacial PV has been extensively used in extraterrestrial applications because of its radiation-capturing capability [47]. For the stratosphere, 14% gain in power was observed as a result of reduced cell temperature in the infrared region of the spectrum [48]. Space missions have also employed bifacial solar cells and have reported an enhancement of 1,020% over monofacial solar cells. They have been investigated for the manufacturing of the wings of solar unmanned aerial vehicles, and 15% energy gain has been reported with cell temperature as low as 264 K [49].

Bifacial solar cells can also be used for load management and peak shifting of load curve [3]. This can be accomplished by optimizing the array configuration and aligning the panel in an east–west direction to yield a broader energy yield profile as shown in Figure 6.4. The usage of vertical bifacial solar

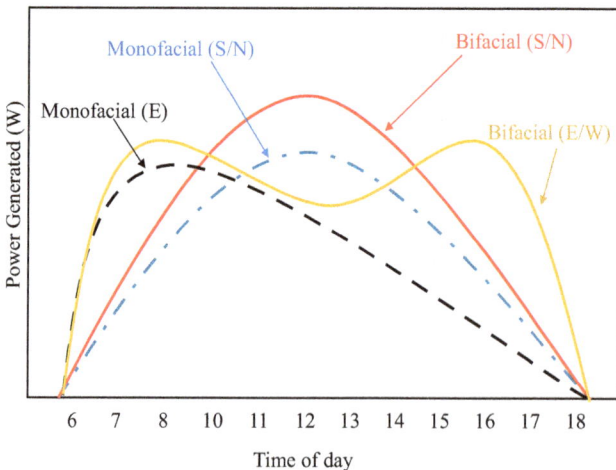

Figure 6.4   Typical specific power profiles of bifacial and monofacial PV installed with vertical east–west (E–W) and north–south (N–S) directions

cell as a noise barrier is reported in the literature and was first installed in Switzerland in 1997 [11].

Bifacial solar cells can be employed on buildings for zero carbon emissions [50]. This can reduce harmful emissions by 40% [11]. Building integrated PV (BIPV) is a key player for making the cities sustainable as it helps to cover its energy aspect, that is, energy management, efficiency, and energy generation from alternative energy resources. Solar thermal applications such as photo-voltaic/thermal (PVT) systems can also utilize the added advantage of bifacial PV. Overall system effectiveness, that is, thermal and electrical efficiency, was improved by 45% for bifacial PVT system [51].

Possible applications for bifacial PV from reviewed literature are shown below:

(i)   Standalone or field systems
(ii) Vertical installations
(iii) Solar park
(iv) Sound barrier
(v)   Building integrated
(vi) Low orbit space applications
(vii) Load management

6.5   Solar cell device optimization

Over the past decade, the field of PV has advanced extraordinarily in many fronts. Many of the stagnated PV technologies were significantly improved. For example, cadmium telluride (CdTe) solar cell efficiency was increased to 22.1% [52] after being pinned around 16.5% for 20 years between 1992 and 2012 [53,54]. This also happened for CIGS solar cells; after 15 years of stagnation, the efficiency around 19% [55,56] was improved in the past 3 years and reached the record of 22.3% [52]. Furthermore, a new family of hybrid PSCs has emerged in 2012 and has been developing exceptionally since then; its efficiency reached 22.1% in just 4 years [52]. As for multijunction cells, Fraunhofer Institute for Solar Energy Systems achieved 46.0% efficiency using a four-junction cell [52]. Actually, the list of recent interesting developments in the field is huge; thus, we refer the reader to the latest comprehensive reviews [57,58]. Basically, there are many reasons for such remarkable developments. The main one—as usual—is economical due to the increased prices and the depletion rates of other fuel sources [59]. Scientifically, nanotechnology and materials sciences have been growing exponentially since the 1990s [60,61]. This, in turn, enriches PV field which relies heavily on the advances in material sciences. Historically, the first practical realization of the solar cell was in the 1950s [62–64]. It was mainly based on crystalline silicon. The space of materials used in solar cells increased considerably in the 1970s as tens of absorbers were considered.

The most prominent output of that era was CdTe and CIGS. The set of explored absorbers has been expanding since then. The general structure of solar cells is shown in Figure 6.5. Besides the absorber, there are two contacts and the electron and hole transport materials (ETM and HTM, respectively). More layers could be used for various purposes. In principle, each absorber shall have a unique set of optimally matching materials to maximize cell efficiency. However, a reduced number of materials is used in different solar cell technologies. For example, cadmium sulfide (CdS) is commonly used for CdTe, CIGS, CuS, and InP cells as ETM. Also, TiO_2 is used as ETM in a very wide set of solar cell technologies [59,65]. There are many other examples as well for the use of a particular "nonabsorbing" material in multiple solar cell technologies. Such coincidences cannot be generally attributed to a device optimization process; but paradigms and experiences play a major role—at least—at the first stage of development. Furthermore, it is practically very arduous to experimentally identify the best matching device materials for a given absorber. Thus, the need for computational device design and optimization.

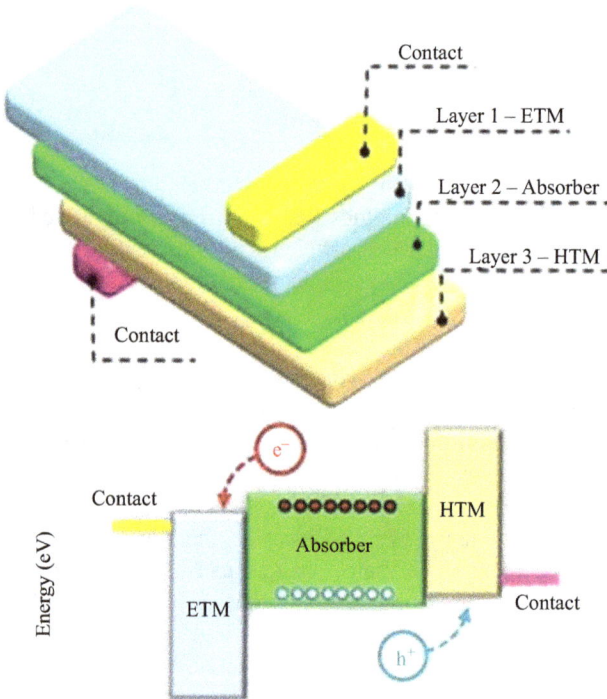

Figure 6.5   The general device structure of solar cells

6.6   Theoretical and practical efficiency limits of solar cells

Despite the considerable progress toward high PCE, which is one of the key performance indicators for solar cells [62], detailed analysis is needed to achieve a better understanding of the actual factors limiting the performance of PSCs and to approach the thermodynamic efficiency limits of around 30–33% for the corresponding band gap of 1.2–1.6 eV [63]. More specifically, this requires further investigations about the nature of the recombination mechanisms at work and the impact on charge transfer dynamics and PCE in the actual PSC devices [64,65]. Previous studies have confirmed that recombination kinetics has served as the basis for technological advancement for other semiconductors such as silicon [66]. The recombination losses arise from band-to-band radiative recombination (U_B) and nonradiative defect-assisted recombination known as Shockley–Read–Hall (U_{SRH}) [67] under low injection level in solar cells. Radiative recombination is a process based on the reciprocity relation for photon emission and absorption, whereas nonradiative trap-assisted recombination is due to the defects in the material [64]. In PSC, the effect of nonradiative recombination is dominant and has a stronger influence, when compared to radiative recombination, thereby it is the mechanism that practically limits the efficiency of PSC [67–71]. Remarkably, the reported rates of defect-assisted recombination for $CH_3NH_3PBI_3$ vary within a very wide range [36] due to the large variety of processes used for perovskite film growth and the complexity of measuring trap-assisted recombination rates.

So far, the PCE limit for PSC has been estimated generally by two methods. In the first, the PCE analysis is carried out for external radiative limits [72,73], while on the other one, the detailed balance approach with alterations for nonradiative recombination and light trapping is investigated [74–76]. Sha *et al.* [76] showed that the detailed balance limit for $CH_3NH_3PBI_3$ perovskite is 31%. However, to have better practical estimators, it is important to consider nonradiative recombination rates as well, which is the dominant recombination effect in PSC. The detailed balance method is not well suited from the perspective of charge transport and extraction [75]. Contrarily, the recently developed full-space optimization method [76] performs an in-depth analysis for PSC devices by incorporating the practical device architecture and charge dynamics such as carrier recombination, mobility and geometry optimization. This can provide better assessment for the limits achievable by a multilayered device. Hence, by considering both radiative and nonradiative recombination pathways for extreme cases (minimum and maximum reported recombination rates) reported in the literature for $CH_3NH_3PbI_3$, realistic limits for PSC can be estimated. This shall provide a guideline to experimentalists for improving multilayered perovskite device performance.

Efficiency limits of solar cell depend on the recombination mechanisms present in the device. They can either arise from intrinsic radiative recombination or extrinsic defect-assisted nonradiative recombination. The recombination losses arise from band-to-band radiative recombination (U_B) and nonradiative defect-assisted recombination known as Shockley–Read–Hall U_{SRH} [67] under low injection level in solar cells. Radiative recombination is a process based on the reciprocity relation for photon emission and absorption, whereas nonradiative trap-assisted recombination is due to the defects in the material [64]. In PSC, the effect of nonradiative recombination is dominant and has a stronger influence, when compared to radiative recombination; thereby it is the mechanism that practically limits the efficiency of PSC [67–71]. Depending on the material quality and process conditions employed for perovskite absorber growth, the reported trap densities are between 10^8 and 10^{16} cm^{-3} [64,72–78].

Table 6.1 illustrates the variation of the band-to-band recombination and product of trap density and capture cross section for perovskite solar absorber reported in the literature. The radiative recombination is found to vary by two orders of magnitude, whereas material quality (quantified by the product of trap density and capture cross section) has a huge range from $\sim 10^1$ to 10^{-4} cm^{-1}.

The extraction or recombination depends collectively on multiple properties such as recombination kinetics, charge dynamics, and device design [79]. Therefore, a comprehensive understanding of the complex relationship between these different factors is needed to estimate the practical limits of PSC and identify the achievable value.

For trap-assisted recombination, a product of trap density (N_T) and capture cross section is employed as a measure of SRH recombination. The reported rates of radiative and nonradiative trap-assisted recombination have a broad range due to the large variety of fabrication practices used for perovskite film and the complexity of measuring trap-assisted recombination rates. Performance wise, despite the vast effort and huge set of developed PSC, the best reported short circuit current J_{sc} (24.2 mA/cm^2 [25]) is considerably smaller than the expected theoretical values of a semiconducting absorber of 1.5 eV. Conceptually, 29 mA/cm^2 should be achievable, and this should result in a better PCE (n%), if open circuit voltage (V_{oc}) and fill factor (FF) are within the theoretical limit range [83]. However, the increase of J_{sc} needs either to have better light management or to increase the thickness of the perovskite layer [74]. The second is mainly limited by material

Table 6.1   A range of band-to-band radiative recombination constant and product of trap density and cross section

Trap density N_T (cm^{-3})	Cross section σ (cm^2)	$N_T \times \sigma$	References	Band to band (max; cm^3 s^{-1})	References
1.60×10^{17}	3.30×10^{-18}	0.528	[77]	1.1×10^{-9}	[80]
8.00×10^{17}	2.97×10^{-19}	0.2376	[77]	0.62×10^{-9}	[33]
1.60×10^{18}	2.74×10^{-19}	0.4384	[77]	9.4×10^{-10}	[68]
7.44×10^{16}	2.00×10^{-17}	1.488	[78]	2.0×10^{-10}	[71]
1.30×10^{15}	2.50×10^{-15}	3.25	[72]	1.7×10^{-10}	[69]
1.40×10^{10}	1.24×10^{-12}	0.01736	[74]	2.6×10^{-10}	[75]
7.6×10^{08}	2.93×10^{-13}	0.000223	[81]	2.4×10^{-11}	[82]

quality, and hence the maximum obtained value of J_{sc} is not fundamental. This is due to the reduced qualities of both the perovskite layer and the interfaces with other layers in the device. The first one limits the possible thickness of the perovskite layer, and hence the photogenerated carriers and J_{sc}, while the second affects the carrier's extraction and results in further reduction in J_{sc}. Optimum thicknesses for device layers will ensure that charge collection is efficient by balancing both absorption and recombination as a whole in the device.

There has been some focus on the practical limitations of J_{sc} as a function of geometry and morphology [84–87]. Correa-Baena *et al.* [92] analyzed the effect of a perovskite capping layer on the generated light current and optical properties. It was found that an absorber layer of 0.5 μm with high-quality crystal size was sufficient to provide champion cell efficiency of 20.8%. Y. Dang *et al.* [96] showed that pinhole-free absorbs layer of 0.8 μm was possible with 16.8% PCE using solvent engineering. Although initial reports have suggested the optimized device thickness in the range of 0.3–0.4 μm [29,87,88], this should increase as the crystalline quality of perovskite has been improving ever since [89]. The first $CH_3NH_3PbI_3$ electrochemical cell showed a J_{sc} of 11 mA/cm^2 [24], which has evolved rapidly by employing all solid device architectures to reach the maximum of 24.2 mA/cm^2 in 2017 and a PCE of 25.7% in 2022 [25]. The question of "How thin is too thin for efficient PSC?" is primarily driven by the fact that the levelized cost of energy is determined by the material cost and the efficiency of the solar cell, which are both dependent on absorber thickness [90]. Thus, it is timely to investigate computationally the practical efficiency limit of PSC utilizing the experimental report values of the diverse quality of $CH_3NH_3PbI_3$.

6.7 Bifacial perovskite silicon tandem

Over the last few years, PV market has seen an incremental growth in terms of establishing industries with big manufacturing companies with large-scale production capabilities to boost the module performance in the range of 14–21% of efficiency using different conventional materials such as silicon (Si), cadmium telluride (CdTe), and copper indium gallium diselenide (CIGS) with a target cost of less than a dollar/watt [91]. Though the momentum of progression is abrupt, a target milestone is to reduce the cost less than $0.5/watt with an increasing output efficiency of more than 25% [92], since till now PV electricity is supporting the entire world electricity need with around 1% only due to its initial installation cost related to the volume of panels. To tackle this issue, it is required to improve the PCE as the cost of installation drops significantly with the reduced number of solar panels. It has become a critical issue due to the shutdown of many PV companies, which eventually failed to compete with the trend of keeping a balance between panel performance and cost of installation—one of the main reasons for securing a vast amount of flowing capital at the beginning, which is essential for any startup company. As a conventional material, Si PVs are saturated with an efficiency of

25% for more than a decade, even though many active PV research groups are working to improve it [93]. A significant approach is to develop tandem solar cell structures using a high band gap material as a top cell and to keep Si as a bottom cell due to favorable optical (1.1 eV) and electronic properties [94,95]. Bifacial tandem solar cells show a practical method to reach beyond the efficiency limits of conventional single-junction solar cells. Bifacial technology harnesses the solar radiation from both the front and the rear side of the solar cells leading to an increase in the photogenerated current, whereas tandem devices increase the PCE by utilizing the multijunction concept with suitable band gaps for photon management. In such structures, shorter wavelength photons can be absorbed by the high band gap material to enhance the voltage by twofolds, whereas Si single junction is unable to compete. To make sure that the top cell is competitive in terms of cost, it becomes a norm to go for an alternative absorber material even with defect states and semicrystallinity. Recently, PSC absorber materials have emerged as one of the best options due to their suitable electronic, optical, and structural properties besides cost-effective processing. A remarkable progress has been achieved over the past years with a rapid increase of efficiency from 3% in 2009 to over 25% in 2023 [96]. This makes PSCs the champion device of thin film-based PV technology. In a tandem structure, a bottom cell is required with a band gap of between 1.1 eV and 1.4 eV to fulfill the ideal condition, where Si is the best choice among other established absorbers because of its capability of capturing photons with low energies to produce electron–hole pairs; however, an ideal band gap is around 1.7 eV for the top cell [97]. Due to the band gap engineering possibility, a band gap shift has been calculated from 1.6 eV to 2.25 eV by only substituting I by Br in the perovskite crystal structure [98]. Due to its perfect matching band gap, perovskite can be a very efficient tandem device with Si solar cell. Practically and due to the high obtained efficiencies for both Si and PSCs, a significant progress has been achieved very recently with a certified power conversion efficiency of 33.7% [96].

The first monolithic perovskite/silicon tandem was made with a diffused silicon p–n junction, a tunnel junction made of n+ hydrogenated amorphous silicon, a Titania electron transport layer, a methylammonium lead iodide absorber, and a Spiro-OMeTAD hole transport layer. The PCE was only 13.7% due to excessive parasitic absorption of light in the hole transport layer, limiting the matched current density to 11.5 mA/cm^2. Switching to a silicon heterojunction (SHJ) bottom cell and carefully tuning layer thicknesses to reduce optical losses increased the current density to 15.9 mA/cm^2 and raised the PCE to a record 21.2%. It is clear from these reports that minimizing parasitic absorption in the window layers is crucial to achieving higher current densities and efficiencies in monolithic tandems. The tandem structure using PSCs has enabled to boost the performance. EPFL has worked on "high-efficiency monolithic perovskite/silicon tandem solar cells," where semi-transparent mixed halide PSC has been coupled with Si device to fabricate four-terminal tandem structure [99]. As it is required to have a transparent electrode for tandem device structure and to keep perovskite as a top cell, a thin layer of MoO$_x$ layer was grown on top of the hole transporting layer using physical vapor deposition (PVD) technique to sputter a thin layer of transparent conductive oxide. Such metal oxide usually helps to resist the surface damage of the perovskite structure occurring from plasma bombardment. Eventually, the team managed to reach up to an efficiency of 10.7%. PVD-grown indium tin oxide (ITO) layer as transparent conductive oxide (TCO) has a significant role in fabricating such high-efficiency devices. Overall efficiency of 25% was presented with encapsulation. However, further optimization is underway for module processing with better device stability. Another generic device structure was presented where atomic layer deposition (ALD) and sputtering were used to fabricate electron transport layer (ETL) and TCO layers, respectively. Such perovskite device was optimized using Anti-Reflection Coating (ARC) from both front and rear to be used for perovskite devices. The fabrication of a high-efficiency cell on textured SHJ solar cell was presented. The steady-state efficiency of 25.24% was certified (the record now is 27.3% by Snaith group). To tackle the stability of the perovskite cell, the team is working toward encapsulation, light soaking tests, and reverse bias stability. The humidity effect on the optical properties was observed for measurements done in summer/winter. The group is also working toward a scalable 4-inch wafer with improved FF by improving the interface as developed at EPFL PV lab. A theoretical study confirms that the performance of a perovskite/Si two-terminal can be enhanced up to 33% using the bifacial module, as the reflection from the earth results in better absorption side through the transparent rear side [95].

6.8   Material characterization model for device assessment

A material characterization model called time-resolved photoluminescence (TRPL) has been used extensively to investigate the carrier extraction at the interfaces of solar cells, particularly PSCs [100,101]. The contactless nature and its application for both complete and noncomplete devices make it an ideal tool for characterizing limitations in PSCs. Careful physical examination of TRPL measurements can provide us with an understanding of the complex attributes of charge dynamics and extraction mechanism in mixed halide PSC. Since their discovery, PSCs have unleashed a great deal of enthusiasm in the PV community because of their technological promises and the open questions raised concerning the physical processes occurring in the device. The nature of the gap (direct/indirect) [102–104], the values of the carrier mobilities and diffusion length [105,106], the charge transport mechanism [107–110], the role of point defects, and the recombination mechanisms responsible for the carrier losses [111–113] have been the subject of intensive debates. In the standard configuration, a PSC consists of a thin perovskite absorbing layer sandwiched between an ETM and a HTM. Under light excitation, the photocarriers are generated in the perovskite absorbing layer and diffuse toward the interfaces. Because of the appropriate band alignment at the interfaces, electrons are extracted at the ETM/perovskite interface while holes are extracted at the perovskite/HTM interface [79]. The carrier extraction at these interfaces occurs within picosecond range after excitation [91,114]. The radiative band-to-band recombination leading to photoluminescence occurs in the nanosecond range [115–125]. The photocarriers generated close to the interfaces are extracted before contributing to the PL signal leading to the PL quenching. The drastic reduction of PL signal in the presence of HTM/perovskite interface has been assigned by many authors to carrier extraction at the selective electrodes [126].

In a typical TRPL experiment, the sample is excited by a very short laser pulse leading to the generation of an initial excess carrier density $n(0)$, which subsequently decays as a result of the recombination and carrier extraction at the interfaces as illustrated in Figure 6.6(a). Figure 6.6 shows the TRPL schematic along with conventional curve fitting and charge dynamics fitting method for PL measurements. Conventionally, TRPL curves are analyzed by performing curve fitting to the PL transients and extracting the associated lifetimes. Due to the complex quasi-exponential shape of the TRPL, one-exponent, two-exponent, and stretched exponential functions for fitting have often been employed as shown in Figure 6.6(b). The main limitation arising from this method is the extraction of arbitrary lifetime constants without consistent

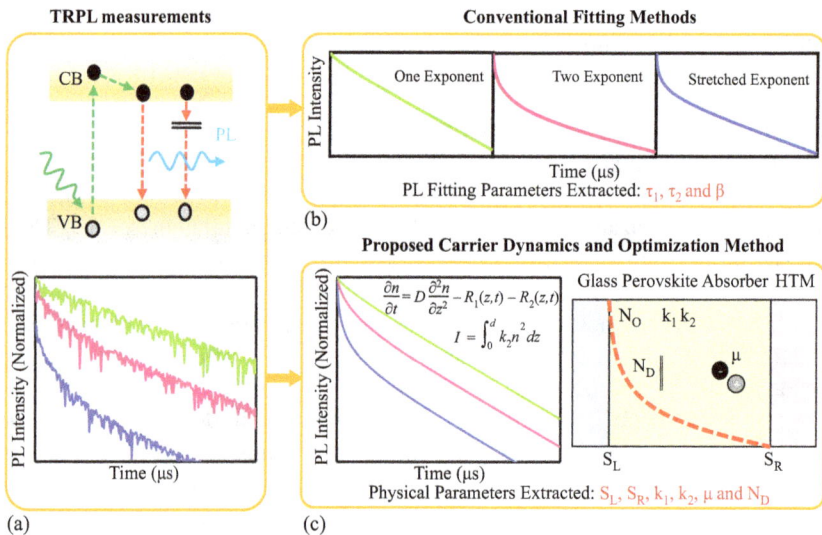

Figure 6.6   *(a) TRPL signals arising from recombination pathways. (b) Conventional fitting method of TRPL measurements for fitting constants (in logarithmic scale). PL curves are often fitted by one-, two-, and stretched exponential functions. Here $\tau_1, \tau_2,$ and β show lifetimes and stretched constant, respectively. (c) The proposed charge dynamics method predicts physical parameters by solving rate equation with appropriate boundary conditions based on the geometry.*

consideration of the physical processes involved in TRPL decay dynamics. Moreover, as the diffusion and varied recombination pathways in bulk are occurring at comparable time scales, there is a high likelihood that the extracted parameters may be convoluted with other properties. To understand the opto-electronic properties and obtain meaningful insights, charge transport for analyzing PL transients must be taken into consideration. This assists in unraveling the physical factors governing the PL decay under different time scales and conditions.

The determination of the PL decay requires solving the continuity equation taking into account the generation, transport, recombination, and extraction of the carriers [127,128]. In the practical experimental setup, the incident light penetrates into the cell through the transparent glass, fluorine-doped tin oxide (FTO), and ETM layers and gets absorbed exponentially in the perovskite layer according to Lambert law. Therefore, a diffusion term needs to be included in the continuity equation, which governs the charge dynamics after a pulsed excitation. The effect of doping concentration, intentional or unintentional, must be considered as well to study the collective impact of doping and radiative recombination on TRPL kinetics [129,130].

More importantly, the addition of excitation density, surface recombination velocity, and diffusion provides detailed insight into the complex kinetics occurring at initial PL decay. This can equip us with the true quality of absorbers coupled with HTM or ETM by incorporating the photophysical processes. We describe, in this work, such an approach along with the boundary conditions and optimization algorithm to consider the carrier extraction and/or recombination at the interfaces. This fundamental work can benefit in understanding the details of the transport and extraction kinetics by dismembering the effect of individual parameters contributing to TRPL. By coupling this comprehensive model with the experimental measurements, the aim is to go beyond the conventional fitting parameters of lifetimes and extract the parameters with physical significance such as surface recombination/interface recombination velocity (S_L and S_R), monomolecular recombination constant (k_1), bimolecular recombination constant (k_2), mobility (μ), and doping density (N_D) as shown in Figure 6.6(c).

6.9   Cell to module to field performance

PV modules are made up of solar cells connected in series and encapsulated between layers. These layers are typically composed of glass, encapsulation, cell, and back sheet material with different optical properties. On the surface of the solar cell, interconnections such as fingers and bus-bars are required to collect generated current and complete the circuit. This encapsulated assembly increases the module's power and a voltage above that of a single solar cell and protects solar cells from the environment [1–4]. The primary goal of the module technology is to provide safe, efficient, and reliable panels extending PV lifetime to around 20–25 years. However, through the process of modularization, the operating conditions for the solar cells change as a result of optical losses due to parasitic absorption and reflection in module sublayers. Furthermore, the solar cell interconnections cause additional power reduction due to the resistance and joule heating, which reduces its electrical performance. As a result, in the modularization process, the power of the module is usually lesser than the summation of the power of all the solar cells connected to assemble the module. This difference in total cells' power/PCE and module power/PCE is called a CTM gap [5].

Technology

A key measure of CTM gap and CTM loss is employed for assessing the effectiveness of the modularization process. A low CTM gap will, therefore, ensure a lower LCOE by enhancing the energy yield and efficiency at the system level [6]. Thus, the aim is to design a module with a CTM gap of 0% such that the module that is installed in the field performs at its best.

Figure 6.7 can explain the importance of lower CTM gap where a general trend of efficiency gap from CTM PCE is revealed. Figure 6.7 shows the state-of-the-art record efficiencies along with their CTM ratio for different solar cell technologies, namely, mono-Si, multi-Si, CIGS, CdTe, GaAs, and perovskite [7]. Currently, the absolute efficiency gap, that is, the CTM gap, is in the range of 2.3–9.3%. Even for matured technology such as mono-Silicon, the CTM gap is 2.3%, and a CTM ratio is 91% [7]. For other technologies, the CTM ratio is in a much lower range showing the potential to enhance the output at the module level.

To understand and minimize various mechanisms responsible for the energy losses/gains from CTM to field, it is vital to know how each component alters the performance. Figure 6.8 shows the efficiency drop mechanisms and the key losses/gains incorporated during the fabrication of a module and installation of these

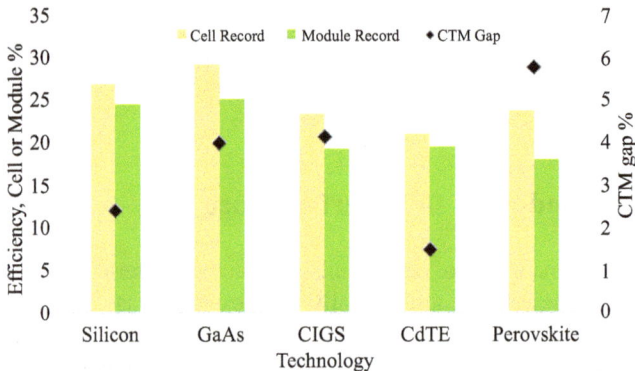

Figure 6.7   Record cell and module PCE with a cell to module gap [7]

CELL TO MODULE TO FIELD PERFORMANCE

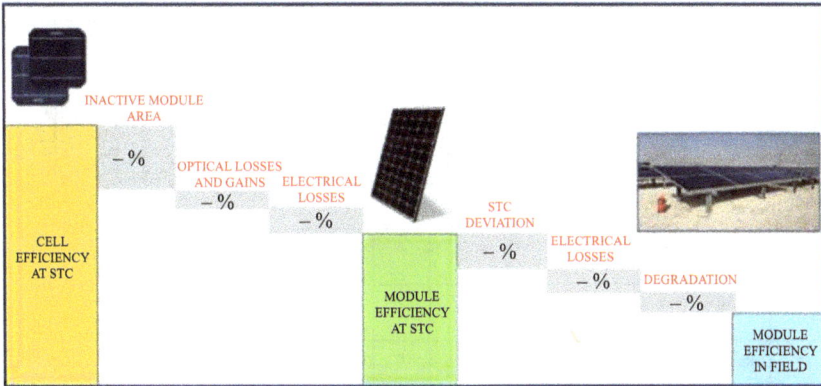

Figure 6.8 *Mechanisms responsible for the losses by upscaling from cell to module to field. STC deviation shows the difference between STC and operating conditions.*

modules in the field. Integrating solar cells into a PV module results in gains and losses that change the module's power and efficiency compared to the original solar cells. The module design and material selection govern the energy output. An optimal module design will require minimum CTM gap, which will result in lower cost/watt. Therefore, to develop such a module it is important to understand the role of an individual component in the module.

It should be noted that these losses are basically evaluated at STC, that is, cell temperature of 25°C, wind velocity of 1 m/s and incident radiation of AM 1.5 spectrum corresponding to 1,000 W/m^2. The nominal operating efficiency of modules and CTM losses are assessed under this STC. Although the focus of module technology is on STC efficiency, it is also vital to understand the dependence of energy yield on environmental parameters differing from STC such as insolation, cell temperature, and spectral variability, which leads to module to field (MTF) losses. The operating conditions of the modules under the desert environment such as high-temperature environment with more incident radiation dose will cause the module losses to be higher due to higher performance degradation. The primary parameter from MTF is varying operating conditions, degradation under field, and FF losses. The FF losses essentially arise from series and shunt resistances from the system.

In principle, the conversion losses and gains in the CTM development can be characterized based on geometry, optics, and electric effects. Figure 6.9 shows a detailed schematic of the gain and loss mechanisms responsible for CTM conversion. They are influenced by:

- **Geometrical**: The losses associated with the geometry of the cell and module are related to the active and inactive areas responsible for photoconversion [2,4].
- **Optical:** This effect can result in both losses and gains because of material selection based on the target function. They include parasitic absorption and

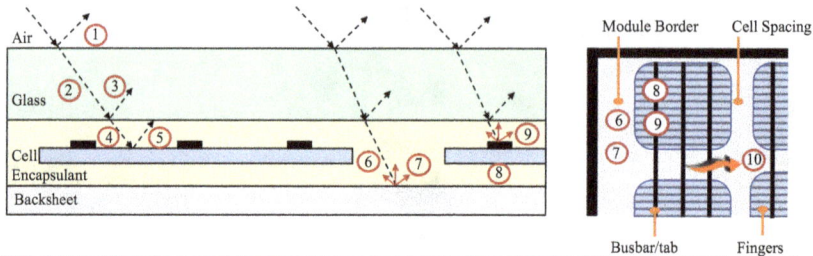

1	Glass reflection loss	6	Geometric loss due to inactive areas
2	Glass absorption loss	7	Reflection gain from backsheet
3	Reflection loss from glass/encapsulation	8	Shadow effect loss from contacts
4	Encapsulation absorption loss	9	Reflection gain from contacts
5	Coupling gain cell/encapsulation	10	Resistive losses due to interconnections

Figure 6.9   Detailed loss analysis involved in evaluating CTM performance

reflection of sublayers. However, there are not only losses: additional light reflection on the solar cell or reduced light reflection from the encapsulated cell results in gains [8,9].

• **Electrical:** The impact is primarily due to resistive losses due to interconnections. Shading effect and reflection from tabs are also essential for the current collection [10].

Therefore, for an optimum PV system, it is vital that CTM to field performance is thoroughly assessed to target any area of improvement at any scale from device to systems. Currently, SHJ technology is the main attraction in the solar community, due to its high PCE and reduced levelized cost of electricity owing to low TC and bifaciality [11,12].

Therefore, SHJ was considered for analysis in this work. SHJ technology is based on low-temperature processing, typically with a maximum process temperature of 200°C, making it less energy intensive [13]. Use of wires to interconnect subsequent cells can drastically reduce the need for silver paste up to 85%. Therefore, the cost of the solar panel can be subsequently reduced or partially transfered to the advanced interconnection technique such as smart wire connection technology (SWCT) [14]. This chapter develops and conducts a comprehensive end-to-end study for the CTM to the field by examining the performance of bifacial heterojunction solar cells under laboratory and in-field. For measuring the effectiveness of the modularization and field installation process, CTM losses and MTF losses are examined for customized bifacial SHJ cells and modules with SWCT.

6.10   Module installation

This chapter presents a comprehensive bifacial PV module parametric study for the hot and sunny desert environment of Qatar. Real-time performance monitoring and complete parametric analysis of bifacial modules are timely needed to support the adoption of this technology. The details of the results from experiment and simulations that were accomplished in this chapter are explained below:

(I)   Bifacial versus monofacial comparison:
   (a)   Experiment: Compared specific average daily energy yield (kWh/day/kW$_p$) for bifacial and monofacial modules installed at test site in-array mounted for a period of 1 year with 2-month cleaning frequency at fixed 22° tilt N–S. Three modules are compared: two bifacial modules of 90% bifaciality (Bi90) and 65% bifaciality (Bi65) with one monofacial module.
   (b)   Simulation: Validated the simulations with real-time field data for 5 months of the bifacial module (Bi90) installed at outdoor test facility (OTF) with fixed 22° tilt N–S.

(II)   Standalone versus in-array mounted module energy yield:
   (a)   Experiment: Fixed 22° tilt N–S Bi90 module compared under standalone and in-array mounting. Both the modules were placed landscape with "1-up" configuration on the mounting frame. The distance between adjacent rows is 4.5 m at OTF.

(III)   Ground albedo:
   (a)   Experiment: Four synthetic and natural ground materials are used for 22° tilt and vertical bifacial panel facing south: desert sand, grass, white cement, and aluminum sheet.
   (b)   Simulation: Albedo coefficient of 0.2, 0.4, and 0.6 used to assess energy yield.

(IV)   Mounting height:
   (a)   Experiment: For 22° tilt N–S, height was varied from 66 cm to 103 cm from the bottom edge of Bi90 module. For vertical facing E–W, height varied from 44 cm to 92 cm from the bottom edge of Bi90 module.
   (b)   Simulation: Height changed from 0.5 to 2.0 m for annual energy yield (AEY) simulations.

(V)   Azimuth orientation:
   (a)   Experiment: Two standard positions of azimuth angle (orientation), that is, N–S (180°) and E–W (90°), were tested for 22° tilt and vertical bifacial (Bi90) module.
   (b)   Simulation: Azimuth angle varied from 90° to 180° for both 22° tilted and vertical modules.

(VI)   Module temperature:
   (a)   Experiment: Nominal operating module temperature (NOMT) for glass/glass monofacial and bifacial modules are compared from

rear-side measurements under 22° tilt N–S module with both of them mounted in-array in the field.

(VII)   Bifaciality:

    (a)   Simulation: From a device perspective, bifaciality was varied from 50% to 100% for vertical and tilted panels to assess the AEY sensitivity.

(VIII)   Temperature coefficient:

    (a)   Simulation: To assess the effect of different TC bifacial modules, values from –0.5%/°C to –0.2%/°C were used with local climatic data.

6.10.1   *Outdoor module performance monitoring*

Figure 6.10(a) and (b) shows the ambient temperature and albedo coefficient measured at the installation site in Qatar for the period of Jan–Dec 2017. Histogram

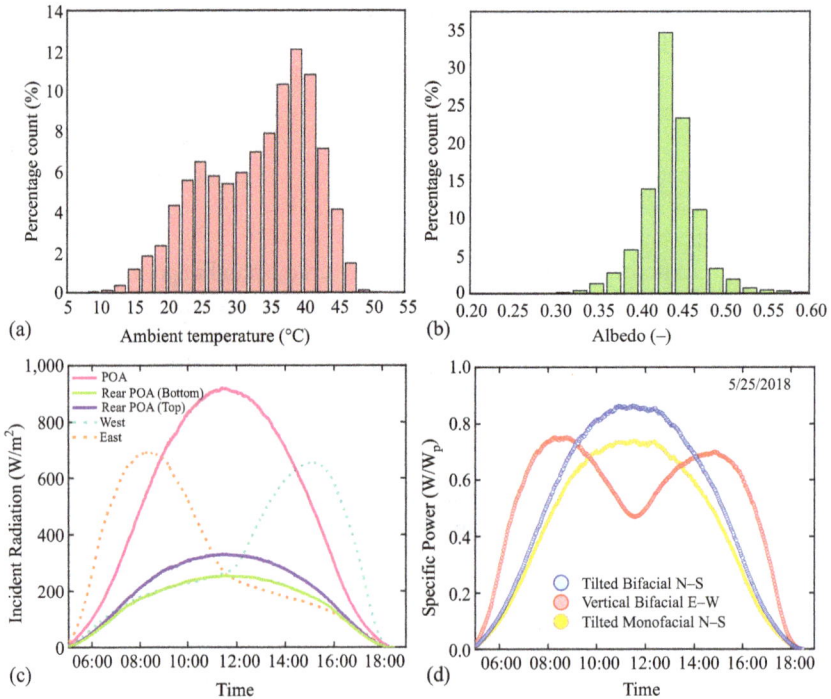

Figure 6.10   *(a) Ambient temperature histogram using an annual data for the location of OTF, Qatar. (b) Histogram of albedo values monitored over a year for local sand at OTF site. (c) Possible radiation sources for bifacial power harnessing from different orientations. Note the difference between the rear-side measurements at the bottom and top pyranometers. (d) Typical specific power profile of bifacial and monofacial PV installed in Qatar with vertical E–W and N–S directions.*

of Figure 6.10(a) shows a higher ambient temperature frequency throughout the year with a mean of 33.3°C and a standard deviation of 7.9°C. A higher frequency of albedo values was found to be in the narrow range around 0.35–0.45 as shown in Figure 6.10 (b) where the mean albedo is found to be 0.43 and a standard deviation of 0.03. The higher standard deviation for ambient temperature shows the seasonal dependence of the temperature values whereas albedo was not found to be significantly different throughout the year. Measured data show the different climatic nature at this region with high environmental temperature and albedo coefficient necessitating the need for local field tests. Figure 6.10(c) provides the general trend of incident radiation and power output of bifacial and monofacial panels aligned in different azimuth orientations installed at OTF, Qatar. It is found that the backside irradiance can be quite nonuniform and this variation can lead to a current mismatch between cell and modules connected in series. Around solar noon, the maximum nonuniformity was observed where the top radiation reference cell showed 330 W/m^2 which was reduced to 254 W/m^2 under the considered 22° tilt orientation for the data of May 2018. Vertical bifacial PV can utilize the radiation in E–W direction throughout the daylight hours as shown in Figure 6.10(d) with the two daily peaks. Two standard positions of azimuth angle (orientation), that is, N–S (180°) and E–W (90°), were tested for 22° tilt and vertical bifacial module. The change in azimuth angle, from south facing to the west, with vertical position offers a unique advantage to the PV community for managing load control and reduce soiling. This way, energy generation profile can be broadened, which shall result in relatively steadier production throughout the daylight hours than the traditional south-facing configuration. The balancing of electrical supply/demand and reduction in storage is, therefore, a possible avenue that may evolve with the optimal system design of hybrid bifacial E–W and N–S panels [9].

6.10.2   *Bifacial versus monofacial energy yield gain and standalone versus in-array mounted losses*

Figure 6.11 shows the potential of bifacial PV installation in the hot desert environment. Measurements are compared for landscaped two bifacial and one monofacial PV panel installed facing south with 22° tilt installed in-array at a height of 100 cm and a row-to-row spacing of 4.5 m.

All the modules tested were mono-crystalline solar cells and followed the same cleaning frequency of 2 months [5]. Bifacial energy yield showed an annual average daily gain of 8.6% for a panel with bifaciality of 65% (Bi65) and 16.3% gain for bifaciality of 90% (Bi90) when compared with monofacial PV as shown in Figure 6.11(a). The energy yield difference is primarily due to the additional bifaciality ratio and rear-side photogeneration from albedo radiation, which was measured to be 40% onsite for a yearly average. Monofacial PV showed an annual average daily specific yield of 4.54 kWh/day/kW$_p$ whereas Bi65 showed 4.93 kWh/day/kW$_p$ and Bi90 increased it to 5.28 kWh/day/kW$_p$. It should be noted that the installation of higher bifaciality panel resulted in a percentage energy gain of 17.85% for absolute bifaciality change of 25%, that is, from 65% to 90%. For exploring the optimal conditions and understanding the fundamentals, standalone

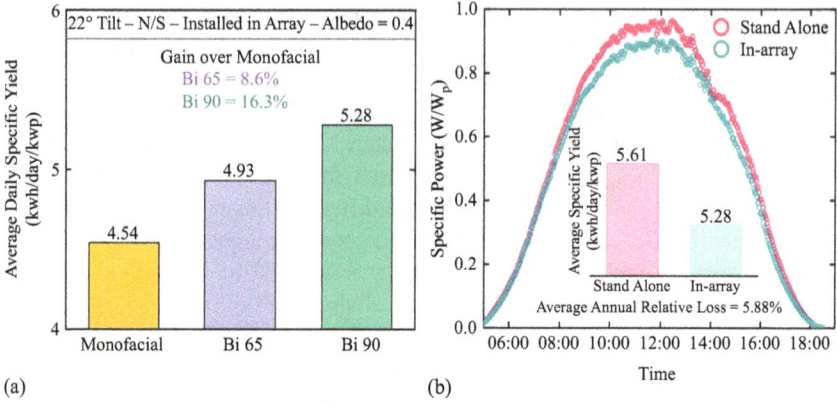

(a)   (b)

Figure 6.11   *(a) Comparison of average daily specific yield for three in-array modules with 2 months cleaning frequency taken over 1 year. Two bifacial PV with different bifaciality (bi) factors of 65% and 90% were tested under natural albedo with standard 22° tilt configuration facing south. (b) Relative loss from a standalone system to realistic in-array bifacial PV modules for a Bi90 module. The graph shows the typical daily power profile for the month of May 2018 with an inset showing the average annual percentage losses from standalone to in-array mounting.*

systems can be employed for different parameter studies. Still, they do not completely represent the in-array installed module. Fixed 22° tilt N–S modules were compared under standalone and in-array mounting. Both the modules were placed landscape with the single module (1-up) on the frame. The distance between adjacent rows is fixed to 4.5 m for in-array mounted configurations. The modules installed in arrays have radiation obstruction from neighboring modules due to their adjacent placement and spacing between rows. On the contrary, the standalone system as portrayed in Figure 6.11(b) is exposed to radiation from the complete hemisphere and only gets restricted with self-shadowing. Therefore, the gaps between standalone and in-array installed modules were tested for the same installation conditions, that is, 40% albedo and 100 cm in height. Figure 6.11(b) shows that by upscaling from standalone to in-array bifacial modules, the average yield was reduced from 5.61 kWh/day/kW$_p$ to 5.28 kWh/day/kW$_p$ for a year's data. Relative loss in average AEY under studied configuration was 5.88% due to additional shadowing from neighboring modules. Considering this analysis, care must be taken when interpreting the data from the installation parameter measurements and the losses must be incorporated for proper system design.

6.10.3   *Effect of mounting height*

One of the most important installation parameters for bifacial PV is the mounting height of the panel from the ground as it determines the amount of irradiation on the rear side.

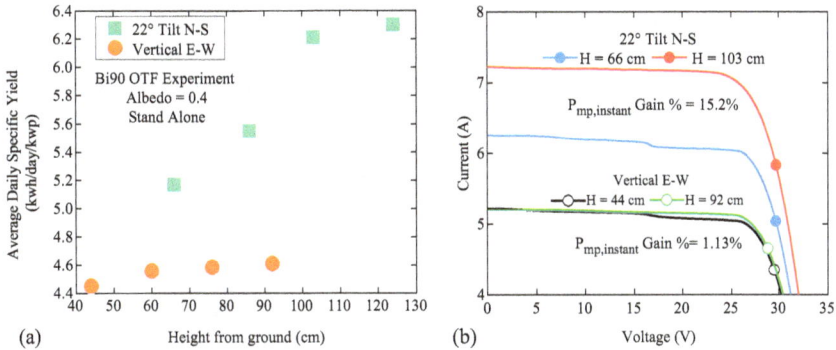

Figure 6.12   *(a) Effect of installation height from the ground on the SEY of 22° tilt N–S and vertical E–W panel for Bi90 module and (b) I–V curves showing kinks due to nonhomogenous irradiation from the rear side of bifacial PV at different heights*

To assess the impact, two standard configurations were tested using standalone systems. For 22° tilt N–S, height was varied from 66 cm to 103 cm whereas for vertical facing E–W, height was varied from 44 cm to 92 cm from the bottom edge of the module. In the experiments for measuring yields, tilted panels were found to be more sensitive to mounting height than vertical bifacial panels as shown in Figure 6.12(a). Overall percentage gain was found to be 3.58% for vertical panels. This, however, was increased to 21.92% for tilted panels by changing height from one extreme to another. From 66 cm to 86 cm, the percentage gain was 7.36%, which was increased to 11.92% from 86 to 104 cm. However, the effect reaches a plateau with only a 1.2% gain from 104 cm to 124 cm for the tilted case.

The outcome is due to self-shadowing of the panels and nonhomogeneity of irradiation at the rear side, which was found to saturate after ~100 cm for tilted panels. This was confirmed by recording the I–V curves shown in Figure 6.12(b). Low rear-side radiation and high nonhomogeneity leading to current mismatching problem can be seen from the kinks in the I–V graphs, which are dominant for tilted panels.

The contribution to rear-side irradiation is primarily a function of solar position, which changes the shadow height of the module throughout the day. This results in less irradiation at the lower portion of the module and higher irradiation at the upper portion, leading to nonhomogenous distribution. However, the influence of albedo and diffuse light on the rear side is complex. Like beam radiation, diffuse light shows higher intensity around the top edge of the module. On the contrary, albedo reflection shows maximum intensity around the bottom edge of the module. Even with this contradictory effect, the collective contribution from diffuse and albedo radiation results in low intensity at the lower edge of the module. Low rear-side radiation and high nonhomogeneity leading to current mismatching problem can be seen from the kinks in the I–V graphs, which are dominant for tilted panels.

To reduce this effect and increase output, higher elevation is required. By elevating the modules, the tilted panel showed an instantaneous gain in maximum power of 15.2% with a smooth I–V curve instead of low irradiation and three kinks (due to three bypass diodes working) present at the lower height of 66 cm. For vertical configuration, the effect was small with only 1.13% increase in the maximum power point. At the cell level, this issue may be tackled by assessing optimal numbers of bypass diodes whereas, at the module level, higher elevation mounting is recommended.

6.10.4   Effect of natural and synthetic ground albedo

With a proportional relation, materials with high albedo coefficient showed better performance by reflecting more light on the rear side of the panel. To explore the optimal installation parameters, four synthetic and natural albedo materials were examined: desert sand, grass, white cement, and aluminum. They were investigated in the standalone configuration for 22° tilt and vertical bifacial panels facing south and the results are shown in Figure 6.13(a) and (b). The effect of local desert sand was encouraging with an average relative gain in power of 18.59% and 20.18% for vertical and tilted, respectively. Grass albedo gained minimum at 8.4% with vertical panels whereas white cement performed best with a maximum bifacial gain of 28.5% for titled systems when compared with monofacial panels under STC. For grass, the relative change was lower for vertical case due to the lower albedo, which was a contributing factor for less irradiation being incident from two sides. This, however, was not the case for the tilted panel where the dominant radiation is captured from the front side.

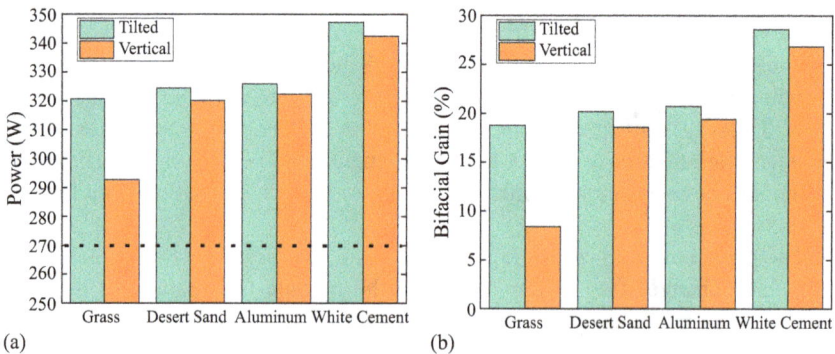

Figure 6.13 (a) Albedo effect on power output, corrected to STC, of bifacial PV (Bi90) modules installed in vertical and 22° tilt position under standalone mounting at OTF, Qatar. The dashed line shows the monofacial front-side rating at 270 W, which is used for comparison, and (b) bifacial gain from monofacial PV module under different ground reflective materials.

Interestingly, aluminum foil with high albedo did not perform better than white cement due to the specular reflection characteristics and not diffusing the light homogenously. Vertical panels were found to be more sensitive to albedo since both the front and rear sides are affected by the coated reflective surface. However, it should be noted that since a tilted panel is optimized for front-side irradiation, in general, it receives more aggregated radiation than vertical bifacial. This is the reason that the average power generated was higher for the tilted bifacial panel as shown in Figure 6.13(b).

6.10.5   Effect of module temperature

NOMT is the temperature when the modules are operating at maximum power point with POA of 800 W/m^2, wind velocity of 1 m/s, and ambient temperature of 20°C based on ASTM [131] and IEC standards [132]. Since these conditions are rarely met in the field, these standards provide guidelines for filtering data above 400 W/m^2 with wind speed under 1.0 ± 0.75 m/s and ambient environment between 5°C and 35°C. Module temperature relative to ambient temperature $(T_m - T_{amb})$ is then plotted against front-side irradiation to develop a regression equation. This can then be employed for future prediction of module considering the environmental parameters and correction factors based on wind speed [133]. Glass/glass monofacial and bifacial modules are compared from rear-side measurements under 22° tilt N–S module with both installed in the array. The effect of module temperature under load was assessed by using two platinum sensors on the rear face of modules with a standard deviation of 1.37°C as plotted in Figure 6.14(a). On average, both glass/glass bifacial and monofacial panels were found to work on same operating temperatures, that is, $T_{bi} = T_{mono}$ with an average $\Delta T = T_{bi} - T_{mono}$ of 0.31 under the same device configuration. However, by careful examination, Figure 6.14(b) shows that ΔT does not stay constant and varies under different irradiation regimes. Bifacial PV module performs thermally better with low module temperature under high irradiances $(T_{mono} > T_{bi})$ whereas monofacial performed well under lower irradiances $(T_{bi} > T_{mono})$. The high module temperature of bifacial than monofacial under low irradiance is related to the relatively higher bifacial gain from the rear side due to the temporal distribution of albedo radiation leading to an increase in the effective heat generation [19]. This nature of albedo is during the early and late hours of the day when the position of the sun is low on the horizon. This suggests that the bifacial power gain determines the cell temperature under different irradiation conditions. The outcome is favorable for sunny climates where there is abundant radiation and clear sky days with long daylight hours.

6.10.6   Optimization of height, tilt, and albedo

Figure 6.15(a) shows the combined effect of the tilt angle, height, and albedo coefficient on the AEY performance of the bifacial panel. This is vital to understand the cumulative effect of height and albedo on the optimal tilt angle for bifacial technology. Unlike standard monofacial PV, bifacial PV needs a more

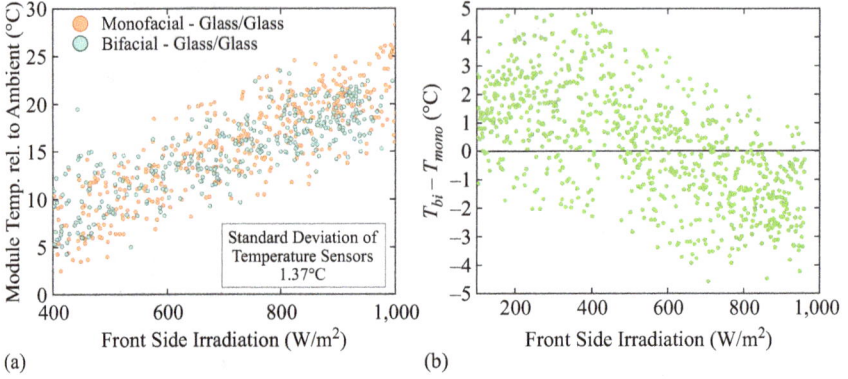

(a)   (b)

Figure 6.14   *(a) Temperature effect of glass/glass monofacial (mono) and bifacial (Bi90) PV module for a range of front-side POA irradiation under 22° tilt N–S module with both mounted in-array. Module temperature relative to ambient temperature is plotted against front-side irradiation with a standard deviation of 1.37°C for two temperature sensors' readings installed at the rear side and (b) difference between bifacial module temperature and monofacial module temperature as a function of front-side irradiation. The line at $T_{bi} - T_{mono} = 0$ separates two regimes $T_{bi} > T_{mono}$ and $T_{mono} > T_{bi}$.*

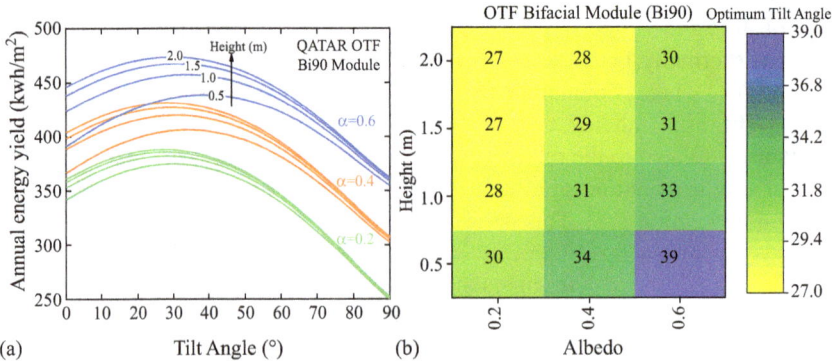

(a)   Tilt Angle (°)   (b)   Albedo

Figure 6.15   *(a) Combined effect of the tilt angle, height, and albedo coefficient on the annual energy yield performance of bifacial modules using simulations and (b) optimal tilt angle for maximum annual energy yield as a function of height and albedo for bifacial PV modules installed at OTF obtained from simulations.*

systematic treatment to calculate the optimal installation conditions because of the direct impact of albedo and mounting settings. In general, higher reflective ground material and height were found to increase the energy yield. The effect of albedo is more dominant among all factors as it has the major contribution to the irradiation received on the rear side. By changing the albedo α from 0.2 to 0.6, with height 2 m, energy yield was enhanced by 21.98%. After carefully selecting site location and albedo, maximum height possible should be selected to increase rear-side radiation component and reduce nonhomogeneity on the rear face of the bifacial module. Bifacial systems require a higher tilt angle for optimum performance so as to collect additional radiation from the surrounding and ground. Another benefit for higher tilt angles is reduced soiling losses. By comparing the maximum energy yield, optimal tilt angle as a function of height and albedo is shown in Figure 6.15(b).

For best performance, larger tilt angles are required than traditional 22° tilt modules for higher albedo coefficient ground. With an increase in albedo, the optimal tilt angle for the bifacial panel was increased whereas with an increase in height, the opposite relationship was found where the optimal tilt angle was decreased. This conflicting nature of tilt angle dependence on the height and albedo shows the importance to analyze all these parameters concurrently. For higher elevation mounting, lower tilt angles are recommended. However, for lower heights, higher tilt angles are required for capturing high radiation with more homogeneity on the rear side of the bifacial module as observed from the peak shift toward left in Figure 6.15(a) and (b). By fixing optimal tilt angles and varying height from 0.5 to 2.0 m, AEY was increased by 3.6%, 6.47%, and 8.6% for α = 0.2, 0.4, and 0.6, respectively.

6.10.7   Effect of azimuth orientation

AEY simulations were performed for azimuth angle variation where the azimuth angle (orientation) was varied from N–S (180°) to E–W (90°) for 22°, 30°, and 90° (vertical) tilt as shown in Figure 6.16(a) and (b). At an elevation of 1 m, vertical bifacial modules facing E–W showed comparable performance with tilted panels with a N–S orientation. Figure 6.16(b) shows the seasonal variation of the same configurations. The maximum AEY achieved for 90° E–W, 22° N–S, and 30° N–S was 390.2, 417.5, and 420.2 kWh/m^2, respectively. Relative change in AEY for vertical E–W was found to be 6.5% and 7.1% with 22° and 30° tilt, respectively. Azimuth angle has a higher influence on vertical panels than tilted panels. Due to their sensitivity, AEY was changed by 27.7% from N–S to E–W facing. On the contrary, by varying azimuth from standard N–S to E–W for tilted bifacial, the relative change was estimated to be 6.23% for 22° and 6.87% for 30°. Since vertical bifacial modules have an advantage of less soiling, the energy losses by changing azimuth angle should be compensated by additional soiling on standard tilted configurations. Additionally, vertical bifacial E–W can be employed for peak shaving of load profile. This way, energy generation profile can be flattened, which shall result in a relatively

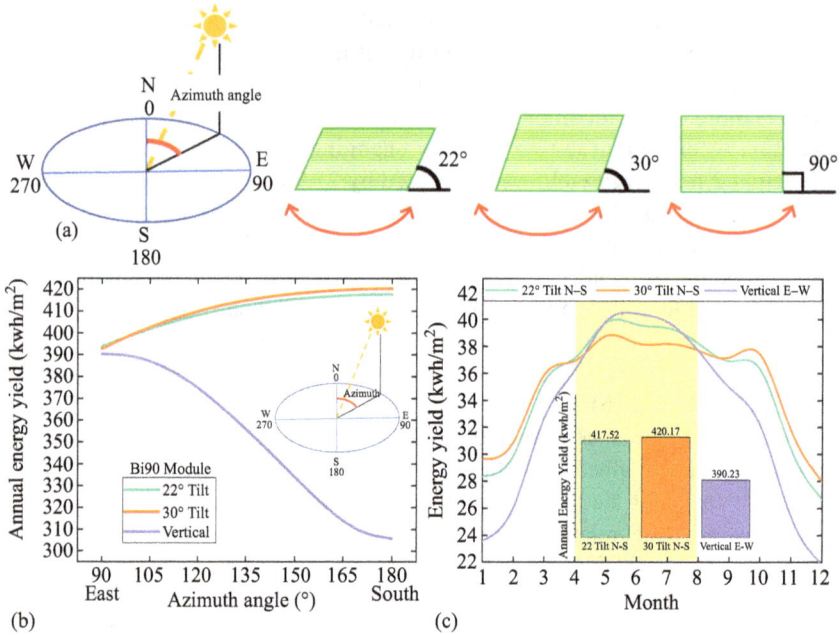

Figure 6.16   (a) Schematic showing the three studied configurations for azimuth
           effect, (b) simulation results for azimuth angle effect on annual
           energy yield of vertical E–W and tilted N–S modules, and (c)
           seasonal variation for vertical E–W and tilted N–S to highlight the
           difference in performance during each month.

steadier production throughout the daylight hours than traditional south-facing
configuration.

6.10.8 Effect of device parameters: bifaciality and temperature coefficient

For better performance, consideration of an appropriate technology with higher
bifaciality and lower power TC is desirable. Both of these cell properties are
vital in the selection of modules for hot and sunny desert environment.
Bifaciality (ratio of rear power to front power at STC) is the main parameter that
characterizes the front- and rear-side response of bifacial solar cells. Bifaciality
factor has a maximum value of 1.0 whereas, generally, it is in the range of
0.6–0.9 for different technologies. This is due to the cell design, especially back
contact and metallization. At the module level, the optical response may be
lower due to additional shadowing from the junction box and different rear-side
(glass or transparent back sheet) properties.

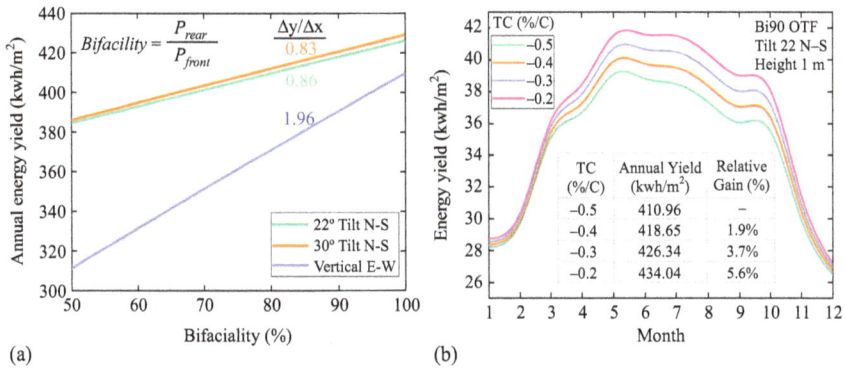

Figure 6.17   *Simulation results for solar cell device-level parameters. (a) Effect of a bifaciality factor on annual energy yield of vertical E–W and tilted N–S modules. Vertical bifacial gains more than tilted modules with higher slope and (b) effect of the power temperature coefficient (TC) on the annual energy yield of bifacial modules.*

Proper utilization of albedo radiation for different mounting configurations can also be carried out by understanding the effect of bifaciality of modules. Qualitatively, AEY showed a direct linear relationship with an increase in bifaciality factor as shown in Figure 6.17(a). The slope, defined by a change in AEY relative to the unit change in bifaciality, was different for each configuration. Vertical bifacial panels gain more from bifaciality than their counterpart tilted panels. For vertical E–W panels, the gradient was the maximum at 1.96 followed by 22° tilt N–S at 0.86 and 30° tilt N–S at 0.83. This implies that for every percent rise in bifaciality, the technology yields this amount of additional energy. The overall change in AEY by changing bifaciality from 50% to 100% was 32.36% for vertical E–W, 10.88% for 22° tilt N–S, and 10.68% for 30° tilt N–S. Therefore, modules of high bifaciality are recommended for vertical configuration as highlighted in Figure 6.17(a).

The power TC is a measure of percentage power drop relative to the unit change in cell temperature. To assess the effect of different TCs, values from –0.5%/C to –0.2%/C for local climatic data of albedo 0.4 is plotted in Figure 6.17 (b). The high average module temperature can be tackled by employing lower TC cells such as SHJ cell (around 0.3%/C) [134]. Currently, most of the commercial modules are in the range of –0.5 to –0.3%/C depending on the device type. Simulation results for TC show that by using the higher temperature resistant device, the maximum gain could be up to 5.6% for modules installed in OTF, Qatar. The best performance TC of –0.2%/C resulted in AEY of 434.04 kWh/m^2, which was reduced to 410.96 kWh/m^2 for a TC of –0.5%/C. Annual energy yield was found to increase linearly for constant system parameters.

Real-time measurements for device and installation parameters are consolidated in Table 6.2.

Table 6.2 Summary of key performance parameters studied for bifacial modules

Parameter examined	Testing conditions	Results
Bifacial versus monofacial (experiment)	Three modules compared at fixed 22° tilt N–S in array mounted under local albedo of 40%: two bifacial modules of 90% bifaciality (Bi90) and 65% bifaciality (Bi65) with one monofacial module	Comparison showed an annual average daily specific yield of Monofacial: 4.54 kWh/day/kW$_p$ Bifacial, Bi65: 4.93 kWh/day/kW$_p$ (relative gain 8.6%) and Bifacial, Bi90: 5.28 kWh/day/kW$_p$ (relative gain 16.3%)
Standalone versus in-array mounted (experiment)	Fixed 22° tilt N–S Bi90 module compared for standalone and in-array with albedo 40%. Landscape and 1-up with 4.5 m distance between adjacent rows	For a particular case, average daily specific yield was reduced from 5.61 kWh/day/kw$_p$ (standalone) to 5.28 kWh/day/kW$_p$ (in-array mounted) Relative loss was 5.88% on yearly average
Ground albedo (experiment)	Four synthetic and natural ground materials used for 22° tilt and vertical bifacial (Bi90) standalone facing south at a height of 1 m: desert sand, grass, white cement, and aluminum sheet	Local desert sand resulted in an average power gain of 18.59% and 20.18% for vertical and tilted, respectively. Grass albedo gained minimum at 8.4% with vertical whereas white cement performed best with a maximum gain of 28.5% for titled when compared with monofacial under STC
Module temperature (experiment)	Nominal operating module temperature for glass/glass monofacial and bifacial modules (Bi90) compared from rear side measurements for in-array mounted 22° tilt N–S	On average, both glass/glass bifacial and monofacial panels were found to work at same operating temperatures, i.e., $T_{bi} = T_{mono}$ with an average $\Delta T = T_{bi} - T_{mono}$ of 0.31 ΔT was found to be a function of bifacial gain, which in turn depends on the temporal distribution of incident and albedo radiation
Mounting height (experiment)	For 22° tilt N–S, height: 66 cm to 103 cm For vertical E–W, height: 44–92 cm Bi90 standalone module tested under local albedo = 40%	Tilted panels were found to be more sensitive to mounting height than vertical bifacial panels. Overall percentage gain in power: Vertical E–W: 3.58% 22° tilt N–S: 21.92%

(Continues)

Azimuth orientation (simulated)	Azimuth angle (orientation) variation From N–S (180°) to E–W(90°) for 22°, 30°, and 90°(vertical) tilt Albedo = 40% and height = 1 m	Azimuth angle showed a higher influence on vertical bifacial module than tilted. Energy yield was changed by 27.7% from N–S to E–W for vertical. For tilted bifacial, the relative change was6.23% for 22° and 6.87% for 30°
Bifaciality (simulated)	50%–100% for vertical E–W and 22° N–S tilted bifacial modules. Albedo = 40% and height = 1 m	Vertical bifacial modules gained more from bifaciality than their counterpart tilted panels. The overall change in energy yield was 32.36% for vertical E–W, 10.88% for 22° N–S, and 10.68% for 30° N–S
Temperature coefficient (TC) (simulated)	TC: –0.5%/C to –0.2%/C	By using the higher temperature resistant device, the maximum gain, varying TC from –0.5%/C to –0.2 %/C, could be up to 5.6% for modules installed in OTF, Qatar
Combined effect of albedo, tilt, and mounting height (simulated)	Albedo: 0.2–0.6 Tilt Angle: 0°–90° Height: 0.5 m–2 m	Higher reflective ground material and height were found to increase the energy yield. The effect of albedo is more dominant among all factors as it has the major contribution to the irradiation received on the rear side. For best performance, larger tilt angles are required than traditional 22° tilt modules. For higher elevation mounting, lower tilt angles are recommended

6.11   Conclusions

- In-array mounted bifacial modules showed an annual AEY gain of 8.6% for a panel with bifaciality of 65% and 16.3% gain for bifaciality of 90% when compared with monofacial PV.
- Comparison between standalone and in-array mounted bifacial module for a particular configuration of fixed tilt 22° N–S showed an AEY reduction of 5.88% due to additional shadowing from neighboring modules for in-array placed module.
- Four synthetic and natural albedo materials were used for 22° tilt and vertical bifacial facing south: desert sand, grass, white cement, and aluminium. Local desert sand showed an increase of 21% in power output relative to monofacial panels under STC for standalone systems. Grass albedo gained a minimum of 8.4% whereas white cement performed best with a bifacial gain of 28.5%.
- One of the most important installation parameters for bifacial PV is the height of the panel as it determines the amount of irradiation on the rear side. In the experiments, tilted panels were found to be more sensitive to mounting height than vertical bifacial panels. This effect is due to self-shadowing of the panels and nonhomogeneity of irradiation at rear side, which was found to saturate after ~100 cm for tilted panels.
- On average, both glass/glass bifacial and monofacial panels were found to work at same operating temperatures with a difference of 0.31°C. However, the temporal nature of albedo and bifacial gain caused the thermal performance of bifacial and monofacial modules to be different during the daylight hours. Overall, the measured higher reflective albedo, the increased gain over monofacial PV, and the matching average module temperature illustrate the promising nature of bifacial PV. The study reveals vital insights that will allow engineers and researchers to optimize bifacial PV performance in hot and sunny desert environments. Future directions for the research include testing the effect of soiling and mounting settings on the performance of bifacial PV arrays.

References

[1]   R. Kopecek and J. Libal, Towards large-scale deployment of bifacial photovoltaics, *Nat. Energy.* 3 (2018) 443–446. doi:10.1038/s41560-018-0178-0.

[2]   J.E. Castillo-Aguilella and P.S. Hauser, Multi-variable bifacial photovoltaic module test results and best-fit annual bifacial energy yield model, *IEEE Access.* 4 (2016) 498–506. doi:10.1109/ACCESS.2016.2518399.

[3]   U.A. Yusufoglu, T.M. Pletzer, L.J. Koduvelikulathu, C. Comparotto, R. Kopecek, and H. Kurz, Analysis of the annual performance of bifacial modules and optimization methods, *IEEE J. Photovoltaics.* 5 (2015) 320–328. doi:10.1109/JPHOTOV.2014.2364406.

[4]   I. Shoukry, J. Libal, R. Kopecek, E. Wefringhaus, and J. Werner, Modelling of bifacial gain for stand-alone and in-field installed bifacial PV modules, *Energy Procedia*. 92 (2016) 600–608. doi:10.1016/j.egypro.2016.07.025.

[5]   B. Figgis, QEERI Solar Test Facility 5-Year Report, 2018. doi:10.13140/RG.2.2.12062.23361.

[6]   E.G. Luque, F. Antonanzas-Torres, and R. Escobar, Effect of soiling in bifacial PV modules and cleaning schedule optimization, *Energy Convers. Manag*. 174 (2018) 615–625. doi:10.1016/j.enconman.2018.08.065.

[7]   A.A.B. Baloch, M. Armoush, B. Hindi, A. Bousselham, and N. Tabet, Performance assessment of stand alone bifacial solar panel under real time conditions, in: *2017 IEEE 44th Photovoltaic Specialist Conference*, Washington DC, 2017.

[8]   J. Rabanal-Arabach, M. Mrcarica, A. Schneider, R. Kopecek, and M. Heckmann, The need of frameless mounting structures for vertical mounting of bifacial PV modules, in: *32nd European Photovoltaic Solar Energy Conference and Exhibition*, 2016, pp. 1630–1634. doi:10.4229/EUPVSEC20162016-5CO.14.5.

[9]   S. Obara, D. Konno, Y. Utsugi, and J. Morel, Analysis of output power and capacity reduction in electrical storage facilities by peak shift control of PV system with bifacial modules, *Appl. Energy*. 128 (2014) 35–48. doi:10.1016/j.apenergy.2014.04.053.

[10]   ITRPV, International Technology Roadmap for Photovoltaic (ITRPV) – Results 2017, ITRPV, 2018, pp. 1–37. doi:http://www.itrs.net/Links/2013ITRS/2013Chapters/2013Litho.pdf.

[11]   R. Guerrero-Lemus, R. Vega, T. Kim, A. Kimm, and L.E. Shephard, Bifacial solar photovoltaics – a technology review, *Renew. Sustain. Energy Rev*. 60 (2016) 1533–1549. doi:10.1016/j.rser.2016.03.041.

[12]   T. Dullweber, H. Schulte-Huxel, S. Blankemeyer, *et al*., Present status and future perspectives of bifacial PERC+ solar cells and modules, *Jpn. J. Appl. Phys*. 6 (2018) 1366–1381.

[13]   M.R. Khan, A. Hanna, X. Sun, and M.A. Alam, Vertical bifacial solar farms: physics, design, and global optimization, *Applied Energy* 206 (2017) 240–248.

[14]   A. Asgharzadeh, B. Marion, C. Deline, C. Hansen, J.S. Stein, and F. Toor, A sensitivity study of the impact of installation parameters and system configuration on the performance of bifacial PV arrays, *IEEE J. Photovolaics* 8 (2018) 798–805. doi:10.1109/JPHOTOV.2018.2819676.

[15]   J.P. Singh, S. Guo, I.M. Peters, A.G. Aberle, and T.M. Walsh, Comparison of glass/glass and glass/backsheet PV modules using bifacial silicon solar cells, *IEEE J. Photovoltaics* 5 (2015) 783–791.

[16]   J.P. Singh, A.G. Aberle, and T.M. Walsh, Electrical characterization method for bifacial photovoltaic modules, *Sol. Energy Mater. Sol. Cells* 127 (2014) 136–142. doi:10.1016/j.solmat.2014.04.017.

[17]   S. Wang, O. Wilkie, J. Lam, *et al*., Bifacial photovoltaic systems energy yield modelling, *Energy Procedia* 77 (2015) 428–433. doi:10.1016/j.egypro.2015.07.060.

[18]   T. Joge, l. Araki, K. Takaku, *et al.*, Advanced applications of bifacial solar modelus, in: *3rd World Conference on Photovoltaic Energy Conversion* 2003, pp. 1–4.

[19]   M.W.P.E. Lamers, E. Özkalay, R.S.R. Gali, *et al.*, Temperature effects of bifacial modules: hotter or cooler?, *Sol. Energy Mater. Sol. Cells.* 185 (2018) 192–197. doi:10.1016/j.solmat.2018.05.033.

[20]   D.S.H. Nussbaumer, F. Baumgartner, and T. Baumann, New opportunities for a better power distribution by the use of bifacial modules in future PV systems, presented in BIFIPV Workshop, Chambéry (2014).

[21]   B. Van Aken, White bifacial modules: monofacial or bifacial: Pmax under STC power rating, in: *European Photovoltaic Solar Energy Conference and Exhibition*, 2016.

[22]   J. Liba and R. Kopecek, Bifaciality: one small step for technology, one giant leap for kWh cost reduction, in: *Cell Processing, Photovoltaics International Papers*, PV Tech., 2015.

[23]   NREL Chart for the Best Solar Cell Efficiencies, 2022, https://www.nrel.gov/pv/cell-efficiency.html

[24]   A. Kojima, K. Teshima, Y. Shirai, and T. Miyasaka, Organometal halide perovskites as visible-light sensitizers for photovoltaic cells, *J. Am. Chem. Soc.* 131 (2009) 6050–6051. doi:10.1021/ja809598r.

[25]   W.S. Yang, B.-W. Park, E.H. Jung, *et al.*, Iodide management in formamidinium-lead-halide–based perovskite layers for efficient solar cells, *Science* 356 (2017) 1376–1379. doi:10.1126/science.aan2301.

[26]   C. Bi, Y. Shao, Y. Yuan, *et al.*, Understanding the formation and evolution of interdiffusion grown organolead halide perovskite thin films by thermal annealing, *J. Mater. Chem. A* 2 (2014) 18508–18514. doi:10.1039/C4TA04007D.

[27]   M.I. Hossain, F.H. Alharbi, and N. Tabet, Copper oxide as inorganic hole transport material for lead halide perovskite based solar cells, *Sol. Energy* 120 (2015) 370–380. doi:10.1016/j.solener.2015.07.040.

[28]   W. Zhang, G.E. Eperon, and H.J. Snaith, Metal halide perovskites for energy applications, *Nat. Energy* 1 (2016) 16048. doi:10.1038/nenergy.2016.48.

[29]   T. Handa, D.M. Tex, A. Shimazaki, A. Wakamiya, and Y. Kanemitsu, Charge injection mechanism at heterointerfaces in $CH_3NH_3PbI_3$ perovskite solar cells revealed by simultaneous time-resolved photoluminescence and photocurrent measurements, *J. Phys. Chem. Lett.* 7 (2017) 3186–3191. doi:10.1021/acs.jpclett.6b02847.

[30]   M.H. Futscher and B. Ehrler, Efficiency limit of perovskite/Si tandem solar cells, *ACS Energy Lett.* 1 (2016) 2–7. doi:10.1021/acsenergylett.6b00405.

[31]   M.R. Filip, G.E. Eperon, H.J. Snaith, and F. Giustino, Steric engineering of metalhalide perovskites with tunable optical band gaps, *Nat. Commun.* 5 (2014) 5757. doi:10.1038/ncomms6757.

[32]   F. El-Mellouhi, E.T. Bentria, A. Marzouk, S.N. Rashkeev, S. Kais, and F.H. Alharbi, Hydrogen bonding: a mechanism for tuning electronic and optical

properties of hybrid organic–inorganic frameworks, *Npj Comput. Mater.* 2 (2016) 16035. doi:10.1038/npjcompumats.2016.35.

[33] F. El-Mellouhi, E.T. Bentria, S.N. Rashkeev, S. Kais, and F.H. Alharbi, Enhancing intrinsic stability of hybrid perovskite solar cell by strong, yet balanced, electronic coupling, *Sci. Rep.* 6 (2016) 30305. doi:10.1038/srep30305.

[34] V. D'Innocenzo, A.R. Srimath Kandada, M. De Bastiani, M. Gandini, and A. Petrozza, Tuning the light emission properties by band gap engineering in hybrid lead halide perovskite, *J. Am. Chem. Soc.* 136 (2014) 17730–17733. doi:10.1021/ja511198f.

[35] H.-S. Kim, C.-R. Lee, J.-H. Im, *et al.*, Lead iodide perovskite sensitized all-solid-state submicron thin film mesoscopic solar cell with efficiency exceeding 9%, *Sci. Rep.* 2 (2012) 591. doi:110.1038/srep00591.

[36] T. Leijtens, G.E. Eperon, A.J. Barker, *et al.*, Carrier trapping and recombination: the role of defect physics in enhancing the open circuit voltage of metal halide perovskite solar cells, *Energy Environ. Sci.* 9 (2016) 3472–3481. doi:10.1039/C6EE01729K.

[37] A. Miyata, A. Mitioglu, P. Plochocka, *et al.*, Direct measurement of the exciton binding energy and effective masses for charge carriers in organic-inorganic tri-halide perovskites, *Nat. Phys.* 11 (2015) 582–587. doi:10.1038/nphys3357.

[38] S.D. Stranks, G.E. Eperon, G. Grancini, *et al.*, Electron-hole diffusion lengths exceeding 1 micrometer in an organometal trihalide perovskite absorber, *Science* 342 (2013) 341–344. doi:10.1126/science.1243982.

[39] M.B. Johnston and L.M. Herz, Hybrid perovskites for photovoltaics: charge-carrier recombination, diffusion, and radiative efficiencies, *Acc. Chem. Res.* 49 (2016) 146–154. doi:10.1021/acs.accounts.5b00411.

[40] G. Xing, N. Mathews, S. Sun, *et al.*, Long-range balanced electron- and hole-transport lengths in organic-inorganic CH3NH3PbI3, *Science* 342 (2013) 344–347. doi:10.1126/science.1243167.

[41] M.E. Madjet, G.R. Berdiyorov, F. El-Mellouhi, F.H. Alharbi, A. V. Akimov, and S. Kais, Cation effect on hot carrier cooling in halide perovskite materials, *J. Phys. Chem. Lett.* 8 (2017) 4439–4445. doi:10.1021/acs.jpclett.7b01732.

[42] D. Bi, W. Tress, M.I. Dar, *et al.*, Efficient luminescent solar cells based on tailored mixed-cation perovskites, *Sci. Adv.* 2 (2016) e1501170–e1501170. doi:10.1126/sciadv.1501170.

[43] F. Fertig, S. Nold, N. Wöhrle, *et al.*, Economic feasibility of bifacial silicon solar cells, *Prog. Photovoltaics Res. Appl.* 24 (2016) 800–817. doi:10.1002/pip.2730.

[44] J. Johnson, W. Hurayb, and Y. Baghzouz, Economic evaluation of energy produced by a bifacial photovoltaic array in the era of time-of-use pricing, in: *4th International Conference on Clean Electrical Power: Renewable Energy Resources Impact, ICCEP 2013*, 2013, pp. 348–352. doi:10.1109/ICCEP.2013.6587013.

[45]   H. Apostoleris, S. Sgouridis, M. Stefancich, and M. Chiesa, Evaluating the factors that led to low-priced solar electricity projects in the Middle East, *Nat. Energy 3* (2018). doi:10.1038/s41560-018-0256-3.

[46]   A. Di Paola, Saudi Arabia Gets Cheapest Bids for Solar Power in Auction, October 2017. (n.d.) https://www.bloomberg.com/news/articles/2017-10-03/saudi-arabia-gets-cheapest-ever-bids-for-solar-power-in-auction#xj4y7vzkg.

[47]   P. Ooshaksaraei, K. Sopian, R. Zulkifli, M.A. Alghoul, and S.H. Zaidi, Characterization of a bifacial photovoltaic panel integrated with external diffuse and semimirror type reflectors, *Int. J. Photoenergy.* 2013 (2013). doi:10.1155/2013/465837.

[48]   J. Frank, M. Rüdiger, S. Fischer, J.C. Goldschmidt, and M. Hermle, Optical simulation of bifacial solar cells, *Energy Procedia.* 27 (2012) 300–305. doi:10.1016/j.egypro.2012.07.067.

[49]   P. Ooshaksaraei, R. Zulkifli, S.H. Zaidi, M.A. Alghoul, A. Zaharim, and K. Sopian, Terrestrial applications of bifacial photovoltaic solar panels, in: *Proceedings of the 10th WSEAS International Conference on System Science and Simulation in Engineering*, 2011, pp. 128–131.

[50]   X. Wang, L. Kurdgelashvili, J. Byrne, and A. Barnett, The value of module efficiency in lowering the levelized cost of energy of photovoltaic systems, *Renew. Sustain. Energy Rev.* 15 (2011) 4248–4254. doi:10.1016/j.rser.2011.07.125.

[51]   S.W. Glunz, and A. Cuevas, Bifacial silicon solar cells – an overview, *Sol. Energy.* (2012) 1–39.

[52]   A. Kylili and P.A. Fokaides, European smart cities: The role of zero energy buildings, *Sustain. Cities Soc. 15* (2015) 86–95. doi:10.1016/j.scs.2014.12.003.

[53]   P. Ooshaksaraei, K. Sopian, R. Zulkifli, and S.H. Zaidi, Characterization of air-based photovoltaic thermal panels with bifacial solar cells, *Int. J. Photoenergy.* 2013 (2013) Article ID 978234. doi:10.1155/2013/978234.

[54]   F. Liu, J. Zhu, J. Wei, *et al.*, Numerical simulation: toward the design of high-efficiency planar perovskite solar cells, *Appl. Phys. Lett.* 104 (2014) 253508. doi:10.1063/1.4885367.

[55]   T. Minemoto and M. Murata, Theoretical analysis on effect of band offsets in perovskite solar cells, *Sol. Energy Mater. Sol. Cells* 133 (2015) 8–14. doi:10.1016/j.solmat.2014.10.036.

[56]   T. Minemoto and M. Murata, Device modeling of perovskite solar cells based on structural similarity with thin film inorganic semiconductor solar cells, *J. Appl. Phys.* 116 (2014) 054505. doi:10.1063/1.4891982.

[57]   P.P. Altermatt, T. Kiesewetter, K. Ellmer, and H. Tributsch, Specifying targets of future research in photovoltaic devices containing pyrite (FeS2) by numerical modelling, *Sol. Energy Mater. Sol. Cells* 71 (2002) 181–195. doi:10.1016/S09270248(01)00053-8.

[58]   M. Burgelman, J. Verschraegen, B. Minnaert, and J. Marlein, Numerical simulation of thin film solar cells: practical exercises with SCAPS, *Numos Work March* (2007) 357–366.

[59] A.A.B. Baloch, M.I. Hossain, N. Tabet, and F.H. Alharbi, Practical efficiency limit of methylammonium lead iodide perovskite (CH 3 NH 3 PbI 3) solar cells, *J. Phys. Chem. Lett.* 9 (2018) 426–434. doi:10.1021/acs.jpclett. 7b03343.

[60] S.H. Song, K. Nagaich, E.S. Aydil, R. Feist, R. Haley, and S.A. Campbell, Structure optimization for a high efficiency CIGS solar cell, in: *2010 35th IEEE Photovoltaic Specialists Conference*, IEEE, 2010, pp. 002488–002492. doi:10.1109/PVSC.2010.5614724.

[61] J. Pettersson, C. Platzer-Björkman, U. Zimmermann, and M. Edoff, Baseline model of graded-absorber Cu(In,Ga)Se2 solar cells applied to cells with Zn1–XMgxO buffer layers, *Thin Solid Films.* 519 (2011) 7476–7480. doi:10.1016/j.tsf.2010.12.141.

[62] S. Heo, J. Chung, H.-I. Lee, *et al.*, Defect visualization of Cu(InGa)(SeS)2 thin films using DLTS measurement, *Sci. Rep.* 6 (2016) 30554. doi:10.1038/srep30554.

[63] S. Sharbati, and J.R. Sites, Impact of the band offset for n-Zn(O,S)/p-Cu(In, Ga)Se2 solar cells, *IEEE J. Photovoltaics.* 4 (2014) 697–702. doi:10.1109/ JPHOTOV.2014.2298093.

[64] D.D. Smith, P. Cousins, S. Westerberg, R. De Jesus-Tabajonda, G. Aniero, and Y.-C. Shen, Toward the practical limits of silicon solar cells, *IEEE J. Photovoltaics* 4 (2014) 1465–1469. doi:10.1109/JPHOTOV.2014.2350695.

[65] W. Shockley and H.J. Queisser, Detailed balance limit of efficiency of p-n junction solar cells, *J. Appl. Phys.* 32 (1961) 510–519. doi:10.1063/ 1.1736034.

[66] J.M. Ball and A. Petrozza, Defects in perovskite-halides and their effects in solar cells, *Nat. Energy.* 1 (2016) 16149. doi:10.1038/nenergy.2016.149.

[67] T.S. Sherkar, C. Momblona, L. Gil-Escrig, *et al.*, Recombination in perovskite solar cells: significance of grain boundaries, interface traps, and defect ions, *ACS Energy Lett.* (2017) 1214–1222. doi:10.1021/acsenergylett.7b00236.

[68] H.J. Queisser, Defects in semiconductors: some fatal, some vital, *Science* 281 (1998) 945–950. doi:10.1126/science.281.5379.945.

[69] W. Shockley and W.T. Read, Statistics of the recombinations of holes and electrons, *Phys. Rev.* 87 (1952) 835–842. doi:10.1103/PhysRev.87.835.

[70] C. Wehrenfennig, G.E. Eperon, M.B. Johnston, H.J. Snaith, and L.M. Herz, High charge carrier mobilities and lifetimes in organolead trihalide perovskites, *Adv. Mater.* 26 (2014) 1584–1589. doi:10.1002/adma.201305172.

[71] P. Ch, N.H. Pbi, C. Applications, *et al.*, Photocarrier recombination dynamics in photocarrier recombination dynamics in perovskite CH 3 NH 3 PbI 3 for solar cell applications, *J. Am. Chem. Soc.* 136 (33) (2014) 11610–11613. doi:10.1021/ja506624n.

[72] F.H. Alharbi and S. Kais, Theoretical limits of photovoltaics efficiency and possible improvements by intuitive approaches learned from photosynthesis and quantum coherence, *Renew. Sustain. Energy Rev.* 43 (2015) 1073–1089. doi:10.1016/j.rser.2014.11.101.

[73] S.D. Stranks, V.M. Burlakov, T. Leijtens, J.M. Ball, A. Goriely, and H.J. Snaith, Recombination kinetics in organic-inorganic perovskites: excitons, free charge, and subgap states, *Phys. Rev. Appl.* 2 (2014) 1–8. doi:10.1103/PhysRevApplied.2.034007.

[74] K. Tvingstedt, O. Malinkiewicz, A. Baumann, *et al.*, Radiative efficiency of lead iodide based perovskite solar cells, *Sci. Rep.* 4 (2015) 6071. doi:10.1038/srep06071.

[75] W. Tress, Perovskite solar cells on the way to their radiative efficiency limit – insights into a success story of high open-circuit voltage and low recombination, *Adv. Energy Mater.* 7 (2017) 1602358. doi:10.1002/aenm.201602358.

[76] W.E.I. Sha, X. Ren, L. Chen, *et al.*, The efficiency limit of CH3NH3PbI3 perovskite solar cells, *Appl. Phys. Lett.* 221104 (2015) 1–6. doi:10.1063/1.4922150.

[77] X. Ren, Z. Wang, W.E.I. Sha, and W.C.H. Choy, Exploring the way to approach the efficiency limit of perovskite solar cells by drift-diffusion model, *ACS Photonics* 4 (2017) 934–942. doi:10.1021/acsphotonics.6b01043.

[78] S. Agarwal and P.R. Nair, Device engineering of perovskite solar cells to achieve near ideal efficiency, *Appl. Phys. Lett.* 107 (2015) 1–31. doi:10.1063/1.4931130.

[79] A.A.B. Baloch, S.P. Aly, M.I. Hossain, F. El-Mellouhi, N. Tabet, and F.H. Alharbi, Full space device optimization for solar cells, *Sci. Rep.* 7 (2017) 11984. doi:10.1038/s41598-017-12158-0.

[80] S. Heo, G. Seo, Y. Lee, *et al.*, Deep level trapped defect analysis in CH 3 NH 3 PbI 3 perovskite solar cells by deep level transient spectroscopy, *Energy Environ. Sci.* 10 (2017) 0–6. doi:10.1039/C7EE00303J.

[81] D. Shi, V. Adinolfi, R. Comin, *et al.*, Low trap-state density and long carrier diffusion in organolead trihalide perovskite single crystals, *Science* 347 (2015) 519–22. doi:10.1126/science.aaa2725.

[82] M.I. Saidaminov, A.L. Abdelhady, B. Murali, *et al.*, High-quality bulk hybrid perovskite single crystals within minutes by inverse temperature crystallization, *Nat. Commun.* 6 (2015) 7586. doi:10.1038/ncomms8586.

[83] M. Saba, M. Cadelano, D. Marongiu, *et al.*, Correlated electron–hole plasma in organometal perovskites, *Nat. Commun.* 5 (2014) 5049. doi:10.1038/ncomms6049.

[84] Y. Bi, E.M. Hutter, Y. Fang, Q. Dong, J. Huang, and T.J. Savenije, Charge carrier lifetimes exceeding 15 μs in methylammonium lead iodide single crystals, *J. Phys. Chem. Lett.* 7 (2016) 923–928. doi:10.1021/acs.jpclett.6b00269.

[85] W. Peng, B. Anand, L. Liu, *et al.*, Influence of growth temperature on bulk and surface defects in hybrid lead halide perovskite films, *Nanoscale* 8 (2016) 1627–1634. doi:10.1039/C5NR06222E.

[86] X. Wen, Y. Feng, S. Huang, *et al.*, Defect trapping states and charge carrier recombination in organic–inorganic halide perovskites, *J. Mater. Chem. C* 4 (2016) 793–800. doi:10.1039/C5TC03109E.

[87]   A.A.B. Baloch, S.P. Aly, M.I. Hossain, F. El-Mellouhi, N. Tabet, and F.H. Alharbi, Full space device optimization for solar cells, *Sci. Rep.* 7 (2017) 11984. doi:10.1038/s41598-017-12158-0.

[88]   M.T. Trinh, X. Wu, D. Niesner, and X.-Y. Zhu, Many-body interactions in photoexcited lead iodide perovskite, *J. Mater. Chem. A* 3 (2015) 9285–9290. doi:10.1039/C5TA01093D.

[89]   Z. Lian, Q. Yan, T. Gao, *et al.*, Perovskite CH3NH3PbI3(Cl) single crystals: rapid solution growth, unparalleled crystalline quality, and low trap density toward 108 cm^{-3}, *J. Am. Chem. Soc.* 138 (2016) 9409–9412. doi:10.1021/jacs.6b05683.

[90]   F.H. Alharbi, S.N. Rashkeev, F. El-Mellouhi, H.P. Lüthi, N. Tabet, and S. Kais, An efficient descriptor model for designing materials for solar cells, *Npj Comput. Mater.* 1 (2015) 15003. doi:10.1038/npjcompumats.2015.3.

[91]   W.E.I. Sha, X. Ren, L. Chen, *et al.*, The efficiency limit of CH3NH3PbI3 perovskite solar cells, *Appl. Phys. Lett.* 221104 (2015) 1–6. doi:10.1063/1.4922150.

[92]   J.-P. Correa-Baena, A. Abate, M. Saliba, *et al.*, The rapid evolution of highly efficient perovskite solar cells, *Energy Environ. Sci.* 10 (2017) 710–727. doi:10.1039/C6EE03397K.

[93]   B. Zhang, M.J. Zhang, S.P. Pang, *et al.*, Carrier transport in CH3NH3PbI3 films with different thickness for perovskite solar cells, *Adv. Mater. Interfaces.* 3 (2016) 1600327. doi:10.1002/admi.201600327.

[94]   J.P. Correa-Baena, M. Anaya, G. Lozano, *et al.*, Unbroken perovskite: interplay of morphology, electro-optical properties, and ionic movement, *Adv. Mater.* 28 (2016) 5031–5037. doi:10.1002/adma.201600624.

[95]   C. Momblona, O. Malinkiewicz, C. Roldan-Carmona, *et al.*, Efficient methylammonium lead iodide perovskite solar cells with active layers from 300 to 900 nm, *APL Mater.* 2 (2014) 081504. doi:10.1063/1.4890056.

[96]   NREL Chart for the Best Solar Cell Efficiencies, 2023, https://www.nrel.gov/pv/assets/pdfs/best-research-cell-efficiencies.pdf.

[97]   Y. Shao, Y. Yuan, and J. Huang, Correlation of energy disorder and open-circuit voltage in hybrid perovskite solar cells, *Nat. Energy* 1 (2016) 15001. doi:10.1038/nenergy.2015.1.

[98]   P. Kowalczewski and L.C. Andreani, Towards the efficiency limits of silicon solar cells: How thin is too thin?, *Sol. Energy Mater. Sol. Cells* 143 (2015) 260–268.

[99]   Q. Chen, N. De Marco, Y. Yang, *et al.*, Under the spotlight: the organic-inorganic hybrid halide perovskite for optoelectronic applications, *Nano Today* 10 (2015) 355–396. doi:10.1016/j.nantod.2015.04.009.

[100] SunShot Vision Study. DOE/GO-102012-3037, 2012.

[101] M.A. Green, The path to 25% silicon solar cell efficiency: history of silicon cell evolution, *Prog. Photovoltaics: Res. Appl.* 17 (2009) 183–189. doi:10.1002/pip.

[102] M.T. Hörantner and H. Snaith, Predicting and optimising the energy yield of perovskite-on-silicon tandem solar cells under real world conditions, *Energy Environ. Sci.* 10 (2017) 0–34. doi:10.1039/C7EE01232B.

[103] R. Asadpour, R.V.K. Chavali, M. Ryyan Khan, and M.A. Alam, Bifacial Si heterojunction-perovskite organic-inorganic tandem to produce highly efficient solar cel, *Appl. Phys. Lett.* 106 (2015) 243902. doi:10.1063/1.4922375.

[104] H.J. Snaith, Perovskites: the emergence of a new era for low-cost, high-efficiency solar cells, *J. Phys. Chem. Lett.* 4 (2013) 3623–3630. doi:10.1021/jz4020162.

[105] P.P. Boix, K. Nonomura, N. Mathews, and S.G. Mhaisalkar, Current progress and future perspectives for organic/inorganic perovskite solar cells, *Mater. Today.* 17 (2014) 16–23. doi:10.1016/j.mattod.2013.12.002.

[106] N.J. Jeon, J.H. Noh, W.S. Yang, *et al.*, Compositional engineering of perovskite materials for high-performance solar cells, *Nature* 517 (2015) 476–480. doi:10.1038/nature14133.

[107] J.P. Mailoa, C.D. Bailie, E.C. Johlin, *et al.*, A 2-terminal perovskite/silicon multijunction solar cell enabled by a silicon tunnel junction, *Appl. Phys. Lett.* 121105 (2015) 30–33. doi:10.1063/1.4914179.

[108] G.E. Eperon, T. Leijtens, K.A. Bush, *et al.*, Perovskite-perovskite tandem photovoltaics with optimized band gaps, *Science* 354 (2016) 861–865. doi:10.1126/science.aaf9717.

[109] P. Löper, S.-J. Moon, S. Martín de Nicolas, *et al.*, Organic–inorganic halide perovskite/crystalline silicon four-terminal tandem solar cells, *Phys. Chem. Chem. Phys.* 17 (2015) 1619–1629. doi:10.1039/C4CP03788J.

[110] C. Motta, F. El-Mellouhi, S. Kais, N. Tabet, F. Alharbi, and S. Sanvito, Revealing the role of organic cations in hybrid halide perovskite $CH_3NH_3PbI_3$, *Nat. Commun.* 6 (2015) 7026. doi:10.1038/ncomms8026.

[111] Y. Zhang, J. Yin, M.R. Parida, *et al.*, Direct-indirect nature of the bandgap in lead-free perovskite nanocrystals, *J. Phys. Chem. Lett.* 8 (2017) 3173–3177. doi:10.1021/acs.jpclett.7b01381.

[112] X. Ke, J. Yan, A. Zhang, B. Zhang, and Y. Chen, Optical band gap transition from direct to indirect induced by organic content of $CH_3NH_3PbI_3$ perovskite films, *Appl. Phys. Lett.* 107 (2015) 091904. doi:10.1063/1.4930070.

[113] W. Rehman, R.L. Milot, G.E. Eperon, *et al.*, Charge carrier dynamics and mobilities in formamidinium lead mixed-halide perovskites, *Adv. Mater.* 27 (2015) 7938–7944. doi:10.1002/adma.201502969.

[114] A.A. Zhumekenov, M.I. Saidaminov, M.A. Haque, *et al.*, Formamidinium lead halide perovskite crystals with unprecedented long carrier dynamics and diffusion length, *ACS Energy Lett.* 1 (2016) 32–37. doi:10.1021/acsenergylett.6b00002.

[115] C. Wehrenfennig, M. Liu, H.J. Snaith, M.B. Johnston, and L.M. Herz, Charge-carrier dynamics in vapour-deposited films of the organolead

halide perovskite CH3NH3PbI3−xClx, *Energy Environ. Sci.* 7 (2014) 2269. doi:10.1039/C4EE01358A.

[116] E.M. Hutter, G.E. Eperon, S.D. Stranks, and T.J. Savenije, Charge carriers in planar and meso-structured organic-inorganic perovskites: mobilities, lifetimes, and concentrations of trap states, *J. Phys. Chem. Lett.* 6 (2015) 3082–3090. doi:10.1021/acs.jpclett.5b01361.

[117] G.R. Berdiyorov, F. El-Mellouhi, M.E. Madjet, F.H. Alharbi, and S.N. Rashkeev, Electronic transport in organometallic perovskite CH 3 NH 3 PbI 3: the role of organic cation orientations, *Appl. Phys. Lett.* 108 (2016) 053901. doi:10.1063/1.4941296.

[118] G.R. Berdiyorov, F. El-Mellouhi, M.E. Madjet, F.H. Alharbi, F.M. Peeters, and S. Kais, Effect of halide-mixing on the electronic transport properties of organometallic perovskites, *Sol. Energy Mater. Sol. Cells* 148 (2016) 2–10. doi:10.1016/j.solmat.2015.11.023.

[119] B. Wang, K.Y. Wong, S. Yang, and T. Chen, Crystallinity and defect state engineering in organo-lead halide perovskite for high efficiency solar cells, *J. Mater. Chem. A* 4 (2016) 3806–3812. doi:10.1039/C5TA09249C.

[120] H.C. Chang, C.J. Huang, P.T. Hsieh, W.C. Mo, S.H. Yu, and C.C. Li, Improvement on industrial n-type bifacial solar cell with >20.6% efficiency, *Energy Proc.* 55 (2014) 643–648. doi:10.1016/j.egypro.2014.08.038.

[121] R.L. Milot, G.E. Eperon, T. Green, H.J. Snaith, M.B. Johnston, and L.M. Herz, Radiative monomolecular recombination boosts amplified spontaneous emission in hc(NH 2) 2 SnI 3 perovskite films, *J. Phys. Chem. Lett.* 7 (2016) 4178–4184. doi:10.1021/acs.jpclett.6b02030.

[122] G.J.A.H. Wetzelaer, M. Scheepers, A.M. Sempere, C. Momblona, J. Ávila, and H.J. Bolink, Trap-assisted non-radiative recombination in organic-inorganic perovskite solar cells, *Adv. Mater.* 27 (2015) 1837–1841. doi:10.1002/adma.201405372.

[123] M. Ye, C. He, J. Iocozzia, *et al.*, Recent advances in interfacial engineering of perovskite solar cells, *J. Phys. D.: Appl. Phys.* 50 (2017) aa7cb0. doi:10.1088/1361-6463/aa7cb0.

[124] D. Webber, C. Clegg, A.W. Mason, S.A. March, I.G. Hill, and K.C. Hall, Carrier diffusion in thin-film CH 3 NH 3 PbI 3 perovskite measured using four-wave mixing, *Appl. Phys. Lett.* 111 (2017) 121905. doi:10.1063/1.4989970.

[125] A. Paulke, S.D. Stranks, J. Kniepert, *et al.*, Charge carrier recombination dynamics in perovskite and polymer solar cells, *Appl. Phys. Lett.* 108 (2016). doi:10.1063/1.4944044.

[126] G. Grancini, D. Viola, Y. Lee, *et al.*, Femtosecond charge injection dynamics at hybrid perovskite interfaces, *ChemPhysChem.* 18 (2017) 2381–2389. doi:10.1002/cphc.201700492.

[127] A. Marchioro, J. Teuscher, D. Friedrich, *et al.*, Unravelling the mechanism of photoinduced charge transfer processes in lead iodide perovskite solar cells, *Nat Phot.* 8 (2014) 250–255. doi:10.1038/nphoton.2013.374.

[128] D.W. de Quilettes, S.M. Vorpahl, S.D. Stranks, *et al.*, Impact of micro-structure on local carrier lifetime in perovskite solar cells, *Science* (80–88). 348 (2015) 683–686. doi:10.1126/science.aaa5333.

[129] P. Piatkowski, B. Cohen, F. Javier Ramos, *et al.*, Direct monitoring of ultrafast electron and hole dynamics in perovskite solar cells, *Phys. Chem. Chem. Phys.* 17 (2015) 14674–14684. doi:10.1039/C5CP01119A.

[130] B. Wenger, P.K. Nayak, X. Wen, S. V. Kesava, N.K. Noel, and H.J. Snaith, Consolidation of the optoelectronic properties of CH3NH3PbBr3 perovskite single crystals, *Nat. Commun.* 8 (2017) 590. doi:10.1038/s41467-017-00567-8.

[131] ASTM-E1036-08, Standard Test Methods for Electrical Performance of Nonconcentrator Terrestrial Photovoltaic Modules and Arrays Using Reference Cells, Annual Book of ASTM Standards, 2008.

[132] IEC61215, Crystalline Silicon Terrestrial Photovoltaic (PV) Modules – Design Qualification and Type Approval. 2005.

[133] T.W. Neises, S.A. Klein, and D.T. Reindl, Development of a thermal model for photovoltaic modules and analysis of NOCT guidelines, *J. Sol. Energy Eng.* 134 (2012) 011009. doi:10.1115/1.4005340.

[134] A. Abdallah, O. El Daif, B. Aïssa, *et al.*, Towards an optimum silicon heterojunction solar cell configuration for high temperature and high light intensity environment, *Energy Procedia* 124 (2017) 331–337. doi:10.1016/j.egypro.2017.09.307.

Chapter 7

Photostatic soiling in desert environment

Brahim Aïssa[1], Benjamin W. Figgis[1] and Klemens Isle[1]

7.1 Introduction

In an increasingly carbon-constrained world, solar-energy technologies represent one of the least carbon-intensive means of electricity generation. Solar power produces no emissions during generation itself, and life-cycle assessments clearly demonstrate that it has a smaller carbon footprint from "cradle-to-grave" than fossil fuels.

Solar energy is inexhaustible, safe, renewable, free, and no toxic or polluting, and its deployment reduces the use of fossil fuels along with its negative impact on climate change. Qatar has set a goal of attaining 20% of its energy from solar power by 2030. However, soiling of solar collectors has been recognized as the main issue for solar-energy systems operating in the Middle East and North Africa (MENA) region, which results in significant losses of solar power generation and an increase of associated plant operational and maintenance (O&M) costs. The state of Qatar, like the other MENA countries, has a large potential for deployment of solar-energy production due to the very high solar irradiations. The average annual global horizontal solar irradiation (GHI) ranges from 1897 kWh m^{-2} to 2286 kWh m^{-2}, and the associated annual photovoltaic (PV) electricity production potential can reach 1,800–2,100 kWh/ kWp for c-Si modules in the MENA region [3]. The first utility-scale PV plant in Qatar completed construction in the summer of 2022 with the capacity of 800 MWp, corresponding to 10% of the country's peak electricity demand. However, the extreme climatic conditions and the dusty environment characterizing this region pose major challenges for solar-energy deployment, performance, and reliability. Therefore, great O&M efforts are proportionally associated with solar power plants in order to ensure the maximum benefits of the systems. Figure 7.1 shows the global dust intensity over the world, and the daily output PV power loss per day in selected countries/cities. More specifically, Figure 7.2 shows the field mapping of the average daily soiling ratio in Qatar (a period of October–December 2021) as measured from the light attenuation on

[1]Qatar Environment and Energy Research Institute (QEERI), Hamad Bin Khalifa University (HBKU), Qatar Foundation, Qatar

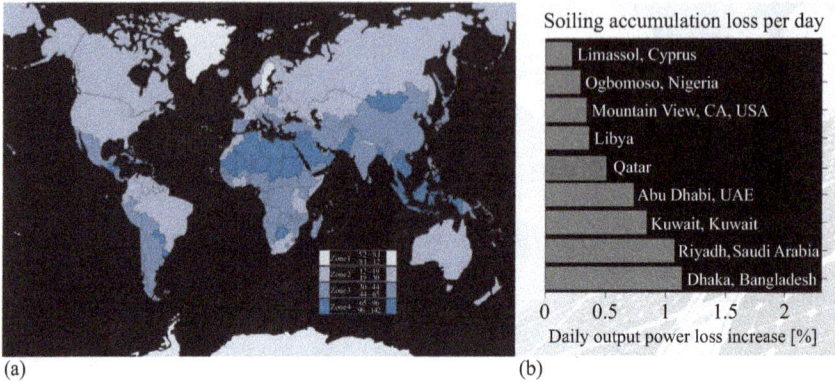

Figure 7.1   (a) Global dust intensity, the darker colors represent a higher $\mu g/m^3$ PM10. (b) Daily output PV power loss per day in selected countries/ cities. Adapted from [1,2] and historical soiling data from Outdoor Test Facility (OTF).

transparent glass coupons distributed over the country. Extreme values up to 0.7%/ day might be noticed. More specifically, from 5-year field measurements performed in Doha, Qatar, the average power loss due to the soiling of PV modules is almost 0.5%/day and is a location-dependent parameter. It means that each month without cleaning, the power of a PV system decreased by approximately 15%. In addition, soiling is a very complex and multidimensional phenomenon, and no "one-solution-fits-all" is applicable to this problem. Thus, PV soiling could be considered a major influence on the performance ratio in a PV power plant and has a serious impact on the return on interments (ROIs) of a PV project.

The term "soiling" may invoke different meanings depending on the context of use. In the aerosol field, it is commonly attributed to the quantity of particles present on a given surface (e.g., surface density which reflects the number (and/or mass) of particles per unit area), while in the PV industry, it is mainly correlated with optical effects due to soiling accumulated on the font glass of PV module (e.g., light transmission, absorption, and reflection losses) as the absorbed light governs directly the associated PV module power. PV soiling has been found to be variable both geographically and seasonally, strongly influenced by prevailing environmental conditions that exhibit marked seasonal variations [4,5]. The influencing environmental factors, such as airborne dust concentration, wind speed (WS), and relative humidity (RH), show significant variations throughout the year. Moreover, the frequency of rainfall and dust storm (DS) events also varies throughout the year [6], leading thereby to a season-dependent performance loss (i.e., due to the soiling of the PV modules), which is worth considering if one wants to predict the soiling behavior, estimate more accurately the power yield, or plan the optimal cleaning schedule. The modeling of soiling is thus necessary to predict the yield of the PV technologies.

Recently, the topic of PV-dust/soiling has recorded a tremendous increase in terms of publications (Figure 7.3). Since 2013, QEERI has started developing a

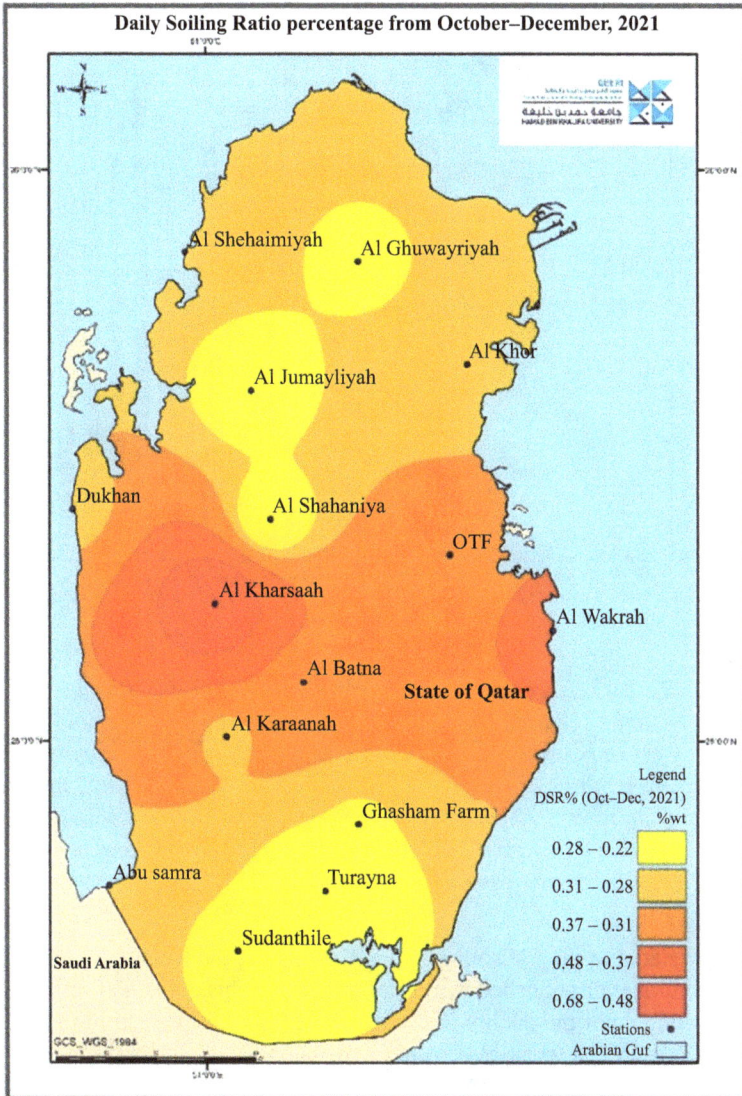

Figure 7.2   Example of the field measurement of the daily soiling ratio percentage in Qatar from October to December 2021

unique regional and worldwide expertise in dealing with the specific problematic of photovoltaic soiling (PV soiling), both from experimental (characteristics, innovative measurements methods developed, etc.) and modeling point of view (predicting of sandstorm, dust deposition, etc.). Based on the multi-year developed expertise, QEERI has become an international reference in this matter and is currently the most active and productive institution in the world in generating

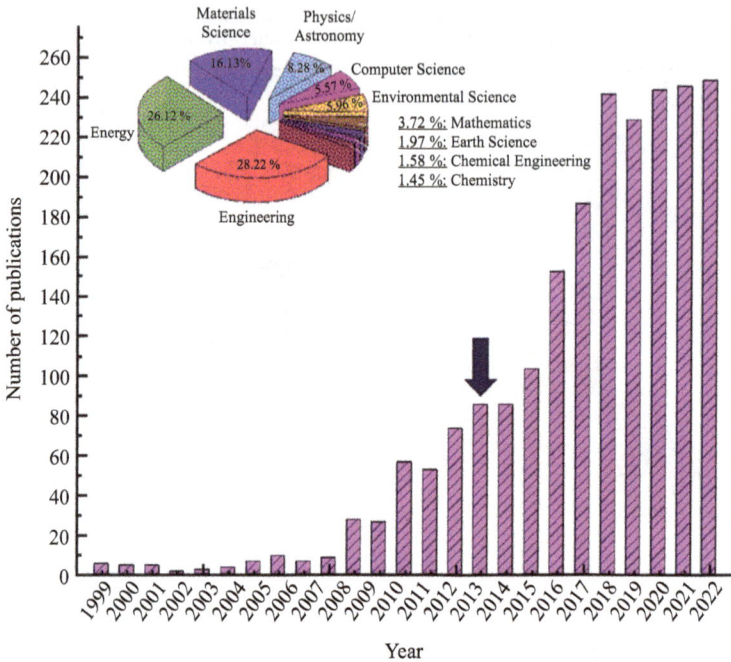

Figure 7.3   Recent refereed publications related to the field of PV-dust/soiling, together with their corresponding distribution of the employed keywords vocabulary and histogram of the type of documents, in the inset. Statistics are available from 1999 to January 2022 inclusively. Data were collected from Scopus® Expertly curated abstract and citation database information service. The arrow shows the year when QEERI started working on PV soiling.

knowledge related to the PV soiling. QEERI is also at the top first in terms of the ratio of the number of published works per scientist in this domain of research. This is mainly to highlight the quality of the expertise being developed in-house.

Figure 7.4 shows a worst-case scenario of soiled PV modules after a single sand storm event at the QEERI OTF in Doha, Qatar. Along this report, the specific terminology is relevant. Indeed, while the meaning of "soiling" is rather ambiguous, we use "soil quantity" and/or "PV soiling loss" for their respective meaning. Moreover, on a PV module, the decrease of light transmission (T%) due to the dust is not perfectly matching the associated PV power loss, because accumulated dust favors the attenuation of short wavelengths, while PV cells have a selective spectral response [7]. However, to avoid additional complexity, light transmission attenuation and PV power loss are handled as equivalent. The relationship between the soiling mass and the PV power loss depends also on the geographical locations. Different dusts have different effects on T%. For the same amount of dust, fine particles have a more dramatic effect on T% because of their higher cross-sectional area-to-volume

*Figure 7.4   Uneven soiling on a PV array following a sandstorm in Doha, Qatar.
            Modules are tilted south; dominant wind direction was from the North
            West; PV arrays facing south.*

ratio. In the same vein, the chemical composition of the dust particles and their shape
is critical on the light scattering and absorption properties. In sum, the quantitative
correlation between soiling mass and PV power loss is linearly proportional for
lightly soiled surface, and nonlinear for heavily soiled ones (i.e., T% loss evolves
slowly than soil mass) [7]. PV soiling depends on various parameters, including:

—   Geographical location of the PV site and the impact of the presence of air-
     borne, particulate matter (PM) size particle, airborne dust concentration, soil-
     ing rate, wind direction (WD), WS, and the occurrence of DS.
—   Installation of the PV system tilt angle, orientation, and height.
—   Meteorological variations (T, RH, WS, WD).
—   Structural and physical properties and size distribution of PV-dust particle.
—   Surface properties of the front glass of PV module, adhesion forces, and
     cementation.

7.2   Description of the OTF

Which solar-energy technologies work best in Qatar's climate? In 2012, this question
led to the establishment of the Solar Test Facility (STF)—updated now to OTF, at
QSTP (Qatar Science & Technology Park). The OTF is a 35,000 m^2 outdoor site with
more than 80 PV systems, a battery storage system, a linear Fresnel collector, solar
trackers, and inverters. Its core purpose is to evaluate the performance of different
solar devices in Qatar's hot, humid, and dusty climate, and determine how to maximize their energy production and reliability. In addition, it serves as a test-bed for
new solar technologies developed in Qatar and internationally.

A summary of typical equipment installed at the OTF [8]:

- 38 crystalline silicon PV systems—mono-Si, multi-Si, bifacial.
- 30 thin-film PV systems—CdTe, CIGS, μSi-aSi (micromorph).
- 4 hybrid PV systems—silicon hetero-junction.
- 7 concentrating PV systems.
- Inverters—central, string, micro-inverters.
- Trackers—1-axis, 2-axis.
- Battery storage system (500 kWh).
- Linear Fresnel thermal collector.
- Anti-soiling coatings.
- Mars^{TM} and DustIQ soiling sensors.
- Meteo station with irradiation [(GHI), Direct Normal Irradiation (DNI), Total Normal Irradiance, Diffuse Horizontal Irradiation (DHI), Plane of Array (POA), UV, albedo].

Full-scale testing commenced at the OTF in March 2013. In 2016, QEERI took over its operation. A large part of the OTF is dedicated to field testing of photo-voltaic (PV) modules and related equipment (Figure 7.5).

The standard rack tilt at the OTF is 22° south, which is optimal for annual irradiation yield (site latitude is 25.3° north). The ground is naturally occurring sand and local gravel, its albedo is approximately 42%. The pitch spacing of the

Figure 7.5   Photo of the QEERI OTF's area

Table 7.1   Systems in OTF measured continuously since 2013

Parameter of interest	PV system	Variations tested
Cleaning frequency	2 kW strings of mc-Si modules, fixed tilt	Weekly cleaning 2-month cleaning 6-month cleaning No cleaning
PV technology	2 kW strings, fixed tilt, 2-month cleaning	mc-Si CdTe CIGS μSi-aSi
Mounting	2 kW strings of mc-Si modules, 2-month cleaning	Fixed tilt 22° Horizontal single-axis tracker

fixed-tilt rows is 6 m, ensuring zero inter-row shading throughout the year. In addition to field measurements at the OTF, modules can also be characterized in QEERI's nearby the certified indoor PV Reliability Lab which has a sun simulator and electroluminescence camera. Qatar is a desert climate with high temperatures, irradiation, dust, and humidity at various times of the year. In particular, PV modules experience very high temperatures and irradiation in summer. The module temperature exceeds 70°C for roughly 50 h per year.

Soiling is the major challenge to PV performance in desert conditions. At the OTF, the average soiling rate (decrease in daily energy yield due to soiling) for fixed-tilt PV modules is 0.49%/day. That is, in an average month without rain or cleaning, the daily energy production of PV modules falls by around 13%. The daily soiling rate is higher in winter, due to more frequent overnight dew. However, there are usually several rainfalls over winter which eventually help cleaning the modules. The standard testing protocol for PV modules at the OTF is strings of approximately 2 kW, connected to grid-tied string inverters, mounted at fixed tilt 22° south, and cleaned once per week by manual dry brushing.

Because soiling has a critical impact on PV energy yield in desert climates, a variety of test stands are available at the OTF to study the effects of cleaning frequency, PV cell technology, and mounting (fixed tilt vs. tracking). A summary of the soiling test stands is provided in Table 7.1. QEERI has operated most of the systems mentioned continuously testing since 2013, providing a long-term historical dataset.

7.3   Impact of the soiling on the attenuation of solar radiation in a desert environment

The study and analysis of solar radiation are crucial when dealing with projects using solar energy as input. More specifically, the assessment of solar resources is needed all through the lifetime of solar power projects, from the feasibility analysis and design to the setup and management of the system. In QEERI, solar radiation measurement, modeling, and forecasting have been developed as main capabilities to help Qatar in deploying solar-energy-based projects. For instance, QEERI runs a network of 13 high-quality solar radiation monitoring stations across the state of Qatar, measuring the three components of solar radiation (direct, global or total, and diffuse), in collaboration with the Qatar Meteorological Department. These measurements are essential and provide comprehensive information on the solar prospects in the country. As mentioned above, since Qatar is a desert area, several research works on solar radiation in QEERI have included the study of atmospheric dust or the accumulation of dust on sensor surfaces.

In order to study the effect of the atmospheric constituents on solar radiation, mainly atmospheric dust and aerosols, we investigated the use of a lidar-ceilometer in the estimation of the extinction of the Beam or DNI [9]. DNI, which is the energy coming directly from the sun disk on a unit area normal to the sun rays, varies along the sun line direction due to changes in the atmospheric contents such as clouds and dust. We studied the relation between the ceilometer signal and DNI under cloud-free conditions, that is, in conditions where clouds are absent and only dust and aerosols are present in the atmosphere.

We studied the daily variation of the integrated backscatter and the hourly averages of DNI for all the selected clear days of 1 year at local noon (12:00). A clear relationship was found between the two measurements, relating to the hourly so-called DNI clearness index. To quantify the effect of aerosols on the attenuation of solar radiation from the high-atmosphere to the ground level, the Aerosol Optical Depth (AOD) parameter is used. In QEERI, for an improved accuracy, we investigated a method of deriving the AOD using spectral ground measurements based on a multi-filter rotating shadow band radiometer (MFRSR) from Yankee Environmental Systems (YES) in two contributions [10].

AOD values were estimated for a number of days. The values were compared with AOD derived from the Copernicus Atmosphere Monitoring Service (CAMS) [11]. The point-by-point bias between AOD-MFRSR and AOD-CAMS shows good results. A subsequent study related to solar radiation and dust accumulation is the degree of soiling on pyranometers in Qatar [12]. Pyranometers are thermopile-based instruments used, as per the recommendation of international organizations, for the GHI measurement, the solar radiation resource used for PV plants. For pyranometers, dust can accumulate on the dome of the sensor, causing lower measured values. The experimental setup consisted of measuring GHI using two pyranometers: one acting as a reference with daily cleaning, and one test sensor with varying frequencies of cleaning. The decreases in the measured irradiances due to dust accumulation on the pyranometer dome are evaluated by analyzing the error rates (relative root mean square errors and relative biases) as well as the ratio of the daily average of GHI measured by both sensors as a function of time since the last cleaning of the test sensor.

The decrease in the measured irradiances due to dust accumulation on the pyranometer dome is evaluated by analyzing the error rates (relative root mean square errors and relative biases) as well as the ratio of the daily average of GHI measured by both sensors as a function of time since the last cleaning of the test sensor. Figure 7.6 shows a sample period of the soiling ratio quantified as the ratio of the daily averages of GHI measured by the soiled sensor to GHI measured by the

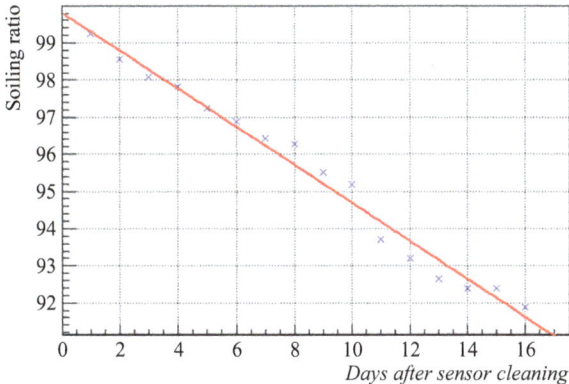

Figure 7.6   Soiling degree of a pyranometer operating in Doha, Qatar, showing the decrease of the measured global horizontal radiation per day due to sensor soiling

reference sensor, as a function of days passed since the last cleaning. The red line is the linear fit function quantifying the correlation between the soiling ratio and time, giving the value of the loss per day in the measured global radiation due to soiling. The daily losses are quantified between 0.3% and 0.5% per day depending on the season, with more losses seen toward the summer, when (in Qatar) dust events are more common and rainfall is practically absent. These values corroborate perfectly the average field-measured data from the PV panels in OTF.

7.4   Soiling rates in different countries

Figure 7.7 shows the global map of modeled annual median airborne PM2.5 dust concentration key parameters for soiling rates are:

- Airborne dust concentration (PM2.5 or PM10) and average length of dry period [13,14].
- Correlation of high irradiation (the global Sun Belt) and high dust loads in the air.
- Dominant global dust sources in the so-called "dust belt" between 15°N and 45°N, which stretches from the west coast of Africa through the Arabian Peninsula and the Middle East to Mongolia/China (Gobi–Taklamakan region) [15–19].
- Further (secondary) sources of mineral dust can be found in North America (Great Basins in Western United States and Mexico), central Australia, South Africa (Namib, Kalahari Desert), and South America (Atacama Desert in Chile).
- Other possible sources for airborne dust including industry (process dusts, exhaust), traffic (carbon particles, soot), agriculture (soil emissions, animal feed dusts, emissions from cattle breeding), and organics (pollen, seeds, bird droppings, leaves, lice, lichen, algae, moss).
- Correlation between reported soiling rates in literature and dust concentration in the air also indicated in Figure 7.8.

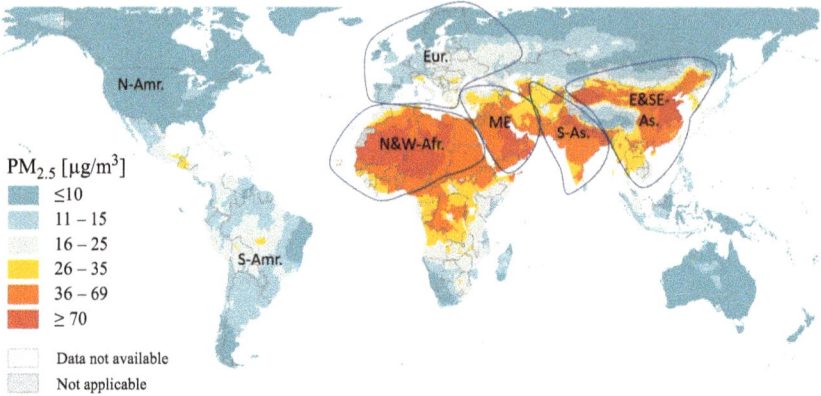

Figure 7.7   *Global map of modeled annual median airborne PM$_{2.5}$ dust concentration (World Health Organization 2016) (modified), selected regions for analysis: Middle East (ME), East- and Southeast Asia (E&SE-As.), North- and West Africa (N&W Afr.), South Asia (S-As.), Europe (Eur.), North America (N-Amr.), South America (S-Amr.). c [20], modified and reprinted with permission from (World Health Organization 2016). Reprinted with permission from Elsevier [18].*

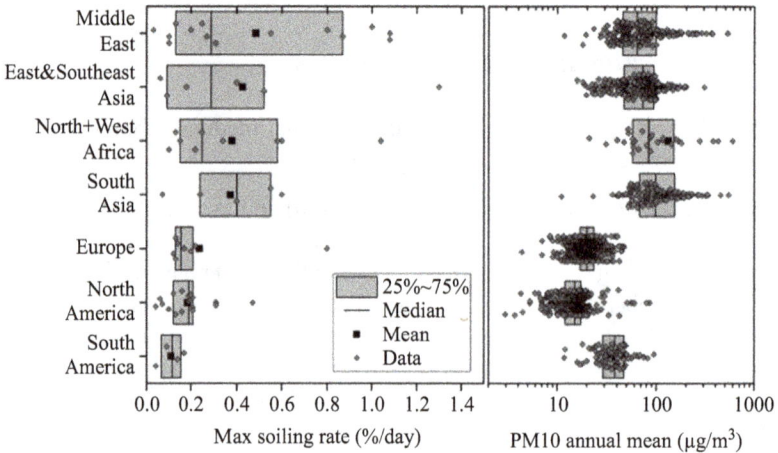

Figure 7.8   *Left: Maximum soiling rates (power/efficiency/transmission loss per day) for different regions of the world, deducted from a review of data from 63 publications on soiling [19]. Right: annual mean airborne dust concentrations of PM$_{10}$ for different regions of the world deducted from the Ambient Air Pollution Database, WHO, May 2016 [20]. Reprinted with permission from Elsevier [18].*

7.5   Dust characteristics in different countries

Figure 7.9 displays the mineral composition of resuspensions of dust collected from over 60 different ground sites throughout the world. It is observed the following:

- Huge variety in local properties of airborne dust, including size distribution, shape, and chemical/mineralogy composition.
- Composition and size distribution of particles strongly influence the optical behavior and corresponding transmission losses [21–26].
- In arid and semi-arid areas, huge fraction of airborne dust composed of minerals such as quartz, illite, kaolinite, smectite, carbonates, gypsum, feldspar, and iron oxides [26].
- Composition of airborne dust expected to be strongly influenced by ground dust composition from local terrain [15].
- Study of Engelbrecht *et al.* [17] on resuspensions of dust collected from over 60 different ground sites throughout the world provides a good overview of expected chemical and mineral compositions.
- Large differences in the mineral dust composition of different locations throughout the world, especially concerning the individual proportions of the main constituents such as quartz, calcite, dolomite, kaolinite, and hematite.
- Similar composition for some regions like the Middle East, but also huge differences, for example, in North and West Africa or North America.

7.5.1   Particle size distribution

Figure 7.10 shows the particle size distribution of dust samples collected from PV module surfaces and analyzed by laser scattering diffraction (wet). It was observed the following:

- For many places, soiling is mainly attributed to dust particles in the silt size range (particle diameter between 2 and 63 μm).
- Lawrence *et al.* [15] reviewed and summarized different size distributions for dust deposits from local, regional, and global sources and found the main fraction of particles to be in the silt size range (50%, 60%, and 70%, respectively) and minor fractions of clay (< 2 μm) and sand.
- Similarity of the results for dust size distributions from different locations agrees with the brittle fragmentation theory, which states that the emitted size distribution of dusts is essentially independent of WS and soil properties and dominant emission is in the range 2–20 μm [16,26].
- Javed *et al.* [27] reported mean and median size ranges on PV modules in Qatar to be between 9 and 25 μm including monthly changes as well as lower size ranges for longer durations of exposure of the surfaces.
- A literature overview of particles accumulating on outdoor surfaces provided by Figgis *et al.* [28] shows dominant particle sizes of 10–32 μm.

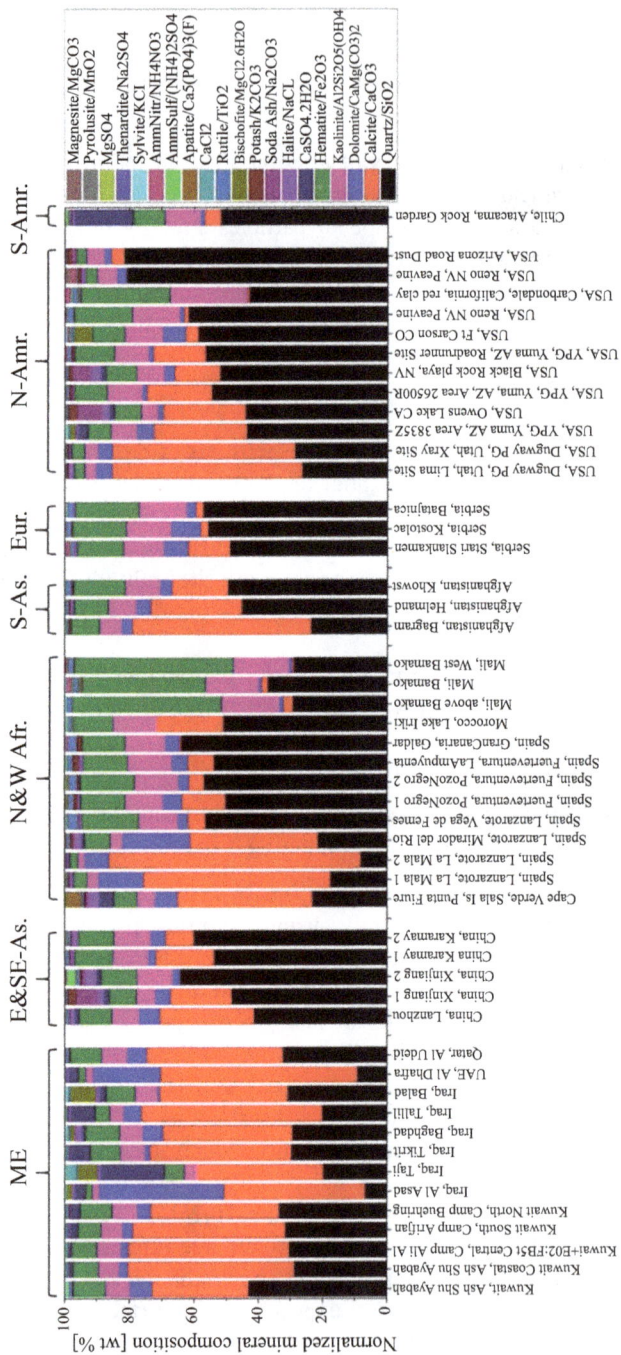

Figure 7.9 Mineral composition of resuspensions of dust collected from over 60 different ground sites throughout the world [17]. Reprinted with permission from Elsevier [18].

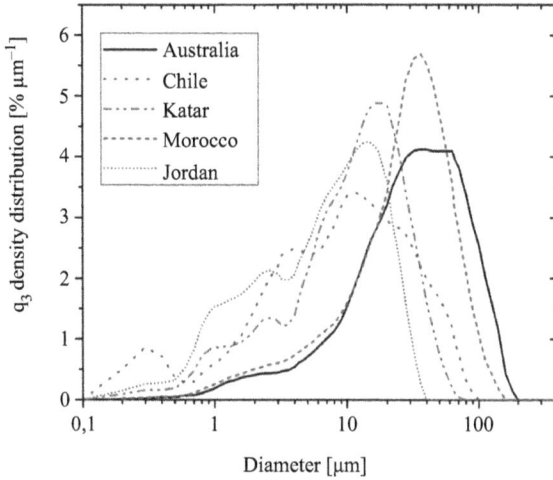

Figure 7.10 Particle size distribution of dust samples collected from PV module surfaces and analyzed by laser scattering diffraction (wet)

- Large literature review of Nayshevsky of 69 data points obtained from 60 articles revealed a median size of 16 μm, http://www.condensationexperiments.com/WorldBubbleMap/mymap.html
- Smaller particle sizes significantly increase losses per accumulated dust mass compared to larger ones [21,25].
- Care should be taken concerning the comparability of results, since different methods could deliver significant differences [29].

7.5.2 *Particle-surface adhesion forces*

Figure 7.11 summarizes an overview of particle size dependency of forces relevant for dust adhesion and removal thresholds for rolling, sliding, or direct lift-off by blowing wind. It was observed the following:

- Dust accumulation and corresponding soiling rate are determined by the interaction of particle deposition, rebound, and resuspension, which are strongly influenced by particle adhesion forces and removal forces.
- Overview of typical particle adhesion forces and theoretical detachment modes by wind for a spherical SiO_2 particle on a glass surface in dependence of the particle size.
- Van der Waals forces for a smooth and a rough glass surface, capillary forces for hydrophilic and hydrophobic glass surface properties, electrostatic forces assuming mean particle charging at the Boltzmann equilibrium as well as gravity (weight of sphere).
- If present, capillary forces are the dominating forces for all particle sizes [30], followed by Van der Waals forces.

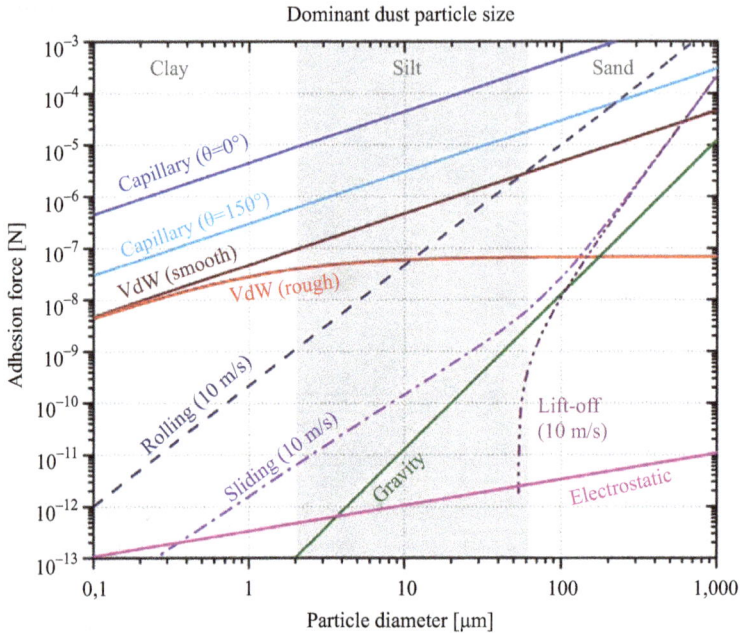

Figure 7.11   *Overview of particle size dependency of forces relevant for dust adhesion and removal thresholds for rolling, sliding, or direct lift-off by blowing wind. Reprinted with permission from Elsevier [18].*

- Gravity and electrostatic forces can be neglected for the relevant particle size range of dust at assumed charge conditions (e.g., no external electrical fields).
- Adhesion forces significantly increase with increasing particle size except for rough surface conditions.

Detailed information and literature reviews are presented in [31,32].

Capillary forces:
- Liquid bridges between particles and the surface.
- Surface tension force and the (dominant) capillary pressure force [31].
- For typical environmental conditions, water can condense and forms nano- or micro-scale menisci already at RH levels well below 100% due to capillary condensation, for example, in surface scratches or small crevices [33].
- Capillary forces can be present already at low humidity levels like 30% or 40% [30–36].
- Capillary forces increase with increasing RH levels since more water will be present at the surface, filling more and more asperities between the (rough) surface and (rough) particles [30–36]. As soon as the RH rises to values close to 100%, the surface may be flooded by water and capillary forces decrease [32]. Besides RH, the surface energy and the corresponding water contact angle (WCA) have a strong impact on the amplitude of capillary forces for a

hydrophilic glass surface (WCA = 0°) and a hydrophobic surface (WCA = 150°). There will be a strong decrease in the capillary adhesion force for hydrophobic, water-repellent surfaces, because no stable meniscus or continuous capillary bridges can form [35]. The same applies to an increased surface roughness, since particle and surface are further apart than for the smooth case and asperities cannot be filled by liquid anymore. For the case of rough surfaces, possible decreases of capillary forces by orders of magnitude are reported in the literature [35].

Van der Waals (VdW) forces:
They are always present between particles and surfaces. They act on very short ranges since they originate from interacting dipoles at two surfaces that are in contact. Consequently, they scale with the effective contact area between the particle and the surface as has been vividly demonstrated by Kazmerski *et al.* [37]. Accordingly, when a surface roughness is introduced, this could significantly reduce the effective contact area and corresponding VdW forces [32,35].

For this, it should be noted that the size scale of roughness is very important. Asperities should be significantly smaller than particle diameter (about three orders of magnitude) to reduce adhesion, since a roughness in comparable size scale can act as a trap for the particles [32]. In addition to roughness, material properties strongly impact VdW adhesion, such as the surface energy, which is strongly correlated to the Hamaker constant, or the material stiffness (Young modulus), which determines the effective contact area between particle and surface. In wet environments, VdW forces are considerably decreased because of the reduction of the Hamaker constant by the interstitial water [30] as well as electrical double layer repulsion as known from DLVO theory [31], which is an important factor regarding wet cleaning processes for soiled surfaces.

Electrostatic forces:
Between particles and surfaces electrostatic forces can be either attractive or repulsive. Its strength depends on many influencing factors, such as the presence of electric fields and their field strength, charge state of particles and surfaces including the charge distribution; the electrical properties of particles and surface (e.g., conductive and non-conductive), the dielectric properties of intervening medium as well as separation distance [31–36]. It can be concluded that for the assumptions made and in the relevant particle size range, the electrostatic forces have much lower impact on particle adhesion than capillary or VdW forces. This conclusion is also valid assuming also a maximum surface-charge density for the particles [30–33]. Furthermore, electrostatic forces due to charged surfaces will further be lowered as soon as thin films of water are present causing charge compensation. Therefore, electrostatic forces are typically not the dominant adhesion forces for particle-surface separations smaller than 20 nm [31]. Nevertheless, if there are electrical fields present above the PV module surfaces, which could be the case for dry environmental conditions (e.g., due to potential differences between solar cells and grounded module frames, which can also cause PID [38–42], the

long range of electrostatic attraction may significantly change the deposition rate of particles which are oppositely charged [43].

Solid bridge bonds:
Solid bridge bonds are including chemical bonds (e.g., covalent, ionic, metallic, or hydrogen bonds), as well as entanglement/mechanical interlocking, but can be assumed to be orders of magnitude higher than the given values for capillary and VdW forces [30]. The solid bridge bonds play an essential role in the phenomenon of cementation involving rather complex processes.

7.6   Dew, cementation, particle caking, capillary aging

Figure 7.12 shows the optical transmittance of atmosphere and black body radiation indicating radiative losses in the atmospheric window. Figure 7.13 displays the mean RH levels at morning hours (4:30–8:00) for different locations throughout the world.

Although rather unexpected at the first glance, dew formation on PV modules is frequently reported for different desert locations. This phenomenon can be attributed to radiative cooling due to the high infrared emissivity of solar glass: at night the front surface of PV modules could cool down below ambient temperature, especially at clear sky conditions [40–57]. Figgis *et al.* [28] provided the example that at the autumn equinox 2016, a PV module in Qatar cooled 5.1 K below ambient temperature during the night. Similar results can be seen in Figure 7.13, which shows an exemplary set of environmental data collected in October 2015 at the OTF in Doha, Qatar, with frequent differences between ambient and module temperatures of more than 6 K.

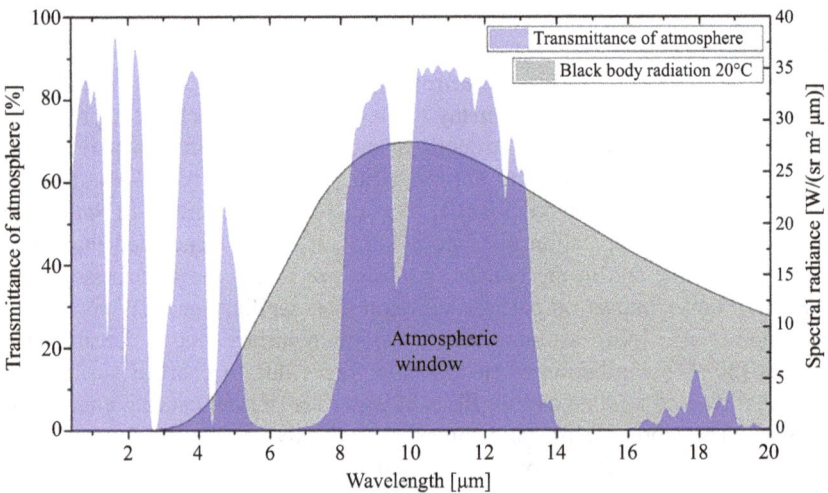

Figure 7.12   Transmittance of atmosphere and black body radiation indicating radiative losses in the atmospheric window

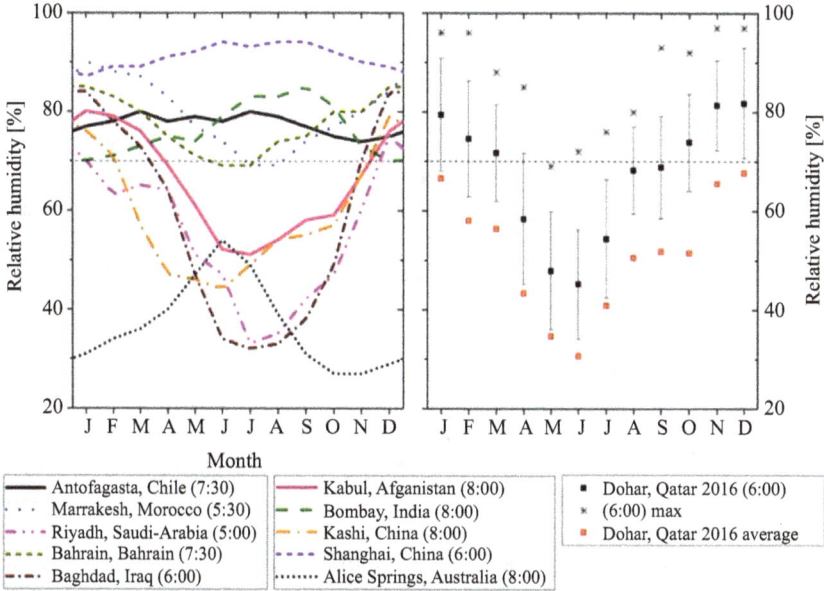

Figure 7.13 *Mean RH levels at morning hours (4:30–8:00) for different locations throughout the world (compiled from Hutchinson World Weather Guide (Pearce und Smith 2000)). For Qatar 2016, monthly mean values at 6:00, monthly maximum values at 6:00, and monthly total average values were deducted from weather station at the OTF, Doha. Reprinted with permission from Elsevier [18].*

Further, for various desert locations high RH is often experienced during late night/early morning hours (see Figure 7.13), especially in proximity to the sea. This is because of the typical decrease of air and surface temperatures, which accompanies an exponential reduction of the saturation concentration of water vapor and is demonstrated by the course of RH and ambient temperature levels in Figure 7.13. In the presence of high RH levels and surface temperatures cooler than the ambient, conditions below dew point are easily achieved, and dew forms at the surface (marked by dotted lines in Figure 7.13). In agreement with our findings, Beysens *et al.* [58] stated that for condenser surfaces with a cooling effect of about 5 K, as a rule of the thumb dew formation can be expected for relative humidity levels in the order of 70% RH with only small variations with air temperature (threshold: 67% at $-10°C$ and 76% at 40°C).

In order to assess the relevance of dew formation for different locations all over the world, a set of data was compiled from Hutchinson World Weather Guide [59], comparing the monthly mean RH during the morning hours, see Figure 7.8. It can be concluded that the threshold condition of about 70% RH is reached frequently for many desert locations throughout the world. For example, for the north of the Negev desert in Israel, dew formation can be expected for about 200 nights in a year [60].

For ease of interpretation, the right part in Figure 7.13 shows the mean values of RH at 6:00 for Doha, Qatar, together with their standard deviation and observed maximum values. Further, the monthly averages of whole-day data are shown which are on average of about 13–22% RH lower than the averages during morning hours. This indicates that for Qatar, even during months with mean values of 50% RH like August or September, dew formation occurs frequently in the late night/morning hours, for example, every second day [61,62]. Furthermore, we detected microscopic condensation already at surface temperature conditions well above dew point due to the presence of hygroscopic dust species and capillary condensation [63–65].

Condensing water can enhance particle deposition rate (Stefan flow) [56]. Furthermore, as soon as dew is present on the PV module surface, the water can interact with dust particles and the glass surface until it is completely evaporated during day when PV module surface temperatures increase to high values (exceeding 70°C at the module rear surface). Besides temporarily increasing capillary forces, the water–particle interaction will promote different processes which permanently increase particle adhesion, such as cementation, particle caking, and capillary aging (see Figure 7.14).

Three different soiling processes result from the formation and subsequent evaporation of dew [18]. A prominent one is the cementation process, which involves the dissolution and precipitation of deposited material (salts such as NaCl and gypsum or palygorskite) forming solid bridge bonds between the particles and the glass surface [64–66]. This also includes the formation of various kinds of

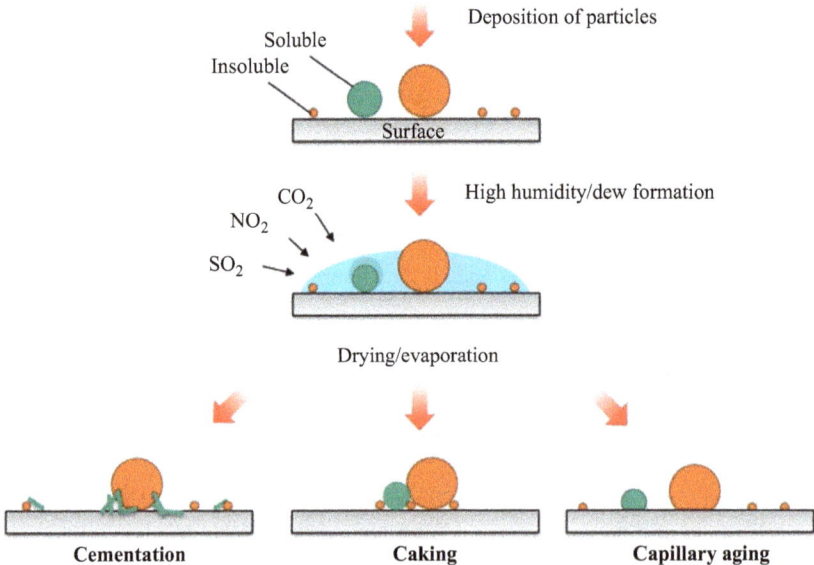

Figure 7.14 Schematic illustration of soiling mechanisms caused by dew formation and subsequent evaporation. Reprinted with permission from Elsevier [18].

chemical bonds including ionic, covalent, metallic, or also hydrogen bonds. Another process is particle caking, caused by rearrangement, agglomeration, and compaction of the dust layer with small particles filling spaces between larger particles and the glass surface [67]. Third, capillary aging describes the increase of adhesion, which is caused by capillary forces pressing the particle against the surface and thereby increasing the contact area. All of these processes can occur simultaneously and significantly increase particle adhesion [68].

7.7   Mechanics of dust accumulation

Dust accumulation on PV modules can be thought of as a competition between deposition and removal of dust particles. The physics underlying each of these processes is complex and notoriously difficult to measure in field conditions. However, a few basic principles help to explain how dust accumulates in natural environments.

Terminology is important, as the same words can be used to mean different things. In particular, here we define "deposition" to mean all dust particles that come into contact with the surface, regardless of whether or not they stick to it (elsewhere, the word has been used to refer only to particles that stick). Once particles are deposited, they can either "rebound" (detach immediately), "resuspend" (detach after being adhered for some time), or "accumulate" (never detach). If we measure these processes as flux rates (mass of dust per surface area per time [g m^{-2} s^{-1}]), then:

Accumulation = Deposition − Rebound − Resuspension

In the following sections, we examine these processes in detail and the factors that influence them. Broadly, the dominant factors are WS and module tilt angle, which greatly affect both deposition and detachment flux rates. Further, deposition is directly correlated to the aerosol concentration, while resuspension is greatly reduced by even one dew cycle.

7.7.1   Particle deposition

Three main mechanisms of particle deposition are gravity settling (sedimentation), inertial or turbulent deposition, and Brownian motion [69–73]. Their relative importance depends on the ratio of viscous to inertial acting forces, that is, particle size and WS. For mineral particles in the order of ∼1 µm diameter and smaller, Brownian motion is significant—they diffuse randomly in space and tend to follow airflow streamlines even as the flow changes direction sharply. Very large particles, around 100 µm and larger, are controlled by sedimentation—gravity dominates and such particles rapidly deposit from the air regardless of the flow behavior. Particles of diameter several to tens of microns experience viscous forces comparable to their own inertia and exhibit inertial deposition. That is, if they are entrained in an air stream that accelerates sharply due to a turbulence eddy or obstruction by an object, the particles may deviate from the stream and impact the object. Thus, in calm (not windy) conditions mid-sized particles deposit by sedimentation, and in windy conditions their deposition rate is enhanced by the inertial mechanism.

To determine which deposition mechanisms are relevant for PV soiling in desert environments, we, therefore, need to know typical particles sizes and WSs. Particles sizes accumulating on outdoor surfaces have been reported in several studies [74], from which it can be summarized that the dominant size is usually between 10 and 30 μm. Wind speeds encountered in deserts are harder to generalize. An atlas of wind speed 10 m above ground in the Middle East showed that it averages less than 4.43 m s^{-1} in all but a few locations [40]. We measured WS at a field station in Doha, Qatar, and found that the median speed was 1.70 m s^{-1} and ninetieth percentile was 4.1 m s^{-1}. Further, we measured the projected area of dust particles depositing onto a horizontal glass surface and found that 96% of deposition occurred when WS was less than 3 m s^{-1}. We conclude that in the Middle East dust deposits on PV modules mainly in low WS conditions, but of course high WS conditions also sometimes occur. Overall then it appears that PV soiling in deserts is usually a case of mineral particles several to tens of microns in diameter settling by gravity, while inertial deposition adds to the deposition rate in windy periods. Brownian motion is negligible because particles smaller than a few microns are a minor component of the surface coverage of soiled modules [74].

The deposition flux rate (deposited dust mass per surface area per time) is given by

$$F = vdep \; C \tag{7.1}$$

where $vdep$ is the dust deposition velocity [m s^{-1}] and C is the airborne dust concentration [g m^{-3}]. The deposition velocity of a particle is influenced by its size, density, and airflow characteristics. For sedimentation (gravity settling), $vdep$ is the terminal velocity reached by the falling particle, which for mid-sized particles is in the Stokes flow regime and is given by the Stokes velocity:

$$v_{Stokes} = (\rho.d^2.g)/(18\,\mu)$$

where ρ is the particle density (kg m^{-3}), d is the particle diameter (m), g is the gravitational acceleration (m s^{-2}), and μ is the dynamic viscosity of air (kg m^{-1} s^{-1}). For the case of a 10 μm mineral dust particle ($\rho = 2{,}700$ kg m^{-3}) in air at 25°C ($\mu = 1.85 \times 10^{-5}$ kg m^{-1} s^{-1}), its deposition velocity is not well defined.

As noted, in windy conditions the deposition velocity of mid-sized particles is further increased by inertial deposition. In particular, the more turbulent the airflow, the greater the likelihood that particles will be ejected from air streams and impact the surface.

Taking again the case of a 10-μm mineral dust particle, and adding the inertial $vdep$ to the sedimentation $vdep$, the total predicted deposition velocity is obtained. This is compared to experimental data obtained from a horizontal glass coupon in Doha, which is seen to fit well. While these models of $vdep$ help to understand the dust deposition process, they are less useful for predicting the soiling rate in natural environments because knowledge is required for the airborne dust concentration of every particle size. That is, (7.1) is more precisely expressed as

$$F = \sum vdep_i C_i \tag{7.2}$$

Yet the C_i for each particle size are not known in practical situations. Instead, an effective vdep and C_i can be assumed which represent the whole aerosol population and produce the "correct" F when multiplied. Although aerosol size distribution of course varies by location and over time, as noted the dominant size has usually been reported to be around 10–30 μm, hence, effective v_{dep} of the same order as those reported here might be expected.

Note that both sedimentation and inertial deposition rates are proportional to the aerosol concentration, C. In practice it is difficult to measure large airborne dust particles, and as a result it is common to use PM_{10} (particulate matter smaller than 10 μm) as a proxy for C. This introduces error to environmental deposition models [75], but dry deposition is so erratic anyway that this approximation is accepted.

The above descriptions of deposition deal with horizontal collectors. For sedimentation, the deposition flux rate simply decreases with the cosine of the collector tilt angle, while for inertial deposition the effect of tilt is unpredictable because the WD can vary. In fact, soiling of solar collectors has been reported to approximately follow the cosine of their tilt in field tests in Israel and India indicating that sedimentation alone controlled deposition in those locations and that wind was inconsequential. Further, in Doha we found that 81% of dust deposition on a horizontal coupon occurred when the WS was less than 3 m s $^{-1}$. It is sometimes perceived that sand storms are the main cause of PV soiling in deserts; however, these results show in fact that it largely occurs in calm conditions.

7.7.2 Particle rebound and resuspension

Recall that deposition refers to all dust particles that contact the PV module. Some of those particles (the "rebound fraction") may immediately bounce from the surface if their kinetic energy exceeds their work of adhesion—the energy required to overcome attractive forces between the particle and surface. Therefore, one would expect the rebound fraction to increase with a particle's size and impact velocity, and decrease in high humidity (which introduces a capillary adhesion force).

These expectations have been generally, but not universally, supported by experimental studies. It has been demonstrated that 1.27 μm latex spheres in vacuum tended to rebound completely (i.e., their rebound velocity approached their impact velocity) when their impact velocity exceeded ~5 m s $^{-1}$. Yet it is difficult to measure a particle's impact velocity; it is more common to use free-stream airflow velocity as a proxy. The fluxes of uranine particles have been measured in a wind tunnel and found that the rebound fraction followed a power law of flow velocity, U. That is, rebound fraction $\propto U^n$, where n is a constant that varies with particle size and experimental conditions. Field studies of dust rebound are also rare. It was found that 14% of dust impacting solar collectors in an Israel desert rebounded. Approximate rebound measurements of various chemical species in a Tennessee forest found that, above the canopy, rebound fraction ranged from 0.41 to 0.70. However, in neither of these field studies was the effect of WS on rebound fraction determined, and rebound may have been confounded with (delayed) particle resuspension.

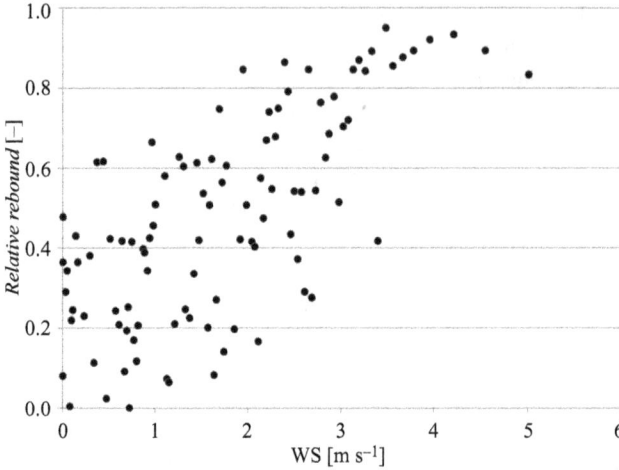

Figure 7.15   Relative rebound as a function of wind speeds

Regarding the effect of RH on the rebound, trajectories study of non-hygroscopic polystyrene particles in a wind tunnel found that the rebound fraction decreased monotonically as RH increased above 60%. An increasing RH inhibited particle rebound. In summary, particle rebound in ambient conditions is expected to be influenced by both WS and RH. Figure 7.15 shows the relative rebound as a function of wind speeds.

Although there is much scatter in the data, a greater proportion of particles rebounded as WS increased, as expected from the kinetic-energy model of rebound. There was a large spread in *Relative Rebound* for a given WS, indicating that other factors also influenced the rebound mechanism. However, the present data (not plotted) showed no correlation between RH and *Relative Rebound*. As noted, previous studies reported that particle rebound was suppressed at high RH due to capillary adhesion. However, when *Relative Rebound* is plotted against RH while holding WS constant there is no discernable consistent influence of RH. It is also seen that during the study a wide range of RH was experienced (~10–80%), which adds confidence to the finding that RH had little effect on rebound fraction. In summary, this study found that WS significantly influenced the degree of particle rebound while RH did not.

Dust particles that deposit on PV modules and do not immediately rebound can later be resuspended by wind. Figure 7.16 displays the relative rebound as a function of RH % for different WSs. The main parameters influencing resuspension are the airflow velocity in the thin boundary layer at the module's surface, the particle/surface adhesion force, and particle size. Adhesion forces are discussed in Section 7.5.2, here we focus on the removal force caused by wind.

In principle particles can detach from a surface via rolling, sliding, or lift-off motions. In practice, rolling is favored because it occurs at lower flow speed than sliding, especially for small particles, and lift-off force is negligible. Considering only static forces on the particle (i.e., ignoring any "rocking" motion it may have), it starts to

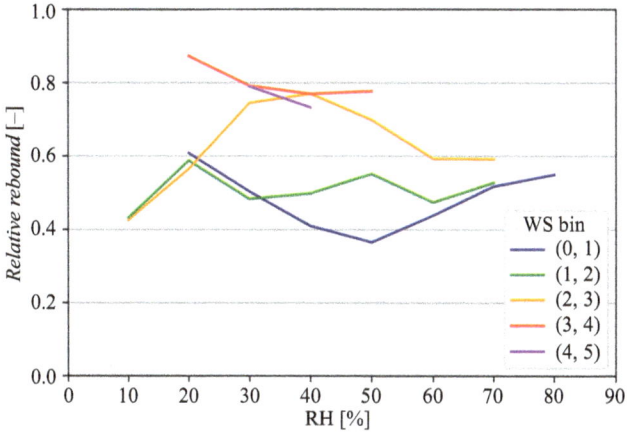

Figure 7.16   Relative rebound as a function of RH % for different wind speeds

roll when the turning moment created by drag exceeds the moment of the adhesion forces. In Stokes flow, which applies to wind speeds and particle sizes typical of PV soiling, drag force acting at the center of a sphere on a surface is given by

$$\text{Fdrag} = \frac{3\pi}{C} f\mu \ \ Dp \ \ v_p \tag{7.3}$$

where f is a correction factor of 1.7009 to account for the presence of the wall and C is the Cunningham slip factor, taken as one for mid-sized particles. A survey of resuspension literature [76] showed that the concentration of resuspending particles, C, had been described as an exponent of wind speed ranging from 1.0 to 6.4, that is, C α Ua where $1.0 \leq a \leq 6.4$. Nevertheless, a key point is that the removal force increases with particle size for two reasons—bigger particles present a larger profile to the wind and they extend further into the boundary layer, where flow velocity and turbulence intensity are greater. As particle size increases, wind removal force increases more rapidly than adhesion forces; hence, large particles resuspend at lower wind speeds than small particles. For this reason, dust accumulating on PV modules tends to be finer than the local airborne PM [77], and small particles are essentially immune to removal by wind [36]. A numerical study by [36] estimated that to dislodge a 1 μm spherical particle from a surface would require a WS in excess of 100 m.s^{-1} and that was excluding any contribution of humidity to the adhesion force. Figure 7.17 highlights the relative resuspension per minute as a function of exposure time in hours.

In locations where dew can occur at night, cycles of condensation and drying will "cement" dust particles to PV modules and virtually eliminate their later resuspension by the wind. Even a single dew cycle has been dramatically increasing particle adhesion force [78]. This increase in particle adhesion force with exposure to humidity was evident in field measurements of dust resuspension flux rate in Doha, Qatar [33]. A glass coupon was cleaned and mounted horizontally,

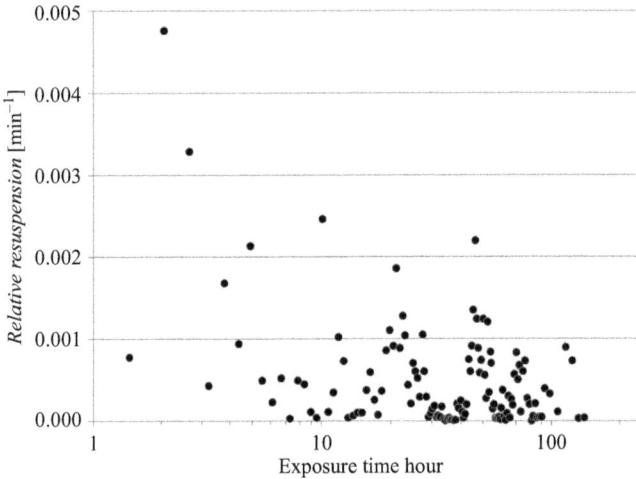

Figure 7.17   Relative resuspension per minute as a function of exposure time in hours

then the flux rate was measured at 10-minute intervals for several days. It was observed that the resuspension rate (flux rate of dust removal as a proportion of surface coverage) started to decrease within hours and reached near zero after several days of field exposure. At that point, it is suspected that the only dust observed to resuspend was "fresh" that had recently been deposited.

7.7.3   Net soiling rate

The preceding sections showed that windy conditions increase both dust deposition and its detachment (rebound and resuspension) from a surface. Therefore, it is not obvious whether wind is "good" or "bad" for PV soiling, and in fact opposing accounts have been published. Yet inconsistent results are not surprising because the balance between deposition and removal (i.e., which is greater) during a field experiment is sensitive to many factors such as module tilt angle, WD, and how the WS evolves over time.

Field tests of soiling tend to measure dust accumulated over days or weeks, which prevents the correlation of the soiling rate to WS. To address this, developed an "outdoor soiling microscope" which could measure dust flux rates continuously. Using small glass coupons in Doha mounted horizontally (so that WD had no effect), we found that the dust accumulation rate clearly decreased with WS [33]. Generally, when WS was less than 2 m s^{-1} the collector became more soiled, and when it exceeded 3 m s^{-1} it became cleaner. Figure 7.18 shows the dust accumulation as a function of the WS (the shown result was independent of aerosol concentration). The significance of this result is that WS *changed* the balance between deposition and resuspension, it did not just magnify some "inherent" imbalance. For the case of horizontal surfaces in

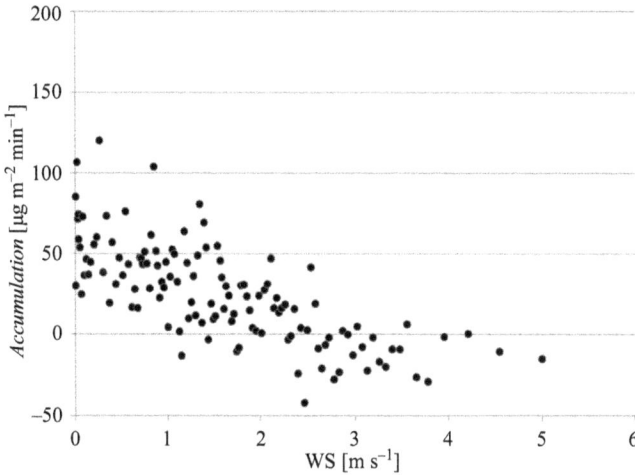

Figure 7.18   Dust accumulation as a function of the WS

desert environments, greater WS increased resuspension more markedly than resuspension, resulting in net cleaning.

7.7.4   Collector geometry

Orientation of PV modules affects their soiling rate. Sedimentation (caused by gravity) depends on the surface's tilt angle, while inertial deposition and resuspension (caused by wind) are influenced by tilt and also azimuth, that is, angle about the vertical axis of a surface exposed to horizontal air flow.

Tilt angle

In calm conditions, dust deposition occurs by sedimentation, and soiling mass per surface area is expected to be proportional to cosine of tilt angle. The desert studies described above confirmed this relation for particles larger than 4 µm. A glass samples were exposed in a desert in India and found that the average transmission loss of those cleaned weekly was 15.1% when horizontal and 9.9% at 45° tilt, which is roughly a cosine relation and indicative of sedimentation. The relation broke down at 90° tilt, at which point the loss was 3.2% rather than zero. Hemispherical domes were constructed to collect soiling on glass samples at different tilts and azimuths in Egypt and found a roughly linear reduction in soiling mass with tilt angle, with a sudden decrease approaching 90°. Figure 7.19 shows the effect of surface tilt angle on deposition and resuspension.

Wind-tunnel studies of soiling on tilted surfaces are a helpful addition to ambient studies because flow velocity, flow direction, humidity, and particulate concentration can be controlled. For deposition, a tilted surface was exposed to airflow dusted with magnetite particles. Maximum deposition was observed at 40° tilt for 1 m s^{-1} flow and 60° for 3 m s^{-1}. Surface tilt caused more variation in deposition at the faster flow speed; at low flow speed tilt was less consequential. For

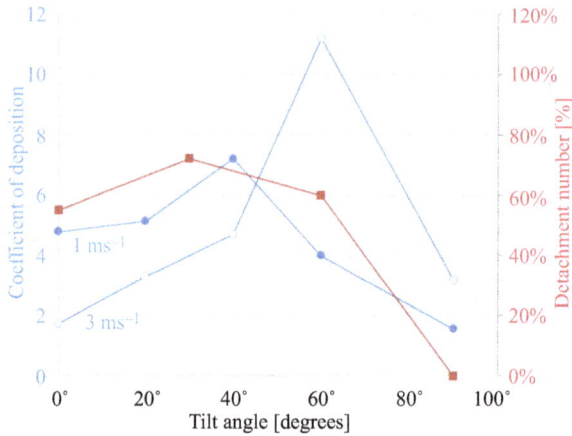

Figure 7.19   Effect of surface tilt angle on deposition and resuspension. Left axis:
coefficient of deposition of magnetite particles, solid blue circles:
1 m s^{-1} flow speed, open circles: 3 m s^{-1} flow speed. Right axis:
detachment number of glass particles from steel surface.

resuspension, a steel surface was dusted with glass spheres, and a detaching force
was applied at different orientations. Like deposition, resuspension increased until
a moderate tilt (30°) was reached and then decreased rapidly as 90° was approached. It is noted that the maximum WS in the deposition experiment was only 3 m
s^{-1}; for understanding of PV soiling it would be useful to have data at higher wind
speeds.

Azimuth angle

Azimuth angle (WD) is also expected to influence PV soiling but in a more complex way than tilt. Flow over a tilted surface whose horizontal edges are perpendicular to the flow can be modeled in two dimensions. However, when the surface
is not perpendicular the flow becomes three-dimensional, sensitive to geometry,
and vertical (especially, flow from behind a PV module may trip over its sharp
edges). Figure 7.20 shows an example of cleaned and un-cleaned PV modules at the
OTF in Doha, Qatar. Reduced soiling in flow recirculation zone downwind of a
step has also been observed in a wind-tunnel study.

Research indicates that soiling is enhanced on surfaces facing the oncoming wind.
Goossens *et al.* [29] performed outdoor experiments with a straight-edged trough
whose sides were angled at 45°. Variable results were reported but notably, during a
period in which wind was predominantly from SW-N, the Eastern side (i.e., tilted
"toward" the wind) accumulated most soiling. Inserting dust into the air upwind of a
cube at several European locations was found that most soiling occurred on the
windward face (average v_{dep} of 0.72 mm s^{-1}) followed by the top face (0.58 mm s^{-1}).
A computational fluid dynamics model of soiling of a PV system produced in-house
with tilted roof showed that although quantitative amounts of soiling on the windward

Figure 7.20   Cleaned and un-cleaned PV modules at the OTF in Doha, Qatar

and leeward roof faces were not reported, stream lines indicate that significantly more particles were predicted to impact the windward face.

It is difficult to isolate the effect of WD on soiling in outdoor tests because WD is not constant. The variation in soiling of vertical samples with different orientations indicated that WD does influence soiling; however, detailed WD data were not reported. To maintain the constant orientation of samples they may be mounted on a wind vane. It was found that soiling was proportional to the cosine of tilt angle, indicating that deposition was controlled by gravity alone (a factor may have been low WS of less than 4 m s^{-1} during the study).

The similarity of deposition and resuspension trends makes it difficult to generalize the overall effect of surface tilt on dust accumulation. The balance between the competing mechanisms will depend on conditions specific to each situation. Outdoor studies generally find that soiling is greatest on horizontal surfaces, whereas wind-tunnel studies (which tend to explore faster flow speeds) find the most soiling on moderately tilted surfaces. For the case of PV soiling in deserts, the literature comprehensively shows that soiling is greatest when the surface is horizontal, and in some cases has even been found to simply follow the cosine of tilt angle. These results indicate that deposition is a more potent process than resuspension in desert PV soiling and that it is largely controlled by sedimentation.

Yet PV modules are not horizontal of course. Regarding tilt angle of the collector, for a surface tilted toward oncoming wind, both particle deposition and resuspension were maximum at some inclined angle (20–60°) and decreased toward both horizontal and vertical positions (and also that the angle for maximum deposition changed with the flow velocity). Again, as for flow velocity, we have the case that tilt angle produces similar trends in both particle deposition and resuspension which makes their net effect (i.e., cleaning vs. soiling) difficult to predict.

From the variation of deposition and resuspension with collector tilt angle, one might postulate that there exists some optimal angle for collectors at which their soiling is minimized. Yet the literature comprehensively shows that the soiling of PV modules in the field is greatest when they are horizontal. In some cases, the amount of soiling has been found to simply follow the cosine of tilt angle. These results indicate that deposition is a more important than resuspension to desert PV soiling, and that deposition is largely due to sedimentation (gravity) rather than wind-induced mechanisms.

This raises the question—which of these is most important when selecting a site for a PV project? Micheli [4] generated 102 variables based on meteorological parameters and compared them to 20 PV soiling stations around the United States and found that the best predictors of soiling rate were particulate matter smaller than 2.5 μm ($PM_{2.5}$) and the length of rain-free periods. At face value, this might seem to contradict our research showing that WS was the dominant parameter in dust accumulation rate. Yet both may be true—WS might best explain minute-to-minute variations in soiling rate on a certain PV module, yet WS varies throughout the day which "dilutes" its long-term influence, so aerosol concentration and rain frequency can still dominate differences between long-term average soiling rates at different locations.

7.8   Field measurement of soiling

There are many techniques for quantifying soiling on modules and coupons, such as mass of soil accumulated, light transmission loss, and output of a PV device. This section aims to summarize and discuss commonly used soiling measurement methods. Further, the effect of sunlight angle of incidence on apparent PV soiling loss in the field is discussed.

7.8.1   Soil mass

The most direct characterization of soiling extent is its mass [74]. The collector surface can either be full-size PV modules and arrays or small surrogate coupons. The advantage of collecting dust from PV arrays is that it is a direct rather than proxy measurement of PV soiling. However, it is labor intensive, difficult to completely collect all dust, and ends the test period. Small coupons, on the other hand, can be easily and accurately weighed before and after soiling, and then returned to the field if desired. Removing soil from the surface allows other properties to be analyzed such as its chemical composition and particle size distribution [75].

Characterizing soiling by its mass directly portrays the quantity of soil accumulated, which is useful for studying physical soiling processes. However, soil mass does not necessarily correspond to PV performance. While in some cases mass has been found to be proportional to PV power output loss and transmission loss, two phenomena can disrupt this linear relation: First, for the same mass of soil, smaller particles have a larger total cross-sectional area and hence occult more light. Second, for surfaces that are already soiled, newly depositing particles may settle on existing particles and so may increase soil mass but not transmission loss [76]. To characterize soiling in the context of PV performance, methods based on light transmission are therefore more appropriate.

7.8.2 Light transmission

Soiling can be characterized by attenuation of light passing through a transparent substrate, which can be easily measured with a pyranometer or spectrometer [9]. Other instruments for analyzing light transmission have also been used to measure the light scattering of soiled samples with an integrating sphere, haze meter, and angular detector. Specular reflectance with an integrating sphere has also been measured. The quantum efficiency meter was also used to measure the transmission of light of different wavelengths.

Pyranometers tend to either be based on a silicon PV cell [9] or else measure total (global) irradiation. Dust tends to preferentially attenuate short light wavelengths, causing a slight red-shift in the transmitted light. An advantage of PV-based pyranometers is that their spectrum response matches that of widely used silicon PV modules, so the measured soiling value mirrors the effect on typical PV performance. Conversely, global pyranometers have a "flat" spectral response and hence provide a universal measure of accumulated soiling. Gostein [75,76] developed a compact CdTe-based pyranometer to achieve both spectral response match with the particular PV technology of interest, and small device size for ease of operation. If the main intent is to measure soil accumulation (rather than simulate the effect of soiling on a PV system), it may be preferable to measure light transmission of samples using an indoor test stand with repeatable, artificial lighting, which eliminates variation in intensity, spectrum, and diffusivity of natural sunlight.

Size of the pyranometer sensor can be important. Small sensors, a few millimeters in diameter, can be sensitive to non-uniform soiling caused by condensation run-off or clustering of particles. That is, moving the sensor to a different point behind the soiled surface may give a markedly different reading. Larger pyranometers [9] mitigate this risk because light transmission is averaged over a greater area. The gap between the back of the work piece and the pyranometer should be kept small and constant. The authors have found that physical contact between the pyranometer and the surface can cause irregular readings, for example, negative transmission loss (light amplification) by a clean piece of glass. Another reason for allowing a small gap is that the rear surface of solar glass may be heavily textured or completely flat; the difference may affect the optical pathway from work piece to the sensor.

Typically, soiling transmission loss (TL) is portrayed as the change in TL of sample when it is clean versus soiled. This approach is adequate if substrates have similar clean TLs and the degree of soiling is moderate, otherwise results may be distorted. For example, substrates A and B have 5% and 20% TL, respectively, when clean. Both are then covered with identical thick layers of dust that block all-light. The apparent soiling loss of substrate A will be 95%, while that of substrate B will be 80%, for the same amount of deposited dust.

Different soiling loss measurements may be obtained under outdoor and indoor test conditions, for example, due to spectrum, light pathways, and pyranometer positioning. It is recommended that these effects be studied further to better understand the comparability of lab and field results.

7.8.3   PV module output

For solar-energy practitioners, the ultimate interest in soiling is its effect on the energy generated by a PV system. This encourages the use of soiling measurements based on the electrical characteristics of an actual PV device, either a cell, module or array. Using PV modules allows the effect of soiling on real-world, commercial PV products to be measured. PV modules may experience uneven soiling over their surface, so another advantage of using full-size modules is that they capture soiling patterns that may not be replicated by smaller work-pieces such as PV cells or glass coupons.

The most straight-forward way of measuring soiling of a PV module in the field is to compare the output of the soiled module to that of a reference module that is kept clean. This raises the question: which measure of module output to use? The three main possibilities are:

• Power of an operating PV module at maximum power point (MPP), P_{mpp} [W].
• Energy yield of an operating module (e.g., per day), E_{mpp} [W h/day].
• Current of a PV module (or cell) operating in short circuit, I_{sc} [A].

In all cases, soiling loss is determined as the proportional difference between the soiled and cleaned modules. Using power measurements, for example, the soiling loss is:

$$Soiling\ loss = 1 - P_{soiled}/P_{clean} \qquad (7.4)$$

The main advantage of using P_{mpp} or E_{mpp} is that they capture the effect of soiling in real module operating conditions, that is, MPP (performance ratio could also be used, but because the irradiation on the soiled and clean modules is presumably identical, it is equivalent to P_{mpp} or E_{mpp}.). MPP operation requires that the modules be connected to an inverter or DC module controller. In contrast, measuring I_{sc} of a module can simply be done with an ammeter while the module is in short circuit. Although it is easier to measure I_{sc}, its response to soiling does not exactly match that of P_{mpp} or E_{mpp}. This is because module voltage also changes slightly with irradiation, also soiled modules tend to run a few degrees cooler than clean ones [73,74]. Further, as demonstrated in a series of papers by Gostein and associates [75,76], non-uniform dust coverage on modules can also cause the I_{sc} response to differ from P_{mpp} and E_{mpp}. This pertains especially to crystalline silicon modules, where shading of some cells can affect the electrical operation of others. The pertinent conclusion is that in order to compare the amounts of soiling on two PV systems I_{sc} is simple and adequate, however, to quantify the actual performance loss of an operating PV system due to soiling then P_{mpp} or E_{mpp} should be used.

The main difference in P_{mpp} and E_{mpp} is angle-of-incidence effects, as discussed in Section 1.4.5. Another question is whether they need to be corrected for module temperature. That would be necessary if the soiled and clean modules operated at different temperatures *and* that temperature difference varied during the test period. Field measurements in Qatar [73,74] showed that soiled glass/back-sheet c-Si modules tended to run 2–3°C cooler than clean ones, which translates to a soiling loss discrepancy of around 1 part in 100, and this temperature difference did vary throughout the day. Thus, it is desirable but not essential to measure the two modules' temperature and adjust their power for any difference.

These attractions of using PV modules for soiling measurement are reflected in a summary of soiling studies presented by Sarver [21]. Of the 73 studies mentioned, 35 used PV modules or arrays, 6 used PV cells, 8 used transparent plates (glass or plastic), 2 used pyranometers, and the other technologies not related to PV, for example, mirrors, heliostats, and solar thermal collectors. Reviews by Costa [78] and Sayyah [79] also showed the dominant use of PV modules for soiling studies.

7.9 Analysis of PV field data

In the case where one has PV modules installed in the field but no regularly cleaned module to compare them to, it is possible to estimate their degree of soiling by computationally modeling the power of a clean module. If one knows the irradiation and module temperature, one can roughly estimate the power that a clean PV module would produce. One then compares this reference power to the actual power of the module to estimate the soiling loss. As noted in the previous section, this difference will vary throughout the day due to angle-of-incidence effects, which can be handled by comparing daily energy (kWh/day) rather than instantaneous power (kW).

Detailed models of PV power output exist that take into account dozens of meteorological parameters and module characteristics. However, for the purpose of estimating the degree of PV soiling, it is sufficient to rely on a simplified model that uses only the major variables in PV power – the amount of received (in-plane) irradiation and the module temperature. Then, the difference between the actual module power and the P_{clean} value gives a measure of the soiling loss.

The advantages of this method of characterizing soiling are that no reference module needs to be kept clean, and the required measurements (irradiation and module power and temperature) are commonly available. On the downside, it does not take into account other factors that influence module power, such as the irradiation spectrum; hence, it gives only an approximation of the soiling. Also, the soiling loss measured in this way depends on the time of day—i.e., angle of incidence of the sun—as discussed in the following section.

7.9.1 Angle-of-incidence effects

A distinction between soiling measurements is whether they are instantaneous, such as TL and I_{sc}, or time-integrated such as energy yield. Light transmission of a soiled surface relative to a clean surface depends on the sun's angle-of-incidence (AOI), primarily because as AOI increases the same mass of surface dust obstructs a narrower beam of direct irradiation. Therefore, apparent soiling loss from instantaneous measurements is affected by time of day, and also time of year: A PV module mounted at latitude tilt experiences AOI of zero at solar noon at equinox but AOI of 23.45° at solar noon at each solstice. Further, the apparent soiling loss is affected by sky conditions: If insolation is diffuse, there is less difference between light transmissions of a soiled and a clean surface at high AOI.

To quantify the effect of AOI variation on soiling measurement, we compared the power output of two PV modules installed at the OTF in Doha, Qatar on April 16, 2015. The modules were 220 Wp poly-crystalline, mounted at 22° tilt (site latitude is 25.33° North). The "clean" module was cleaned shortly before the day of study, and the "soiled" one had been exposed for the preceding 6 weeks, which included a DS. Solar noon on the day of study was 11:34 am (AOIs are calculated using the spherical law of cosines with sun angles from www.esrl.noaa.gov/gmd/grad/solcalc/). At this instant, power of the soiled module relative to that of the clean one (P_{soiled}/P_{clean}) was measured as 76.8% (Figure 7.21). It is seen that the ratio P_{soiled}/P_{clean} decreased before and after solar noon as AOI increased, then the ratio increased again toward dawn and dusk as insolation became overwhelmingly diffuse. For approximately 1½ hours either side of solar noon P_{soiled}/P_{clean} was within two percentage points of its solar noon value, that is, above 74.8%. During these 1½ hours AOI changed by 22.1°. Given that the absolute value of AOI at solar noon varies by 23.45° throughout the year, the apparent soiling loss for a heavily soiled module may change by up to a couple of percentage points over year, even when measured at exactly solar noon. This effect is negligible for short test periods, but it is recommended it be considered when comparing soiling measurements taken at different times of the year.

Distortions in measured soiling loss caused by sun angle and light diffusivity can be eliminated by indoor measurement, where lighting is invariant. Yet for real PV systems sun angle, diffusivity, and spectrum do change throughout the day and

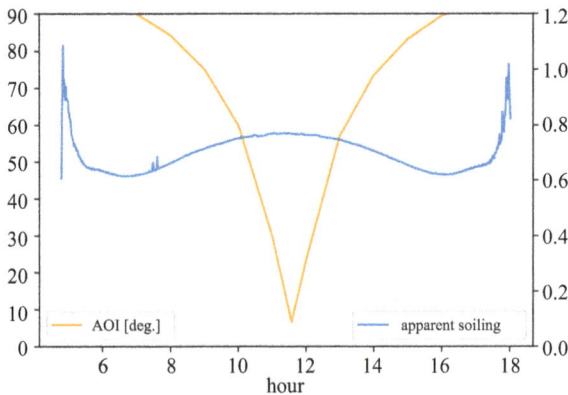

Figure 7.21   Apparent soiling vs. AOI with respect to time

year. For this reason, time-integrated soiling measurements, such daily energy yield (kWh/day) or performance ratio (kWh/day/kWh-irradiation), better convey the impact of soiling on PV system performance [79]. The downside of these approaches is the need to continually measure PV electrical parameters in outdoor conditions over a long period, and the soiling loss measured is specific to that particular test period rather than being a universal result.

The power of a clean module can be estimated as

$$P_{clean} = \beta G_{POA}(1 + k(T_{\text{mod}} - 25)) \tag{7.5}$$

where G_{POA} is the in-plane irradiation [W/m^2], T_{mod} is the module temperature [°C], k is the temperature coefficient of power obtained from the module's datasheet (usually a negative number) [%/°C], and β is an empirical calibration constant obtained by measuring the module power when clean.

7.9.2   Surface imaging

Photographic and micrographic image analysis are emerging tools for studying PV soiling, as camera and computing costs decline. Yap [40] evaluated five algorithms for analyzing photographs of soiled PV modules and found the most accurate was that which matched them to a database of photographs of known PV soiling. It was proposed that such a tool could be deployed by a drone aircraft. Microscope images convey surface coverage of soiling, that is, the projected area of dust particles as a proportion of the surface area [80], and can also be used for particle sizing.

Common to all soiling measurement methods mentioned so far is that their measurement interval is at least 24 h—their responses do not change appreciably with the amount of dust that typically settles in one day (even in deserts the average daily soiling loss rarely exceeds 1% [80]; in Qatar we found it to be roughly 0.5%). To study soiling mechanics, it would be desirable to measure soiling in the field in time intervals as short as a few minutes. This goal prompted some of us to develop an "outdoor soiling microscope." The device was able to measure the deposition and resuspension of individual dust particles larger than 10 μm^2 in real time and was used to quantify the effect of WS, RH, and airborne PM on soil accumulation at a Qatar PV site.

The different methods appear suited to different tasks. For example, to compare anti-soiling effect of many different coatings, samples might be exposed in the field then brought indoors and characterized by J_{sc} or light transmission under controlled lighting. To confirm the effect of a coating on actual PV energy production, modules with and without the treatment might be installed in the field, and their daily energy yields (kWh/day) compared over a long period.

7.10   Cleaning and soiling mitigation

Up to date, there is no passive anti-soiling technology (based on surface coatings) that has demonstrated a complete elimination of the cleaning event. Moreover, there is not a universally recommended cleaning process, as it depends on the economics and availability of local resources and cleaning frequencies.

Generally, cleaning methods can be categorized into manual, semi-automatic, and fully automatic methods (Figures 7.22 and 7.23). A further distinction can be made between:

(i)   Dry cleaning technologies that are currently only available for PV and not concentrated solar power (CSP) and mostly applied in regions with water scarcity such as desert environments, and

(ii) Wet cleaning technologies, which are generally preferred due to their increased cleaning efficiency and lower damage potential [80].

In the current global solar capacity, and despite all this, the fully autonomous cleaning market represents only 0.13% of the current global solar capacity and is expected to grow from about 1.9 GW today to 6.1 GW by the end of 2022, [81]

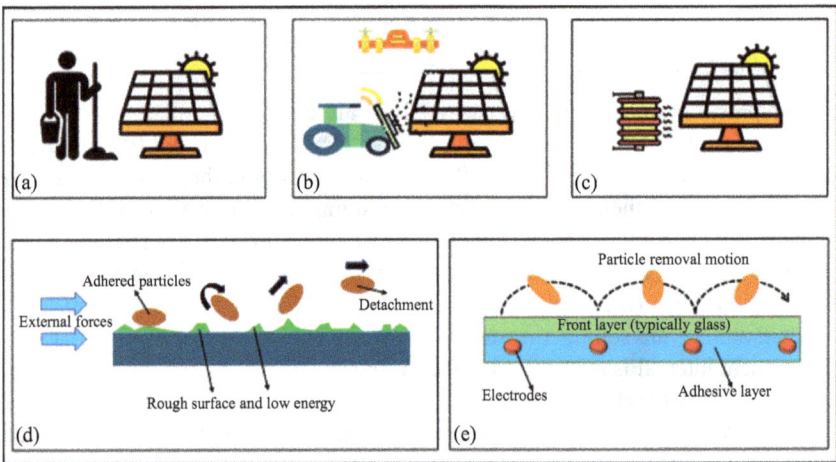

Figure 7.22   Soiling mitigation technologies (a) manual cleaning, (b) automated cleaning, (c) dew mitigation for cementation control using heating, (d) anti-soiling coating, and e) electrodynamic screen or shields for repelling dust particles

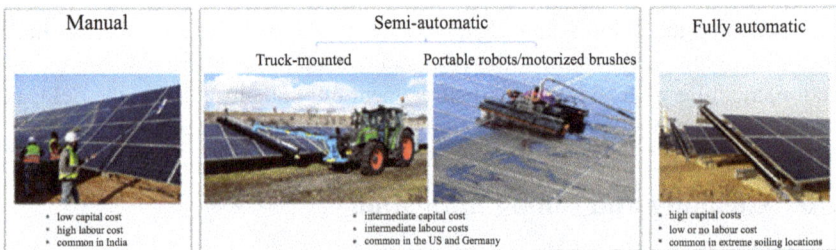

Figure 7.23   Overview of different cleaning technologies sorted by category: manual, semi-automatic (including truck-mounted solutions and portable robots), and fully automatic. Reproduced from [94].

thanks to the recent developments of dry, fully automated robots, which can be already integrated into the PV plant design. A minimum of eight factors are influencing the decision on which optimal cleaning technology to adopt. This includes the soiling type, the deposition rates, the water availability, the accessibility to the site, the system configuration (e.g., tracking vs. fixed-tilt angle, roof vs. ground mounted), the labor cost, the equipment required, and the feed-in contract conditions. Our ongoing efforts are focused now to identify an optimal cleaning schedule for Qatar based on soiling rate detection and weather conditions and dust forecasts. Table 7.2 summarizes the soiling reduction potential and costs for selected soiling mitigation technologies.

Table 7.2   Soiling reduction potential and costs for selected soiling mitigation technologies. Adapted from Refs. [82,83].

Mitigation technology	Potential optimum reduction of soiling rates	Costs	Potential limitations	Application scenario
Fully automated cleaning	>95%	2.4–8.2 €/m^2 [82,83]	Integration in plant design	PV utility scale, ground mounted
Anti-soiling coatings—Applied by glassmanufacturer-Retrofit	<<80% (literature review), <20–50% (authors estimate) 32% reported forcommercial coating	<2 €/m^2	Performance dependent on location and season, degradation by cleaning and environmental stresses	Utility scale, residential, ground mounted and rooftop, BiPV, CSP+ extra benefit from anti-reflective (AR) property
Tracking	<40%–60%	N.A.	Integration in plant planning, additional costs	Utility scale, ground mounted, state-of-the-art in CSP
Electrodynamicscreen/ shield (EDS)	<<98% (laboratory) 32% reported for 2-year study in Saudi Arabia	<30 €/m^2	Expensive, large-scale application needs to be proven	BiPV, island systems, street lighting, rooftop, CSP
Heating- PCM— Active cell heating— PVT	<20–60%	<80 €/m^2 (PCM) N.A.	Expensive, large-scale application needs to be proven	BiPV, island systems, street lighting, rooftop installations + extra benefit from cooling during day for PCM + PVT
Optimized PVmodule designand orientation	<65%	%0 €/Wp	Integration into mass production	Utility scale, rooftop installations
Site adaption	Unknown, site specific	N.A.	Little experience, research needed	Utility-scale PV and CSP

7.10.1   Cleaning economics

The 5-year dataset of actual daily soiling amounts and rain events was used to simulate the effectiveness of different cleaning strategies. Three scenarios were modeled, spanning "high cleaning frequency" assumptions (20% module efficiency, 7 ¢/kWh electricity price) to "low frequency" (14% and 3 ¢/kWh). Labor and water costs for cleaning were estimated. Simulations were held for different cleaning "trigger points," that is, how much soiling was allowed to accumulate before cleaning. The trigger-point approach is more efficient than cleaning at a fixed schedule, because it avoids cleaning soon after rain and instantly cleans the PV after dust storms. Figure 7.24 shows the economic gain in $/kWp year as a function of various the cleaning rates.

It was found that the economically optimal strategy was to clean when the soiling loss reached 10%, except for the low-frequency case when it was 15%). On average, the strategy to "clean at 10% soiling loss" would have triggered 13 cleanings per year. The cost of cleaning rapidly increases as the trigger-point decreases, for example, cleaning at 5% soiling loss would require 25 cleanings per year on average. The exact optimal trigger-point depends on project-specific economics but appears likely to be in the range 10–15% and trigger 13–8 cleanings per year.

7.10.2   Abrasion effects

Another important consideration of PV cleaning and exposure to desert environments is abrasion. Although wind-blown sand does not usually scratch PV modules, even in deserts, the action of manually or mechanically cleaning them with brushes risks abrading their anti-reflective coating (ARC). In the early days of the PV, industry modules tended not to use bare glass; however, today most employ ARC often made of silica nanoparticles in a porous structure. ARCs are weaker than glass and are typically not designed to last for the 20–30 years life of the module.

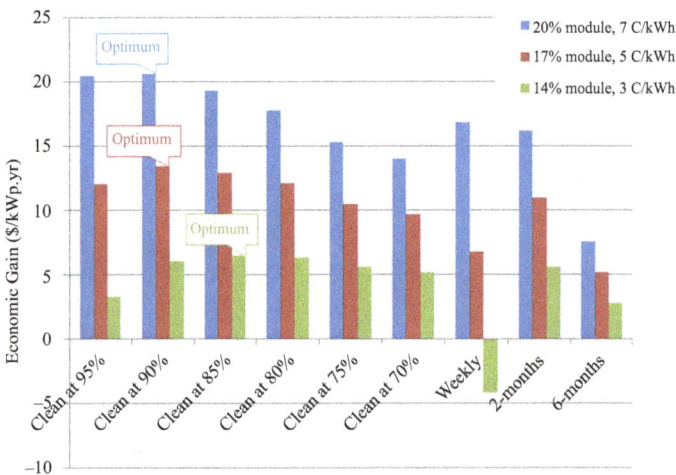

Figure 7.24   Economic gain in $/kWp year as a function of various cleaning rates

Hence, two pertinent questions regarding PV cleaning are: How much does it scratch the ARC, and does the scratching actually reduce the module performance?

Currently, there are no long-term field studies that decisively answer these questions. A comprehensive effort is underway by the National Renewable Energy Laboratory (NREL) using coupons tested with various coatings, cleaning mechanisms, and test locations. Important findings so far are that cleaning does scratch the coatings (dry-brush being harsher than wet cleaning, as expected), yet the scratches tend to forward-scatter sunlight rather than absorb or reflect it, resulting in little adverse effect on module performance. Assessment of actual PV modules at a test field in Doha, Qatar, yielded consistent results [73,74]. Arrays that were manually cleaned every week for 5 years showed the same change in performance ratio as arrays that were cleaned only twice per year. Although both studies indicate that some abrasion of the ARC does not affect module performance, one should still consider that frequent cleaning might accelerate the total removal of the ARC, which could well cause a noticeable performance decrease.

At the moment, there is no standard way of quantifying a module's resistance to abrasion, or the harshness of a cleaning machine; however, an international test standard is in development. In the meantime, it can be assumed that for modules with ARC, cleaning will accelerate the degradation or removal of their coating. In contrast, uncoated modules are expected to be relatively unaffected by cleaning abrasion. Of course, the benefit of ARCs may still favor their use in deserts, even if dry-brush cleaning does hasten their deterioration. It should also be considered that exceptionally harsh cleaning techniques could in principle void a module's warranty.

7.10.3 *Development of anti-soiling coating*

Anti-soiling coatings (ASC) is one of the leading soiling mitigation techniques thoroughly researched in the community. It is applied at the front glass of PV modules or CSP mirrors, with the goal to minimize the soiling and decrease the need for frequent cleaning. Ideally, ASCs have to be optically transparent, with anti-reflection and self-cleaning ability, non-toxic, stable (/durable), cost-effective, and deployable at an industrial level. This rends ASC somehow the "Holy Grail" of the soiling community. This passive cleaning strategy has seen minimal consumer adoption as it does not remove the necessity of cleaning but rather provides prolonged cleaning periods. Soiling level decrease as high as 80% has been monitored in the field using ASC. For extended times, nevertheless, typical anti-soiling efficiency is usually much weaker (i.e., 20%–50%) and may get worse than bare front cover depending on the quality of coating, regional weather, and deterioration level. The major motivation, however, for anti-soiling costs comes from its potential cost which is less than 2€/m^2 [84–94]. Figure 7.25 shows a schematic illustration of Soiling Mitigation Technologies.

Five dry and wet soiling mechanisms were identified through outdoor and indoor laboratory testing [71–75], and are of particular importance for ASC performance in desert regions. They are:

(i) rebound (particles bouncing off the surface upon impact),
(ii) resuspension (delayed removal of particles by wind),

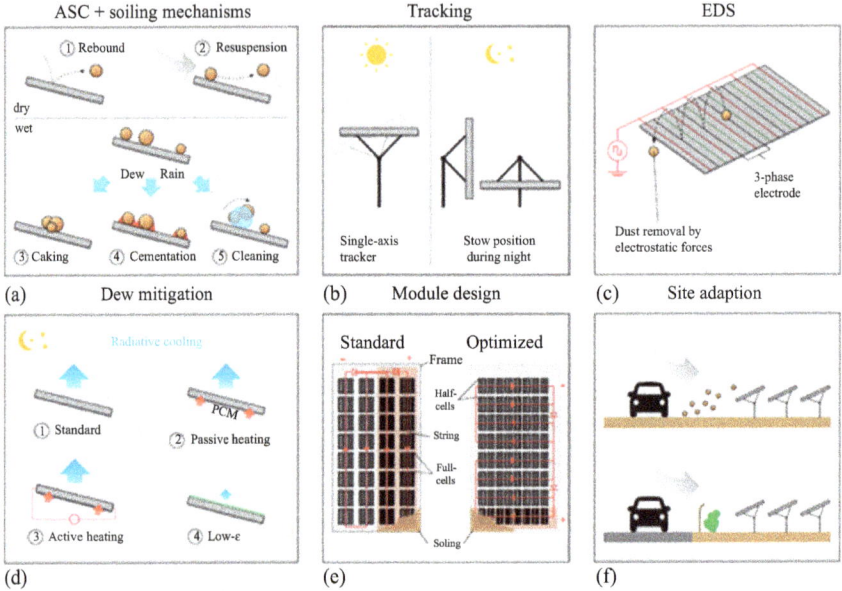

Figure 7.25   Schematic illustration of Soiling Mitigation Technologies. Reproduced from [137]. (A) Important soiling mechanisms which could be addressed by ASCs. (B) Single-axis tracking and optimization of night stowing position. (C) Working principle of EDS (standing wave version). (D) Dew mitigation by low-ε coatings and active and passive heating. (E) PV module design approaches for soiling loss reduction: the red overlay indicates lost cell strings dew to soiling. (F) Site adaption.

(iii)   caking (rearrangement and compaction of particles during dew events),

(iv)   cementation (formation of chemical and/or solid bridges between particles and surfaces after dew cycles), and

(v)   water cleaning (particles washed off by rain or strong dew).

Rebound, resuspension and water cleaning should be enhanced to reduce soiling. Cementation should be avoided. The optical losses (projected area) of caked particles should be also minimized via hydrophobic coating [72,73].

However, limitations arising from physical phenomena must be considered:

— **Geographical and seasonal variation:** Factors affecting the soiling rate and anti-dust coating performance change dramatically with the geographical location. They change also seasonally as the weather conditions change. Consequently, coatings need to be tailored to specific site conditions.

— **Particle adhesion physics:** The particle size distribution differs by location, with a volume fraction of particles <20 μm generally in the range of 35–90% [74]. Dust particles smaller than 10–20 μm diameter are irremovable by the wind as they are immersed in a thin viscous part of the boundary layer [74]. Thus, during the dew time, these particles typically become more strongly adhered to the surface [75], and over time, comprise an increasing fraction of the soiling layer and optical losses.

— **Durability/stability:** Coating performance degrades over time either permanently or temporally. The degradation might be induced by rain or condensed water [76], temperature cycles, UV irradiation, particle settlement and cementation [77] , and abrasion (by cleaning or sand storms) [78,79].

A permanent degradation is due to irreversible physical damage leading to the removal of the ASC coating. A temporal degradation is due to an external contamination of the coating surface which obstructs the anti-soiling properties. This latter might become permanent if not addressed timely and properly [80–83]. The long-term stability of coatings is actually difficult to predict, and the PV-community has attempted to identify standard methodologies to test them in advance, for example, IEC 62788-7-3 or VDI 3956-1.

ASCs have witnessed so far a limited market deployment, since they do not eliminate permanently the need for cleaning, however offering longer periods between successive cleaning events. Despite this, passive anti-soiling solutions are still very attractive and their development continues to grow [80–84]. For instance, more than 80% of soiling rate reduction is reported with anti dust coating (ADC) even at outdoor exposure; however, the issue is with the longer periods, where the average performances decrease down to 20%–50%, which could even be worse with type of coatings, and local climatic conditions [58–90].

7.10.4   Development of a TiO_2-based self-cleaning coating

Inconsistent results regarding the performance of anti-soiling coating tested in the field are found in the literature. Indeed, one may expect that hydrophobic coating would reduce soiling because it prevents water uptake by the soiling layer, and the dust particles undergo less adhesion on hydrophobic surfaces at all RH levels. However, this has not been established systematically in hydrophobic-coated PV modules in real conditions. Conversely, hydrophilic coating could promote condensation, leading to more water run-off and self-cleaning. Thus, ASC should be tested at different ranges of RH and tilt angles to distinguish between coating performance in dry and humid conditions. Variation in RH, tilt angles, and surface roughness may partly explain inconsistent results from anti-soiling coating in the literature. Studying the parameters affecting the degree of condensation on soiled surfaces; how that condensation in turn affects the various mechanisms of dust accumulation including capillary adhesion, cementation, droplet run-off, remains to be performed in depth.

Isaifan *et al.* [91] reported on the preparation of TiO_2 nanocolloids by polyol method and their application as coating films onto borosilicate glass substrates via adsorptive self-assembly process for self-cleaning and anti-dust application. Nanoparticles size is ranging from 2 to 5 nm with an average particle size around 2.60 nm. The self-cleaning properties were investigated in OTF after 7 days of natural soiling. The optical images demonstrated a 56% reduction of dust

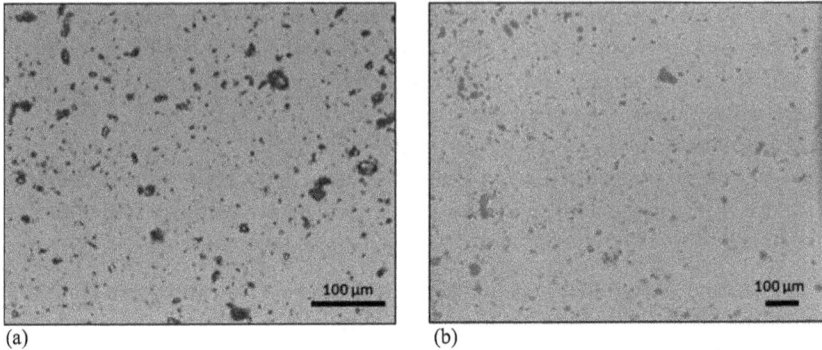

*Figure 7.26   Optical images of glass substrate samples (a) uncoated and (b)
coated with TiO₂ after soiling for 7 days in Doha for all samples
(reproduced from [91])*

accumulation over the coated surfaces compared with bare glass ones and an
average reduction of only 6% of T% in the visible region compared with clean
uncoated surface, the latter is attributed to the increased surface roughness.
Figure 7.26 shows optical images of glass substrate samples uncoated, and TiO2
coated after 7 days in OTF.

In our study of Ref. [91], we have demonstrated that the improved performance
of the coated samples was due to the following factors:

– High photocatalytic activity of TiO_2 anatase with the well-known associated
self-cleaning and anti-reflection properties.
– The small nanoparticles (with an average particle size of 2.61 nm) have shown
higher activity than larger particles.
– The low refractive index of the coated samples enhanced T% [92,93] by at
least 6%.

7.10.5   *Electrodynamic shield/screen*

Translucent electrodynamic shields (EDSs), sometimes known as electrodynamic
screen, neutralizes soil deposits by generating a transient electrical charge throughout
a substrate. Electrodes enclosed in a supportive film produce the charges, provided
with alternate voltages. They have been effectively tested in the laboratory and are
frequently suggested as a future anti-soil technique. However, its implementation in a
harsh environment and moisture conditions poses a challenge to its electronics. Using
EDS, a new industrial project in Kingdom of Saudi Arabia (KSA) registered a 32%
decrease in the soiling level; however, wide-scale deployment has still not taken place
owing to its reasonably high budget of cleaning cost of about € 30/m². For comparison, PV module costs typically vary from 30 €/m² to 90 €/m² [94]. Attempts are
being made to wide-scale produce EDS devices to reduce their price. Nevertheless,
the opportunities for economies of scale and performance in a range of climates and
reliability would need to be proven for consumer acceptance.

To conclude, it is anticipated that the automatic cleaning solutions, ASC, and customized PV module installation layout meets a feasible cost range on a grid-scale PV plant. The decreased soiling levels will result in significantly smaller cleaning costs for such technologies, such that the projected investment costs become rational, particularly in regions with elevated soiling deposition levels. Still, the financial circumstances are very demanding and complex, because, for instance, a 50% decrease in the soiling level may require increased expenses in the range of 2€/m^2 [94]. Electrodynamic shields and thermal PV systems, in contrast, seem too costly and the design is also not advanced yet. Further work is required on the site-dependent effectiveness of these proposed solutions, their future environmental impacts and the long-term durability of PV modules, as well as the operational procedures to ensure efficient and safe implementation. Finally, the soiling mitigation plan should begin at the stage of location selection and project design, along with the technical solutions. There is generally a lack of research on this topic, indicating that additional efforts are required on soiling management and analysis, and incorporation into energy yield models. Table 7.3 summarizes the state of the art of the PV soiling research and market.

Table 7.3   PV soiling research and market—state-of-the-art

(A) International and regional efforts		
	Institution	**Detail of the product**
Regional		
[95]	KISR (Kuwait)	Field Testing
[96]	KAUST (KSA)	Field Testing/Product Development—Cleaning Robot NOMADD
[97]	DEWA (UAE)	Field Testing
[98]	QEERI (Qatar)	Field Testing/Dust characterization/Fundamental Research/Anti-soiling coatings/Statistical Models
International		
[99]	Fraunhofer	PV Soiling and Degradation
[100]	NREL	Photovoltaic Module Soiling Map, Forecasting Tools, Fundamental Research, Abrasion, Soiling, etc.
[101] ISC Konstanz, Fraunhofer Chile, SERC and French CEA, INES	AtaMoS-Tec—Chile project: Photovoltaic Module Soiling study, Forecasting Tools, Fundamental Research	
[102]	SANDIA LABS/ Arizona State University	Soiling Loss Research, PV Reliability
[103]	University of Colorado/Pon-	Soiling Science and Technol-

(Continues)

Table 7.3   (Continued)

(A) International and regional efforts

	Institution	Detail of the product
	tifícia Universidade Católica de Minas Gerais (Brasil)	ogy, Coatings and Films
[104]	Institute of Solar Research of the DLR (German Aerospace Center)	Airborne soiling measurements and product development
[105]	TÜV Rheinland (Germany)	ASC
[106]	International PV Quality Assurance Task Force (PVQAT)	Sensors and Monitoring, Cleaning Solutions and Anti-Reflective and ASC, Standardization
[107]	European Cooperation in Science and Technology (COST) "inDust" program	International effort by WHO, WMO, ECMWF - Dust monitoring and forecasting models
(B) Cleaning solutions (automatic, robotic)		
[96]	NOMADD	Desert, Utility Scale, Dry Brush
[108]	Eccopia	Desert, Utility Scale, Dry Brush
[109]	Washpanel	Moderate Climate, Rooftops, Wet cleaning
[110]	Greenbotics/ SunPower	General Utility, Wet Cleaning
[111]	First Solar/ DEWA	Desert, Utility Scale, Dry Brush
[112]	Serbot Gekko	Moderate Climate, Rooftops, Wet cleaning
[113]	SOLRIDER	General Utility, Wet Cleaning
[114]	Enerwhere	Desert, Rooftops,
[115]	BladeRanger	General Utility, Dry Cleaning
(C) Soiling sensors		
[116]	German Aerospace Center	Qfly (Airborne soiling measurement of entire solar fields)
[117]	Campbell Scientific	Soiling Index Measurement Solution
[118]	Kipp and Zonen	DustIQ Soiling Monitoring System
[119]	Nor-Cal Controls	MaxSun Soiling Station
[110]	NRG	Soiling Measurement Kit
[111]	Ammonit	Soiling Measurement Kit
[120–122]	Atonometrics	Mars Optical Soiling Sensor
[123]	Kintech Engineering	Soiling Measurement Kit
(D) Smart PV monitoring systems (Internet of Things (IoT)/data analytics)		
[124]	Alternative Energy Solutions	AES PIT (Uses machine learning/advanced data analysis platform)
[125]	InnoEnergy	Solar Energy 3.0 (Smart PV

(Continues)

Table 7.3   *(Continued)*

(A) International and regional efforts

	Institution	Detail of the product
		monitoring esp. for detecting degradations)
[126]	Solar IoT platform	TrackSo (Data-driven predictive and condition monitoring)
(E) ASC		
[127]	CSD Nano	MoreSun Multi-Function Coating (Electrodynamic Dust Shield, EDS)
[128]	Anti-Soiling (AS) coating	DSM (Surface Modified Anti-Soiling Coating)
[129]	Hydrophil AS coating	Lotus Leaf Coatings (HydroPhillic Coatings)

7.11   Impact on global solar power production and energy costs

There is obviously a clear competition between the cleaning cost and revenue losses originating from the soiling between two cleaning events. In order to estimate with a descent accuracy the global impact and the associated cost of the soiling process, we have determined for the top twenty dominant PV and CSP markets the optimum between these two competitive processes (i.e., soiling vs. cleaning) [130–145]. The associated dataset was compiled both from the relevant literature and from our exchanges with stakeholders. This study has included the soiling rates, cleaning costs, and simulated local energy yields, presented in Figure 7.27(B, C, and D), respectively. Furthermore, the optimum number of the cleaning cycles per year was calculated for each country, by considering the reported installed capacity and the local feed-in-tariffs from 2017 to 2018 (Figure 7.27(E)), in addition to the medium growth scenario and the average electricity price of about 0.03 €/kWh for 2023. Moreover, we determined the total costs of soiling as the sum of optimized annual cleaning costs and the remaining revenue losses (Figure 7.27(F)). According to our data analysis, the global solar power production reduction to the soiling is estimated to be at least of 3–4% in 2018, causing global revenue losses of at least 3–5 billion €. Note that additional costs related to non-optimized PV cleaning schedules (e.g., in residential application that accounted for about 30% of global installations in 2018) and cleaning rooftop installations which are 3–8 times costlier than cleaning ground-mounted PV might be added to this estimation [145,146]. Also, collateral effects such as increases in loan rates could also have a financial impact. This study would increase the optimum cleaning frequency and the related cleaning expenses. Thus, global soiling losses could rise significantly to 4–7% of annual power production,

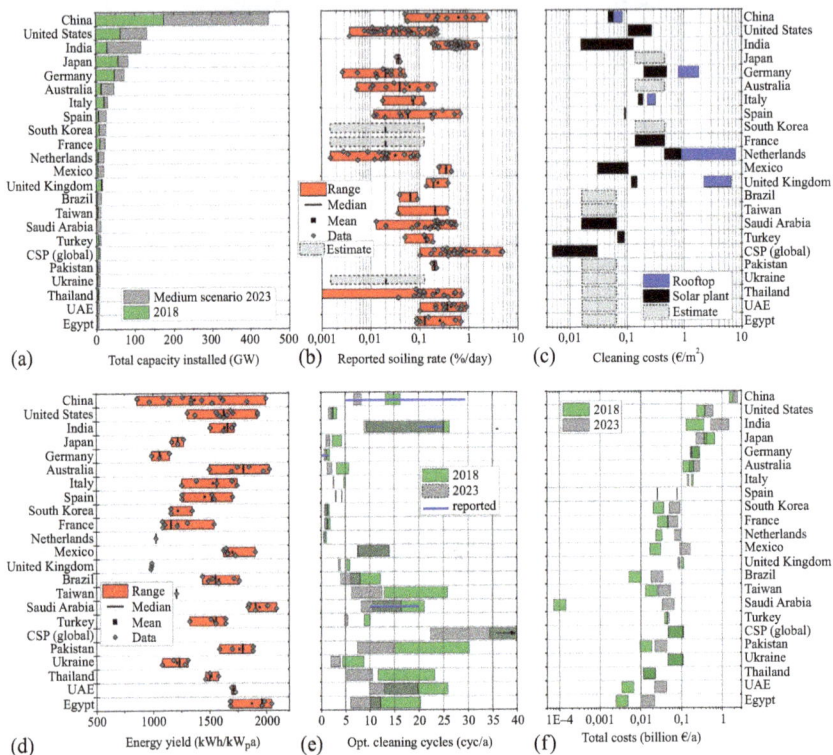

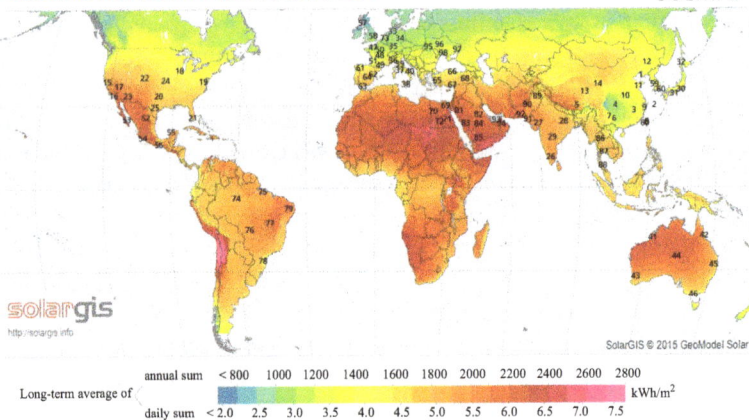

(g)

Figure 7.27　Impact of soiling on solar power generation. Reproduced from [94]. (A)
PV capacity installed by 2018 and medium estimate for 2023, sorted by
country for the top 22, and global CSP capacity. (B) Corresponding
soiling rates reported in literature. (C) Reported cleaning costs per
cleaning and square meter. (D) Typical energy yield in kWh/kWp for
representative locations. (E) Calculated range of optimal number of
yearly cleaning cycles (bars) and actual range of typical yearly cleaning
cycles reported in literature (blue lines, see Model Validation). The
arrow indicates that for CSP, the numbers are out of range (up to 85 in
2018 and 55 in 2023). (F) Minimum expected financial losses due to
soiling calculated from optimum cleaning cycles. (G) Overview of
locations used for estimation of average specific yield.

causing more than 4–7 billion € economic losses by 2023 [147,148]. Additional factors such as improved air quality in some parts of the world [149–153] could reduce anthropogenic sources of soiling, although air-quality policies typically operate over long time scales.

7.12 Renewable energy and soiling in the Gulf Cooperation Council (GCC) context

The development of large PV installation programs in dust-abundant environment (such as the US, South America, Middle East, North Africa, China, and India) has raised increasing concerns about their O&M. Although the GCC region has high PV generation potential due to its abundant irradiance, it is also subject to harsh climate conditions, with DSs and high concentrations of airborne particulates. In the GCC, where soiling of PV module front glass considerably decreases the irradiation received by the PV cells and thus the power output of the PV system.

Despite this harsh context, many utility-scale PV projects have been already implemented, such as Noor Abu Dhabi Solar PV power plant, Siraj 1 (Qatar) which is expected to generate 800 MWp and offset one million metric tons a year of carbon emissions. A recent IRENA report [154] has highlighted that developing just 1% of the most suitable area in the GCC region could result in more than 600 GW of solar PV capacity. Figure 7.28(a) shows the status of the PV farms installed and planned to be installed in the GCC. Electricity generated from large-scale solar PV is currently even cost-effective than that produced from natural gas produced locally or even that purchased as liquefied natural gas (LNG). As a matter of fact, the 2.34 US cents/kWh bid for Sakaka in Saudi Arabia is lower than the lowest gas-based generation costs, estimated at 3 US cents/kWh (assuming a gas cost of USD 2 per MMbtu) [155]. On the other hand, CSP (including storage) is competitive as well with gas-fired electricity generation (i.e., 7.3 US cents/kWh vs. USD 5–6/MMBtu), which is even lower than the opportunity cost of gas in the majority of GCC countries [156,157]. Table 7.4 shows the estimated renewable energy capacity in GCC by 2030. Figure 7.28(b) summarizes the current targets, where UAE is leading, followed by KSA. Figure 7.28(d) summarizes the renewable power planned additions by country. Looking forward, Qatar is set to see bright with a major acceleration in renewable energy deployment. As a matter of fact, the Siraj-1 solar power plant, also known as the Al Kharsaah solar power plant, is an 800MW PV solar power facility, is being developed in Al Kharsaah region, in Qatar, and is considered as the first large-scale solar power project in Qatar. Led by the UAE, GCC region has a total of about 7 GW in renewable power generation capacity as planned by 2020s. Solar PV remains the dominant technology with a share of over 75%, followed by CSP (around 10%), and a 9% share for wind projects.

As part of the global effort to increase the use of renewable energy, GCC countries are rapidly building utility-scale PV plants, including the very large Noor Abu Dhabi (1.2 GW) and Al Dhafra (2 GW) plants. While solar irradiance

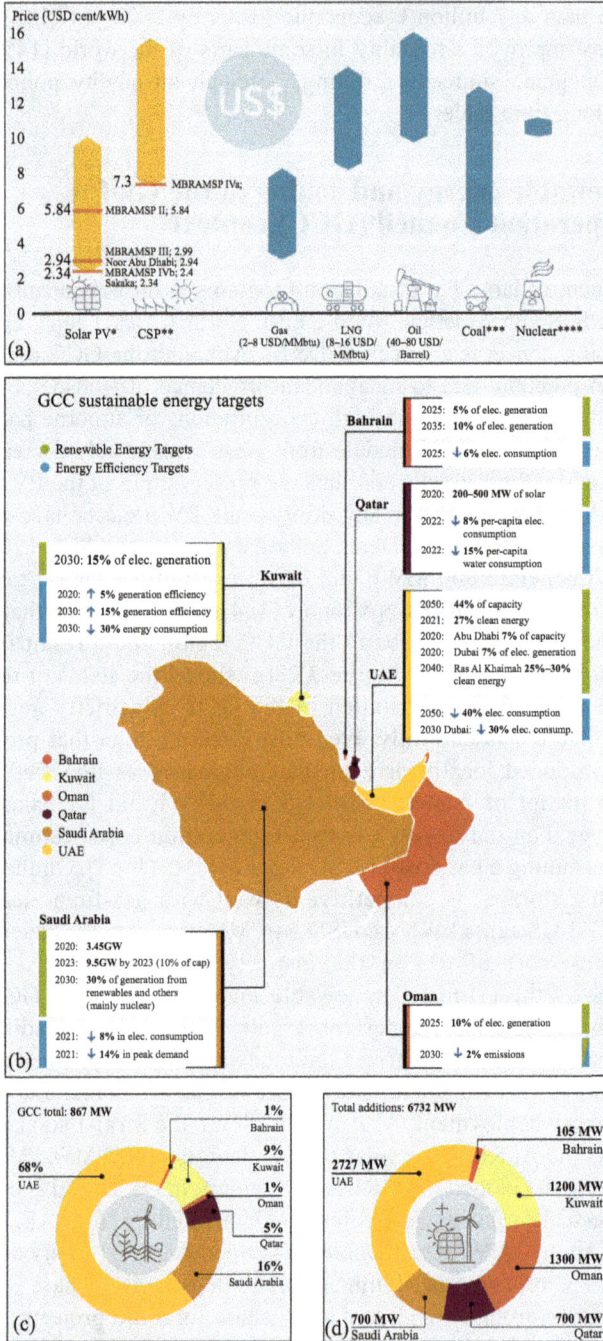

Figure 7.28   (a) Price of utility-scale electricity generation technologies in the GCC. (b) Sustainable energy targets. (c) Installed renewable energy capacity in GCC countries as a share of the total by 2018. (d) Renewable power planned additions by country. Reproduced with permission from [154].

Table 7.4   *Predicted renewable energy capacity in GCC countries by 2030.*   AQ2
Adapted from Ref. [154].

	Wind	PV rooftop	PV utility scale	CSP	Waste to energy	Total	Source
	Capacity in 2030 (MW)						
Oman	1,210	990	2,420	770	110	5,500	Target: 2.6 GW (~2,025) +0.6 GW every year up to 2030
Bahrain	20	70	520	70	20	700	IRENA expectation
Kuwait	200	1,000	5,800	1,000	–	8,000	Inputs from country
Qatar	–	150	2,250	600	100	3,100	IRENA expectation
United Arab Emirates	300	4,200	18,900	6,000	600	3,0000	Based on Masdar Institute/ IRENA
KSA	3,500	750	10,500	9,500	750	25,000	Target: 9.5 GW (~ 2,023) +2 GW every year up to 2030

in the GCC is among the highest in the world, the high soiling rate of PV panels in this region can severely reduce their efficiency and which impacts the maintenance expense and cost of electricity generated by commercial PV projects.

To the brief:

- A typical monthly decrease of 10% of PV power output is observed due to soiling in the GCC region. The amount of soiling depends on weather conditions (i.e., humidity, wind, airborne dust concentration) and the time of accumulation, while a single dust storm can reduce the PV module's power output by 20%.
- Characterization of PV soiling in these GCC countries has shown that its main components are the minerals calcite and quartz, with oxygen, calcium, magnesium, silicon, and iron being the dominant elements. Particles size ranging between few and tens of microns. Dust composition and size also depend on the local environment, for example, proximity to roads, industry, or farms, and can vary with timescales from hours (e.g., day versus night) to months (e.g., winter versus summer).

7.13   Conclusions

This comprehensive review summarizes our experience with our field-measurements key findings and challenges in addressing soiling research obtained from our last 7 years of testing at the OTF. The PV soiling has been demonstrated to be a complex phenomenon, with high degree of freedom and various interplayed factors, including environmental ones, but also physico-chemical, structural, morphological, etc.

The soiling problem is far from being solved, and further research efforts are needed to understand and to tackle the issue. In the short term, commercially available PV cleaning machines and real-time scheduling of the cleaning based on atmospheric forecasts are promising approaches. In the long term, ASC may further reduce the cleaning frequency and cost. Encouraging results were obtained using different textured thin films (silica, metal oxides, and fluorides); further improvement and demonstration of the effectiveness and durability of the coatings will support their deployment in commercial PV projects. By learning lessons from the exposure of PV panels to the harsh environment of this region, Qatar has a unique opportunity to develop mitigation strategies that will benefit other arid regions of the GCC and the world. Finally, AS coating may serve as an efficient solution that complements the active cleaning process. Indeed, innovative solutions, cleaning concepts, and novel coatings are continually developed, addressing new functionalities, including self-healing, condensation run-off, and retrofit application.

References

[1] Sayyah, A., Horenstein, M., and Mazumder, M. Yield loss of photovoltaic panels caused by depositions. *Sol. Energy* 107, 576–604 (2014).
[2] Maghami, M., Hizam, H., Gomes, C., *et al.* Power loss due to soiling on solar panel. *Renew. Sustain. Energy Rev.* 59, 1307–1316 (2016).
[3] Betak, J., Súri, M., Cebecauer, T., and Skoczek, A. Solar resource and photovoltaic electricity potential in EU-MENA region. In *Proceedings of the 27th European Photovoltaic Solar Energy Conference and Exhibition*, 2012, pp. 2–5.
[4] Micheli, L. and Muller, M. An investigation of the key parameters for predicting PV soiling losses. *Prog. Photovolt.: Res. Appl.* 25(4), 291–307 (2017).
[5] Micheli, L., Ruth, D., and Muller, M. Seasonal trends of soiling on photovoltaic systems. In *2017 IEEE 44th Photovoltaic Specialist Conference (PVSC)*. IEEE, 2017, pp. 2301–2306.
[6] Jung, Y., Lee, H., Kim, J., Cho, Y., Kim, J., and Lee, Y. Spatio-temporal characteristics in the clearness index derived from global solar radiation observations in Korea. *Atmosphere* 7 (4), 55 (2016).
[7] Figgis, B., Ennaoui, A., Ahzi, S., and Rémond, Y. Review of PV soiling particle mechanics in desert environments. *Renew. Sustain. Energy Rev.* 76, 872–881 (2017).
[8] Figgis, B. QEERI Outdoor Test Facility Technical Details. doi:10.13140/RG.2.2.29594.98244.   file:///C:/Users/baissa/Downloads/QEERIOTFTechnicalDetailsv2021-03.pdf
[9] Bachour, D. and Perez-Astudillo, D. Deriving solar direct normal irradiance using lidar-ceilometer. *Sol. Energy* 110, 316–324 (2014).

[10]   Martín Pomares, L., Polo, J., Perez-Astudillo, D., and Bachour, D. Assessment of daily atmospheric turbidity databases using aerosol optical depth and direct normal irradiance measurements. In *Conference Proceedings – Solar World Congress*, 2015. doi:10.18086/swc.2015.07.16.

[11]   Bachour, D. Dust impact on shortwave solar radiation: case study of dust events in Doha, Qatar. In *6th International Conference Energy & Meteorology*, Denmark, 2019.

[12]   Barchour, D., Perez-Astudillo, D., and Martín Pomares, L. Study of soiling on pyranometers in desert conditions. In *Conference: Eurosun 2016*, Palma de Mallorca, Spain, 2016. doi:10.18086/eurosun.2016.09.08.

[13]   Micheli, L. and Muller, M. An investigation of the key parameters for predicting PV soiling losses. *Prog. Photovolt.: Res. Appl.* 25 (4), 291–307 (2017).

[14]   Micheli, L., Deceglie, M.G., and Muller, M. Mapping photovoltaic soiling using spatial interpolation techniques. *IEEE J. Photovolt.* 9(1), 272–277 (2019).

[15]   Lawrence, C. and Neff, J. The contemporary physical and chemical flux of Aeolian dust: A synthesis of direct measurements of dust deposition. *Chem. Geol.* 267, 46–63 (2009).

[16]   Kok, J.F., Parteli, E.J.R., Michaels, T.I., and Karam, D.B. The physics of wind-blown sand and dust. *Rep. Prog. Phys.* 75(10), 106901 (2012). https://doi.org/10.1088/0034-4885/75/10/106901.

[17]   Engelbrecht, J., Moosmüller, H., Pincock, S., Jayanty, R.K.M., Lersch, T., and Casuccio, G. Technical note: Mineralogical, chemical, morphological, and optical interrelationships of mineral dust re-suspensions. *Atmos. Chem. Phys. Discuss.* 16, 1–33 (2016).

[18]   Ilse, K., Figgis, B., Naumann, V., Hagendorf, C., and Bagdahn, J. Fundamentals of soiling processes on photovoltaic modules. *Renew. Sustain. Energy Rev.* 98, 239–254 (2018).

[19]   John, J.J., Warade, S., Tamizhmani, G., and Kottantharayil, A. Study of soiling loss on photovoltaic modules with artificially deposited dust of different gravimetric densities and compositions collected from different locations in India. *IEEE J. Photovolt.* 6, 1–8 (2015).

[20]   World Health Organization. Ambient Air Pollution: A Global Assessment of Exposure and Burden of Disease, WHO Labrary Cataloguing-in-Publication Data, Geneva, Switzerland, 2016.

[21]   Sarver, T., Al-Qaraghuli, A., and Kazmerski, L.L. A comprehensive review of the impact of dust on the use of solar energy: history, investigations, results, literature, and mitigation approaches. *Renew. Sustain. Energy Rev.* 22, 698–733 (2013).

[22]   Burton, P.D. and King, B.H. Artificial soiling of photovoltaic module surfaces using traceable soil components. In *Proceedings of the IEEE 39th Photovoltaics Specialists Conference*, Tampa, FL, 2013. pp. 1542–1545.

[23]   Burton, P.D. and King, B.H. Application and characterization of an artificial grime for photovoltaic soiling studies. *IEEE J. Photovolt.* 4(1), 299–303 (2014).

[24] Piedra, P. and Moosmüller, H. Optical losses of photovoltaic cells due to aerosol deposition: role of particle refractive index and size. *Sol. Energy* 155, 637–646 (2017).

[25] Zaihidee, F.M., Mekhilef, S., Seyedmahmoudian, M., and Horan, B. Dust as an unalterable deteriorative factor affecting PV panel's efficiency: why and how. *Renew. Sustain. Energy Rev.* 65, 1267–1278 (2016).

[26] Perlwitz, J.P., Pérez García-Pando, C., and Miller, R.L. Predicting the mineral composition of dust aerosols – Part 1: representing key processes. *Atmos. Chem. Phys.* 15(20), 11593–11627 (2015).

[27] Javed, W., Guo, B., and Figgis, B. Modeling of photovoltaic soiling loss as a function of environmental variables. *Sol. Energy* 157, 397–407 (2017).

[28] Figgis, B., Ennaoui, A., Ahzi, S., and Rémond, Y. Review of PV soiling particle mechanics in desert environments. *Renew. Sustain. Energy Rev.* 76, 872–881 (2017).

[29] Goossens, D. Wind tunnel protocol to study the effects of anti-soiling and anti-reflective coatings on deposition, removal, and accumulation of dust on photovoltaic surfaces and consequences for optical transmittance. *Sol. Energy* 163, 131–139 (2018).

[30] Tomas, J. Mechanics of Particle Adhesion: Review Paper of Chisa Event, Prague, 2004. http://www.hydrosimulation.com/wp-content/uploads/2015/03/Mechanics _Particle_Adhesion _full.pdf

[31] Mittal, K.L. and Jaiswal, R. (eds.). *Particle Adhesion and Removal*. Hoboken, NJ: Wiley, 2015.

[32] Picotti, G., Borghesani, P., Cholette, M.E., and Manzolini, G. Soiling of solar collectors – modelling approaches for airborne dust and its interactions with surfaces. *Renew. Sustain. Energy Rev.* 81, 2343–2357 (2018).

[33] Figgis, B., Guo, B., Javed, W., Ahzi, S., and Remond, Y. Dominant environmental parameters for dust deposition and resuspension in desert climates. *Aerosol Sci. Technol.* 52 788–798 (2018).

[34] Moutinho, H.R., Jiang, C.-S., To, B., *et al.*, Adhesion mechanisms on solar glass: effects of relative humidity, surface roughness, and particle shape and size. *Sol. Energy Mater. Sol. Cells* 172, 145–153 (2017).

[35] Brambilla, S., Speckart, S., and Brown, M.J. Adhesion and aerodynamic forces for the resuspension of non-spherical particles in outdoor environments. *J. Aerosol Sci.* 112, 52–67 (2017).

[36] Moutinho, H.R., Jiang, C., To, B., *et al.* Adhesion mechanisms on solar glass: effects of relative humidity, surface roughness, and particle shape and size. *Sol. Energy Mater. Sol. Cells* 172, 145–153 (2017).

[37] Kazmerski, L.L., Diniz, A.S. A.C., Maia, C.B., *et al.* Soiling particle interactions on PV modules: surface and inter-particle adhesion and chemistry effects. In *Proceedings of the IEEE 43th Photovoltaics Specialists Conference*, Portland, OR, pp. 1714–1717.

[38] Masuda, H., Yoshida, H., and Higashitani, K. (eds.), *Powder Technology Handbook*, 3rd ed. Boca Raton, FL: Taylor & Francis, 2006.

[39] Schubert, H. Grundlagen des Agglomerierens. *Chem. Ing. Tech.* 51(4), 266–277 (1979).

[40] Hinds, W.C. (ed.). *Aerosol Technology: Properties, Behavior, and Measurement of Airborne Particles*, 2nd ed., Wiley-Interscience, 2012.

[41] Weiler, C., Egen, M., Trunk, M., and Langguth, P. Force control and powder dispersibility of spray dried particles for inhalation. *J. Pharm. Sci.* 99(1), 303–316 (2010). https://doi.org/10.1002/jps.21849.

[42] Luo, W., Khoo, Y.S., Hacke, P., *et al.* Potential-induced degradation in photovoltaic modules: a critical review. *Energy. Environ. Sci.* 10(1), 43–68 (2017). https://doi.org/10.1039/C6EE02271E.

[43] Bethea, R.M., Barriger, M.T., Williams, P.F., and Chin, S. Environmental effects on solar concentrator mirrors. *Sol. Energy* 27(6), 497–511 (1981).

[44] Rivera, J.L., Sutherland, J.W., and Allen, J.S. Lift-off behavior of micro and nanoparticles in contact with a flat surface. *J. Fluids Eng.* 135(10), 101205 (2013).

[45] Ibrahim, A., Dunn, P., and Brach, R. Microparticle detachment from surfaces exposed to turbulent air flow: controlled experiments and modeling. *J. Aerosol. Sci.* 34(6), 765–782 (2003).

[46] Ahmadi, G., Guo, S., and Zhang, X. Particle adhesion and detachment in turbulent flows including capillary forces. *Part Sci. Technol.* 25(1), 59–76 (2007).

[47] Rabinovich, Y.I., Adler, J.J., Ata, A., Singh, R.K., and Moudgil, B.M. Adhesion between nanoscale rough surfaces. *J. Colloid Interface Sci.* 232 (1), 17–24 (2000).

[48] Engineering Toolbox. Accessed on 1 June 2018. https://www.Engineering toolbox.com/.

[49] Benito, J.G., Uñac, R.O., Vidales, A.M., and Ippolito, I. Validation of the Monte Carlo model for resuspension phenomena. *J. Aerosol. Sci.* 100, 26–37 (2016).

[50] García, M., Marroyo, L., Lorenzo, E., and Pérez, M. Soiling and other optical losses in solartracking PV plants in navarra. *Prog. Photo: Res. Appl.* 19(2), 211–217 (2011).

[51] Kimber, A., Mitchell, L., Nogradi, S., and Wenger, H. The effect of soiling on large gridconnected photovoltaic systems in California and the Southwest Region of the United States. In: *Proceedings of the 2006 IEEE 4th World Conference on Photovoltaic Energy Conference*, IEEE, 2006. pp. 2391–2395.

[52] Quan, Y., Zhang, L, Qi, R., and Cai, R. Self-cleaning of surfaces: the role of surface wettability and dust types. *Sci. Rep.* 6, 38239 (2016). https://doi.org/10.1038/srep38239.

[53] Ilse, K.K., Rabanal, J., Schonleber, L., *et al.* Comparing indoor and outdoor soiling experiments for different glass coatings and microstructural analysis of particle caking processes. *IEEE J. Photovolt.* 8(1), 203–209 (2018).

[54] Nayshevsky, I., Xu, Q., Barahman, G., and Lyons, A. Anti-reflective and anti-soiling properties of Klean Boost^{TM}, a superhydrophobic nano-textured coating for solar glass. In *Proceedings of the IEEE 44th Photovoltaics Specialist Conference*, Washington, DC, 2017.

[55]   Hülsmann, P., Heck, M., and Köhl, M. Simulation of water vapor ingress into PV-modules under different climatic conditions. *J. Mater.* 2013(1–2), 1–7 (2013).

[56]   Lombardo, T., Ionescu, A., Chabas, A., Lefèvre, R., Ausset, P., and Candau, Y. Dose-response function for the soiling of silica-soda-lime glass due to dry deposition. *Sci. Total Environ.* 408(4), 976–984 (2010).

[57]   Ilse, K.K., Figgis, B.W., Khan, M.Z., Naumann, V., and Hagendorf, C. Dew as a detrimental influencing factor for soiling of PV modules. *IEEE J. Photovolt.* 10, 100084 (2018).

[58]   Beysens, D., Muselli, M., Nikolayev, V., Narhe, R., and Milimouk I. Measurement and modelling of dew in island, coastal and alpine areas. *Atmos. Res.* 73(1–2), 1–22 (2005).

[59]   Pearce, E.A. and Smith, C.G. *The Hutchinson World Weather Guide*, 4th ed. Oxford: Helicon, 2000.

[60]   Beysens, D., Muselli, M., Milimouk, I., *et al.* Application of passive radiative cooling for dew condensation. *Energy* 31(13), 2303–2315 (2006). https://doi.org/10.1016/j.energy.2006.01.006.

[61]   Figgis, B.W. Solar Test Facility in Doha, Qatar. http://dx.doi.org/10.13140/RG. 2.2.21049.77925.

[62]   Ilse, K.K., Figgis, B.W., Werner, M., *et al.* Comprehensive analysis of soiling and cementation processes on PV modules in Qatar. *Sol. Energy Mater. Sol. Cells* 186, 309–323 (2018).

[63]   Mazumder, M.K., Horenstein, M.N., Heiling, C., *et al.* Environmental degradation of the optical surface of PV modules and solar mirrors by soiling and high RH and mitigation methods for minimizing energy yield losses. In *Proceedings of the IEEE 42nd Photovoltaic Specialist Conference (PVSC)*, 2015, pp. 1–6.

[64]   Kazmerski, L.L., Diniz, A.S.A.C., Brasil, M.C, *et al.* Fundamental studies of the adhesion of dust to PV module chemical and physical relationships at the microscale. In *Proceedings of the IEEE 42nd Photovoltaic Specialist Conference (PVSC)*, 2015, pp. 1–7.

[65]   Ilse, K.K., Werner, M., Naumann, V., Figgis, B.W., Hagendorf, C., and Bagdahn, J. Microstructural analysis of the cementation process during soiling on glass surfaces in arid and semi-arid climates. *Phys. Status Solidi RRL* 10(7), 525–529 (2016).

[66]   Perkins, C., Muller, M., and Simpson, L. Laboratory studies of particle cementation and PV module soiling. In *Proceedings of the IEEE 44th Photovoltaics Specialist Conference*, Washington, DC, 2017.

[67]   Yilbas, B.S., Ali, H., Khaled, M.M., Al-Aqeeli, N., Abu-Dheir, N., and Varanasi, K.K. Influence of dust and mud on the optical, chemical, and mechanical properties of a PV protective glass. *Sci. Rep.* 5, 15833 (2015).

[68]   Chabas, A., Fouqueau, A., Attoui, M., *et al.* Characterisation of CIME, an experimental chamber for simulating interactions between materials of the cultural heritage and the environment. *Environ. Sci. Pollut. Res. Int.* 22(23), 19170–19183 (2015).

[69] Melcher, M., Wiesinger, R., and Schreiner, M. Degradation of glass artifacts: application of modern surface analytical techniques. *Acc. Chem. Res.* 43(6), 916–926 (2010).

[70] Leaper, M.C., Prime, D.C., Taylor, P.M., and Leach, V. Solid bridge formation between spraydried sodium carbonate particles. *Dry Technol.* 30(9), 1008–1013 (2012).

[71] Gong, G., Zhou, C., Wu, J., Jin, X., and Jiang, L. Nanofibrous adhesion: the twin of gecko adhesion. *ACS Nano* 9(4), 3721–3727 (2015).

[72] Stokes, M., Charman, J., Epps, R.J., and Griffiths, J.S. *Soil and Rock Description and Characteristics 25*. Geological Society, London: Engineering Geology Special Publications, 2012, pp. 143–57 (Chapter 5).

[73] Figgis, B., Guo, B., Javed, W., Ilse, K., Ahzi, S., and Remond, Y. *Time-of-day and Exposure Influences on PV Soiling 5th International Renewable and Sustainable Energy Conference (IRSEC'17)*, Tangier, Morocco, IEEE, 2017, pp. 1–4.

[74] Figgis, B., Guo, B., Javed, W., Ahzi, S., and Remond, Y. Dominant environmental parameters for dust deposition and resuspension in desert climates. *Aerosol Sci. Technol.* 52, 788–798 (2018).

[75] Gostein, M., Passow, K., Deceglie, M.G., Micheli, L., and Stueve, B. Local variability in PV soiling rate. In *Proceedings of the World Conference on Photovoltaic Energy Conversion (WCPEC-7)*, Hawai, 2018.

[76] Gostein, M., Faullin, S., Miller, K., Schneider, J., and Stueve, B. Mars_soiling sensor. In *35th European Photovoltaic Solar Energy Conference and Exhibition*, 2018, pp. 1523–1527.

[77] Sarver, T., Al-Qaraghuli, A. and Kazmerski, L.L. A comprehensive review of the impact of dust on the use of solar energy: history, investigations, results, literature, and mitigation approaches. *Renew. Sustain. Energy Rev.* 22, 698–733 (2013). doi: 10.1016/j.rser.2012.12.065

[78] Costa, S., Diniz, A., and Kazmerski, L.L. Dust and soiling issues and impacts relating to solar energy systems: literature review update for 2012–2015. *Renew. Sustain. Energy Rev.*, 63, 33–61 (2016). doi: 10.1016/j.rser.2016.04.059

[79] Sayyah, A., Horenstein, M.N., and Mazumder, M.K. Energy yield loss caused by dust deposition on photovoltaic panels. *Sol. Energy* 107, 576–604 (2014).

[80] Einhorn, A., Micheli, L., Miller, D.C., *et al.* Evaluation of soiling and potential mitigation approaches on photovoltaic glass. *IEEE J. Photovoltaics*, 1–7 (2018).

[81] Ben Gallagher. Rise of the Machines: Solar Module-Washing Robots. https://www.woodmac.com/our-expertise/focus/Power–Renewables/rise-of-the-machines-solar-module-washing-robots/, 2018.

[82] Ferretti, N. PV Module Cleaning Market Overview and Basics, 2018.

[83] Choori, G.. Robotic module cleaning. In *Indian Context International PV Soiling Workshop*, Denver, CO, 2018.

[84] Ilse, K., Figgis, B., Naumann, V., Hagendorf, C., and Bagdahn, J. Fundamentals of soiling processes on photovoltaic modules. *Renew. Sustain. Energy Rev.* 98, 239–254 (2018).

[85] Ilse, K., Figgis, B.W., Werner, M., *et al.* Comprehensive analysis of soiling and cementation processes on PV modules in Qatar. *Sol. Energy Mater. Sol. Cells* 186, 309–323 (2018).

[86] Figgis, B., Guo, B., Javed, W., Ahzi, S., and Rémond, Y. Dominant environmental parameters for dust deposition and resuspension in desert climates. In *Aerosol Science and Technology*, vol. 52, Philadelphia, PA: Taylor & Francis Inc., 2018.

[87] Ilse, K., Khan, M., Naumann, V., and Hagendorf, C. Dew as a detrimental influencing factor for soiling of PV modules. *IEEE J. Photovoltaics* 9, 287–294 (2018). doi:10.1109/JPHOTOV.2018.2882649.

[88] Figgis, B., Nouviaire, A., Wubulikasimu, Y., *et al.* Investigation of factors affecting condensation on soiled PV modules. *Sol. Energy* 159, 488–500 (2018).

[89] Ilse, K., Khan, M.Z., Voicu, N., *et al.* Advanced performance testing of anti-soiling coatings – Part I: sequential laboratory test methodology covering the physics of natural soiling processes. *Sol. Energy Mater. Sol. Cells* 202, 110048 (2019).

[90] Ilse, K., Khan, M.Z., Voicu, N., *et al.* Advanced performance testing of anti-soiling coatings – Part II: particle-size dependent analysis for physical understanding of dust removal processes and determination of adhesion forces. *Sol. Energy Mater. Sol. Cells* 202, 110049 (2019).

[91] Isaifan, R., Johnson, D.J., Mansour, S., *et al.* Theoretical and experimental characterization of efficient anti-dust coatings under desert conditions. *J. Thin Film. Res.* 2, 25–29 (2018).

[92] Zhang, R., Ai, X., Wan, Y., *et al.* Surface corrosion resistance in turning of titanium alloy. *Int. J. Corros.* 2015, 1–8 (2015).

[93] Shi, E., Zhang, L., Li, Z., *et al.* TiO_2-coated carbon nanotube-silicon solar cells with efficiency of 15%. *Sci. Rep.* 2, 884 (2012).

[94] Ilse, K., Micheli, L., Figgis, B.W., *et al.* Techno-economic assessment of soiling losses and mitigation strategies for solar power generation. *Joule* 3, 2303–2321 (2019). doi:10.1016/j.joule.2019.08.019.

[95] Steensma, G., Román, R., Marshall, C., *et al.* Shagaya renewable energy park project. In *AIP Conference Proceedings*, vol. 2126, 2019. Melville, NY: AIP Publishing LLC.

[96] PV Cleaning Solution Company NOMADD Desert Solar Solutions Secures Funding Round from Saudi Arabian Investor CEPCO, 2018. http://www.nomaddesertsolar.com/news

[97] Albadwawi, O., John, J., Elhassan, Y., *et al.* Quantification of spectral losses of Natural soiling and detailed microstructural analysis of dust collected from different locations in Dubai, UAE. In *2019 IEEE 46th Photovoltaic Specialists Conference (PVSC)*, 2019. doi:10.1109/PVSC40753.2019.8980937.

[98] Figgis, B. *QEERI Solar Test Facility 5-Year Report*, 2018. doi:10.13140/RG.2.2.12062.23361.

[99] Nitz, P. *Investigations into Soiling and Degradation*. https://www.ise.fraun-hofer.de/en/business-areas/photovoltaics/iii-v-and-concentrator-photovoltaics/concentrator-optics/investigations-into-soiling-and-degradation.html.

[100] Simpson, L.J., Muller, M., Deceglie, M., *et al.* NREL efforts to address soiling on PV modules. In *2017 IEEE 44th Photovoltaic Specialist Conference (PVSC)*, pp. 2789–2793, 2017. doi:10.1109/PVSC.2017.8366040

[101] Cabrera, E., Araya, F., Schneider, A., *et al.* AtaMoS-TeC Project: The Bifacial Institute for Desert PV, 2018.

[102] King, B. *Overview of Sandia's Soiling Program: Experimental Methods and Framework for a Quantitative Soiling Model*, 15 (2015). https://www.osti.gov/servlets/purl/1331410

[103] Welcome to PVReliability.org. Our Collaborative R&D Program Aimed at Ensuring the Reliability of Solar PV Systems. http://www.pvreliability.org/

[104] Institute of Solar Research in the EU-Funded Project WASCOP: Soiling Prediction Model, Tests of Mirror Coatings and Innovative Cleaning Methods for Solar Mirrors, 2016. https://www.dlr.de/sf/en/desktopdefault.aspx/tabid-10436/20174_read-47350/

[105] Herrmann, W. Impact of soiling on pv module performance for various climates. In *4th PV Performance Modelling and Monitoring Workshop*, 2015.

[106] Soiling: Task Group 12. https://www.pvqat.org/project-status/task-group-12.html.

[107] https://cost-indust.eu/the-action/working-groups

[108] Empowering Solar World Leader in Robotic Solar Cleaning, https://www.ecoppia.com/

[109] Automatic and Semiautomatic Portable Washing Systems for PV Plants. http://www.washpanel.com/en/

[110] ERIC WESOFF, H.T. SunPower Cleans Up Solar with Acquisition of Greenbotics, 2013. https://www.greentechmedia.com/articles/read/sun-power-cleans-up-solar-with-acquisiton-of-greenbotics

[111] First Solar Rooftop in the UAE with 100% Robotic Maintenance is Completed, 2019. http://www.mesia.com/2019/02/18/first-solar-rooftop-in-the-uae-with-100-robotic-maintenance-is-completed/

[112] https://www.serbot.ch/en/

[113] Solar Panel Automatic Cleaning Robot with Self-Propelling Guide Rail (The World's First), 2014. https://www.skyrobot.co.jp/en/solrider.html

[114] Solar Hybrid Technology. https://www.enerwhere.com/technology

[115] BladeRanger Autonomous Cleaning and Inspection of Solar Power Installations: Let the Sunshine In! https://bladeranger.com/.

[116] Wolfertstetter, F., Fonk, R. Prahl, C., *et al.* Airborne soiling measurement of entire solar fields with Qfly. In AIP Conference Proceedings, vol. 2303, p. 100008, 2019.

[117] SMP100 Solar-Module Performance Monitoring System. https://www.campbellsci.com/smp100

[118] DustIQ Soiling Monitoring System. https://www.kippzonen.com/Product/419/DustIQ-Soiling-Monitoring-System

[119]   https://norcalcontrols.net/

[120]   Soiling Measurement Kit. https://www.nrgsystems.com/products/solar/detail/soiling-measurement-kit0

[121]   Solar Measurement for Solar Site Assessment. https://www.ammonit.com/en/component/content/article/165-ammonit-messsysteme/solar-resource-assessment/407-solar-resource-assessment

[122]   Mars Optical Soiling Sensor – Revolutionary New Product. http://www.atonometrics.com/mars-optical-soiling-sensor-revolutionary-new-product/

[123]   Soiling Measurement Kit. https://www.kintech-engineering.com/catalogue/soiling-measurement-kit/soiling-measurement-kit/

[124]   Quantifying Solar O&M'S Dirty Side. https://europe.solar-asset.management/new-updates-source/2017/10/19/quantifying-solar-oms-dirty-side

[125]   Solar Energy 3.0. https://www.innoenergy.com/discover-innovative-solutions/sustainable-products-services/electricity-heat-production/solar-pv/solar-energy-30/

[126]   https://trackso.in/

[127]   CHRIS COWELL. (2019) Anti-soiling Coating That Can Increase Solar Panel Output Heads to Commercial Deployment. https://solarbuildermag.com/products/anti-soiling-coating-that-can-increase-solar-panel-output-heads-to-commercial-deployment/

[128]   Anti-Soiling Coatings. https://www.dsm.com/dsm-in-solar/en_US/technologies/anti-soiling-coatings-for-solar-glass.html

[129]   Nano-coatings for Water Control Moisture Condensation Wetting. https://lotusleafcoatings.com/

[130]   Castillo Aguilella, J. and Hauser, P. Multi-variable bifacial photovoltaic module test results and best-fit annual bifacial energy yield model. *IEEE Access* 4, 1 (2016).

[131]   Comello, S., Reichelstein, S., and Sahoo, A. The road ahead for solar PV power. *Renew. Sustain. Energy Rev.* 92, 744–756 (2018).

[132]   Daßler, D., Malik, S., Figgis, B., Bagdahn, J., and Ebert, M. Bifacial gain simulations of modules and systems under desert conditions. In *bifi PV Workshop*, 2017.

[133]   Figgis, B. and Abdallah, A. Investigation of PV yield differences in a desert climate. *Sol. Energy* 194, 136–140 (2019).

[134]   Figgis, B., Ennaoui, A., Ahzi, S., and Rémond, Y. Review of PV soiling particle mechanics in desert environments. *Renew. Sustain. Energy Rev.* 76, 872–881 (2017).

[135]   Figgis, B. Solar Test Facility in Doha, *Qatar*, 2016. doi:10.13140/RG.2.2.21049.77925.

[136]   Fuentealba Vidal, E., Ferrada, P., Araya, F., *et al.* Photovoltaic performance and LCoE comparison at the coastal zone of the Atacama Desert, *Chile. Energy Convers. Manag.* 95, 181–186 (2015).

[137]   Ilse, K., Micheli, L., Figgis, B.W., *et al.* Techno-economic assessment of soiling losses and mitigation strategies for solar power generation. *Joule*, 3, 2303–2321 (2019) doi:10.1016/j.joule.2019.08.019.

[138] IRENA. *International Renewable Energy Agency. Renewable Power Generation Costs in 2017*. International Renewable Energy Agency, 2018.

[139] Vest, B. Levelized cost and levelized avoided cost of new generation resources in the annual energy outlook 2016. *Us Eia Lcoe* 1–20 (2016).

[140] Khan, M., Hanna, A., Sun, X., and Alam, M. Vertical bifacial solar farms: physics, design, and global optimization. *Appl. Energy* 206, 1–9 (2017).

[141] Tian Shen, L., Pravettoni, M., Deline, C., *et al.* A review of crystalline silicon bifacial photovoltaic performance characterisation and simulation. *Energy Environ. Sci.* 12, (2019).

[142] Luque, E., Antonanzas, F., and Escobar, R. Effect of soiling in bifacial PV modules and cleaning schedule optimization. *Energy Convers. Manag.* 174, 615–625 (2018).

[143] Figgis, B. *QEERI Solar Test Facility 5-Year Report*, 2018. doi:10.13140/RG.2.2.12062.23361.

[144] Sayyah, A., Horenstein, M.N., and Mazumder, M. K. Energy yield loss caused by dust deposition on photovoltaic panels. *Sol. Energy* 107, 576–604 (2014).

[145] Schmela, M., Beauvais, A., Chevillard, N., Paredes, M.G., Heisz, M., and Rossi, R. Global market outlook for solar power. *Glob. Mark. Outlook* 92 (2018).

[146] Jäger-Waldau, A. *PV Status Report* 2017. doi:10.2760/452611.

[147] Jäger-Waldau, A. Snapshot of photovoltaics – February 2018. *EPJ Photovoltaics* 13, 9, (2018).

[148] Apostoleris, H., Sgouridis, S., Stefancich, M., and Chiesa, M. Evaluating the factors that led to low-priced solar electricity projects in the Middle East. *Nat. Energy* 3, 1109–1114 (2018)

[149] Labordena, M., Neubauer, D., Folini, D., Patt, A., and Lilliestam, J. Blue skies over China: The effect of pollution-control on solar power generation and revenues. *PLoS One* 13, e0207028 (2018).

[150] Boys, B., Cooper, M.J., Hsu N.C., *et al.* Fifteen-year global time series of satellite-derived fine particulate matter. *Environ. Sci. Technol.* 48, 11109–11118 (2014).

[151] Yang, X., Jiang, L., Zhao, W., *et al.* Comparison of ground-based PM2.5 and PM10 concentrations in China, India, and the U.S. *Int. J. Environ. Res. Public Health* 15, 1382 (2018).

[152] Butt, E., Turnock, S.T., Rigby, R., *et al.* Global and regional trends in particulate air pollution and attributable health burden over the past 50 years. *Environ. Res. Lett.* 12, 104017 (2017).

[153] Weagle, C., Snider, G., Li, C., *et al.* Global sources of fine particulate matter: interpretation of PM 2.5 chemical composition observed by SPARTAN using a global chemical transport model. *Environ. Sci. Technol.* 52, 11670–11681 (2018).

[154] IRENA. Renewable Energy Market Analysis: GCC 2019, Irena, 2019.

[155] Graves L. *Abu Dhabi Plant to Produce Region's Cheapest Electricity from Solar*, 2017. https://www.thenationalnews.com/business/abu-dhabi-plant-to-produce-region-s-cheapest-electricity-from-solar-1.29977

[156]   Robin, M. *UAE's Push on Solar Should Open Eyes Across Wider World*, 2017.

[157]   Sgouridis, S., Abdullah, A, Griffiths, S., *et al.* RE-mapping the UAE's energy transition: an economy-wide assessment of renewable energy options and their policy implications. *Renew. Sustain. Energy Rev.* 55, 1166–1180 (2015)

Chapter 8

Desert PV applications

Ali Elrayyah[1] and Mohd Zamri Che Wanik[1]

8.1 Introduction

Agricultural farms form a major power consumer in a rural desert environment. It is estimated that there are more than 1,300 farms in Qatar [1] and most of them are off-grid based farms in respect to electrical supply. The farms in desert areas are projected to surge in the near future to meet the increasing need for food supply as limited open trade situation due to regional political situation. Besides that, the energy demand in these farms is expected to increase over time as farmers are moving more toward advance farming technologies such as greenhouses which can extend the farming season, allows introducing new types and plant, and thus increases power consumption accordingly. For this isolated farm, the electricity is mostly supplied by a diesel generator. Some of these farms, however, are already having excess to the electricity supply from an electrical grid.

The energy consumption in the remote farm is mainly the water pump for pumping underground water to be used for irrigation, lighting and domestic users. The energy is also needed for lighting and other house appliances for the house or building that co-exist with the farm. In a typical farm that exist in the desert, there are plantation area, animal barns buildings and a small housing area. Electrical energy is needed to power the farm for water pumping, lighting, air conditioning and other commercial or human activities.

In the desert environment, solar resources are known abundantly available and can be utilized for PV application in generating electrical energy. It is forecasted that the PV generation will be peaked during the midday with the bell-shaped curve. During this peaking time, it is forecasted that the energy generated by the PV surpassing the farm load consumption. This extra energy has the potential to be stored and used at later of the day to energize the farm.

This chapter deals with some of the main desert PV applications in the state of Qatar. The climate of the desert is described first including the availability of solar resources and the performance of PV in the desert. Then follow by the discussion

[1]Qatar Environment and Energy Research Institute (QEERI), Hamad Bin Khalifa University (HBKU), Qatar Foundation, Qatar

on the electricity consumption in the desert farm describing the type of typical load available. The chapter also discussing the various option of system design, control, and operation for the PV system. Finally, with the different PV energy system options applicable, the techno-economic analysis for the optimal energy system for the farm is presented and discussed.

8.2   Desert climate and solar resource

8.2.1   General weather profile of a desert

Desert climate (often called arid climate) is a type of climate where precipitation is generally below than 250 mm a year. Low rainfall is the main feature of deserts such as the Arabian, central Australia, and the Sahara [2,3]. A "true dessert" is referring to the one where plant cover is very sparse, and rainfall is very infrequent. Due to this, desert air is very dry where there is a little moisture to hold onto the heat of the day. Desert nights are usually very cold. This wide temperature band-width can make it quite challenging to live in.

Some deserts get more than 250 mm of rainfall yearly but are still arid areas. For example, the Kalahari Desert gets up to 640 mm (25 in) a year. It has great sand dunes like the Sahara, but they do not shift in the winds. They have plants that anchor the sand and help to keep their shape. Other areas can also have more than 250 mm of precipitation but lose more water via evapotranspiration that falls as precipitation. Tucson and Alice Springs are good examples of this. Tucson gets an average of 303 mm (12 in) of rainfall a year. Alice Springs receives about 540 mm (21 in) yearly. Some scientists do not consider these as true deserts.

In terms of temperature, there are three types of desert climates: a hot desert climate, a cold desert climate and a mild desert climate. Hot deserts have very hot summers that have temperatures that can even reach 45°C (113°F). Temperatures can even be very warm during winter. Cold deserts can have hot summers as well, but the winters are usually very cold. They are usually at high altitudes and can be drier than hot deserts.

8.2.2   Arabian and Qatar desert climate

The Arabian Desert has a subtropical, hot desert climate, close to the climate of the Sahara Desert, the world's largest hot desert. In fact, the Arabian Desert is an extension of the Sahara Desert over the Arabian Peninsula. The climate is mainly hot and dry with plenty of sunshine throughout the year. The rainfall amount is generally around 100 mm, and the driest areas can receive between 30 and 40 mm of annual rain. Such dryness remains very rare throughout the desert, however. There are hardly any hyper-arid areas in the Arabian Desert, in contrast with the Sahara Desert, where more than half of the area is hyper-arid (annual rainfall below 50 mm). The sunshine duration is very high by global standards in the Arabian Desert, between 2,900 h (66.2% of the daylight hours) and 3,600 h (82.1% of the daylight hours) but is typically around 3,400 h (77.6% of the daylight hours), which clearly indicates clear-sky conditions prevail over the region and cloudy periods are just intermittent. Even though the sun and moon are bright, the dust and humidity have lower visibility for the traveler. The temperatures remain high all year round. Average high temperatures in summer are generally over 40°C (104°F) at low elevations and can even soar to 48°C (114.8°F) at extremely low elevations, especially along the Persian Gulf near the sea level. Average low temperatures in summer remain high, over 20°C (68°F) and sometimes over 30°C (77°F) in the southernmost regions. Record high temperatures are above 50°C (122°F) in much of the desert, due in part to very low elevation.

The Desert in Qatar is part of the Arabian desert that stretches from Yemen to the Persian Gulf and Oman to Jordan and Iraq. It occupies most of the Arabian Peninsula with an area of 2,330,00 square kilometers. The Arabian desert is the fourth largest desert in the world and the largest in Asia. Occupying a small land on the eastern coast of the Arabian Peninsula, the sovereign state of Qatar has a dry, subtropical desert climate with low annual rainfall and intensely hot and humid summers. The weather in Qatar can be broadly grouped into two seasons: hot (May to October) and cool (December to February). Being a small and flat country, it has a uniform climate throughout the territory. Qatar is mainly a desert; Qatar's largest area of the sand desert lies to the south west of the capital.

8.2.3   Solar resource in the Qatar desert

For PV applications, the most important climate data required is solar radiation, temperature and wind speed. These climate data for six different locations in Qatar are tabulated in Tables 8.1–8.5 and Figures 8.1–8.11. The data from Abu Samrah, Ar-Ruwais, Dukhan, and Messaid is satellite data provided by NASA. The data from Doha International Airport, however, is ground measurement data. The data provided is a daily average data that can be used for designing a PV system.

Table 8.1   Temperature, GHI, and wind speed at Abu Samrah [4]

	Abu Samrah at 10 m			
Month	**Air temperature (Celsius)**	**Daily solar radiation – horizontal (kWh/m²/d)**	**Wind speed (m/s)**	**Earth temperature (°C)**
Jan	17.2	4.02	3.6	18.5
Feb	19.3	4.88	4.1	21.1
Mar	23.1	5.61	3.9	25.4
April	28.7	6.4	3.6	31.9
May	34	7.31	3.9	37.5
Jun	36.1	7.93	4.4	39.6
Jul	37.8	7.31	4.1	41.5
Aug	37.3	7.18	4	41.1
Sep	34.1	6.69	3.6	37.6
Oct	29.5	5.74	3.3	32.2
Nov	24.1	4.59	3.2	26
Dec	19.4	3.79	3.6	20.6
Annual average	28.4	5.96	3.8	31.1

Table 8.2   Temperature, GHI, and wind speed at Ar-Ruwais [4]

Ar-Ruwais at 10 m				
Month	Air temperature (Celsius)	Daily solar radiation – horizontal (kWh/m²/d)	Wind speed (m/s)	Earth temperature (°C)
Jan	19.9	3.88	4	22
Feb	20	4.82	4.5	20.9
Mar	21.6	5.6	4.3	21.4
April	25.2	6.75	4	23.7
May	29.7	7.51	4.5	27.1
Jun	32.2	8.09	4.6	29.9
Jul	33.5	7.77	4.2	31.9
Aug	33.9	7.23	4.1	33.2
Sep	32.6	6.72	3.7	32.8
Oct	29.9	5.76	3.4	31.1
Nov	26.3	4.43	3.5	28.1
Dec	22.4	3.59	3.9	24.6
Annual average	27.3	6.02	4.1	27.3

Table 8.3   Temperature, GHI, and wind speed at Doha International Airport [4]

Doha International Airport				
Month	Air temperature (Celsius)	Daily solar radiation – horizontal (kWh/m²/d)	Wind speed (m/s)	Earth temperature (°C)
Jan	17.5	3.7	3.9	21.1
Feb	18.5	4.4	4.5	21.3
Mar	21.5	4.9	4.6	23.2
April	26.4	5.7	4.4	27.1
May	32.1	6.2	4.7	31.4
Jun	34.5	6.5	4.9	33.9
Jul	35.4	6	4.2	35.8
Aug	35	5.8	3.8	36.5
Sep	32.9	5.5	3.5	34.9
Oct	29.7	4.8	3.3	31.8
Nov	24.8	4.1	3.6	27.7
Dec	20	3.5	3.8	23.6
Annual average	27.4	5.09	4.1	29.1

8.2.4   Daily PV profile in the desert environment

PV power generation is directly related to the irradiance level of the sun. Being a small and flat country, the climate condition is uniform throughout the region. The profile of sun radiation measured in one region can be generalized for the

Table 8.4   Temperature, GHI, and wind speed at Dukhan [4]

		Dukhan		
Month	**Air temperature (Celsius)**	**Daily solar radiation – horizontal (kWh/m²/d)**	**Wind speed (m/s)**	**Earth temperature (°C)**
Jan	17.6	3.69	3.7	18.9
Feb	19.1	4.65	4.2	20.7
Mar	22.4	5.34	4	24.2
April	27.6	6.24	3.7	29.9
May	32.9	7.22	4.1	35.3
Jun	35.3	8.01	4.4	37.6
Jul	36.8	7.47	4.1	39.5
Aug	36.6	7.15	4	39.5
Sep	33.7	6.62	3.6	36.5
Oct	29.6	5.52	3.3	32
Nov	24.5	4.17	3.2	26.3
Dec	20	3.35	3.6	21.2
Annual average	28	5.79	3.8	30.2

Table 8.5   Temperature, GHI, and wind speed at Messaeid [4]

		Messaeid		
Month	**Air temperature (Celsius)**	**Daily solar radiation – horizontal (kWh/m²/d)**	**Wind speed (m/s)**	**Earth temperature (°C)**
Jan	18.5	3.71	3.6	19.9
Feb	20.1	4.66	4.2	21.8
Mar	23.5	5.24	3.9	25.6
April	28.7	6.15	3.7	31.5
May	33.7	7.28	4	36.6
Jun	35.7	7.86	4.3	38.8
Jul	37.4	7.19	4	40.8
Aug	37.2	6.75	3.9	40.6
Sep	34.3	6.24	3.5	37.6
Oct	30	5.26	3.3	32.7
Nov	25.1	4.18	3.1	27
Dec	20.8	3.36	3.6	22.1
Annual average	28.8	5.66	3.8	31.3

rest of the region. The typical irradiance profile measured on certain days throughout the year is as shown in Figures 8.4–8.11. Each figure represents a different type of day [5].

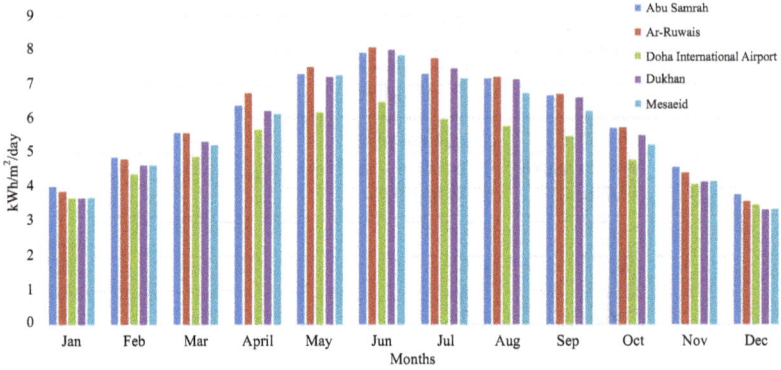

Figure 8.1   Daily average solar radiation – horizontal at five different locations in Qatar [4]

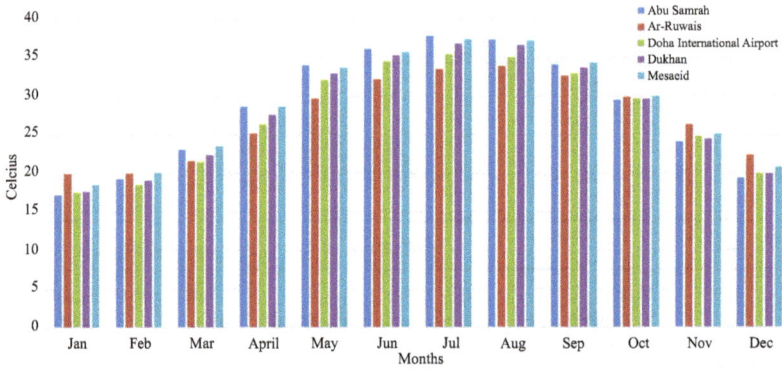

Figure 8.2   Average air temperature at five different locations in Qatar [4]

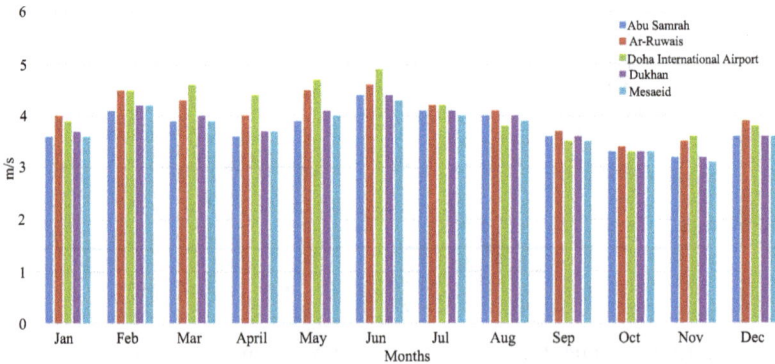

Figure 8.3   Average wind speed at five different locations in Qatar [4]

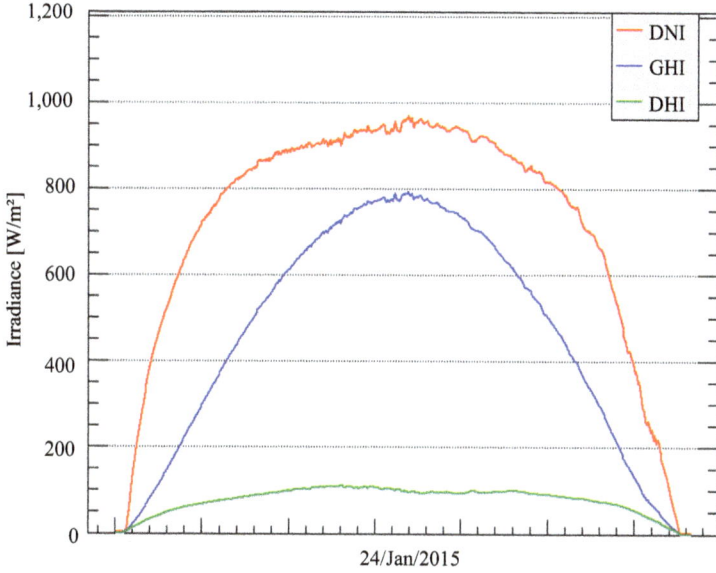

Figure 8.4   A clear day in winter

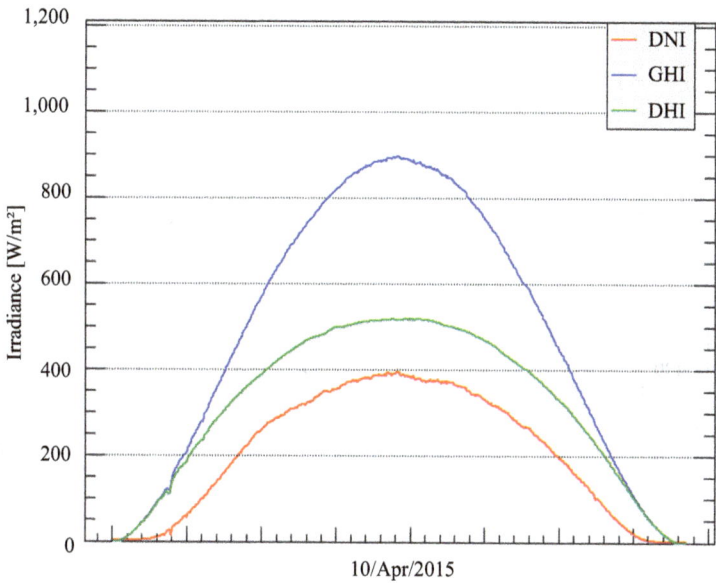

Figure 8.5   Clear day with haze/dust

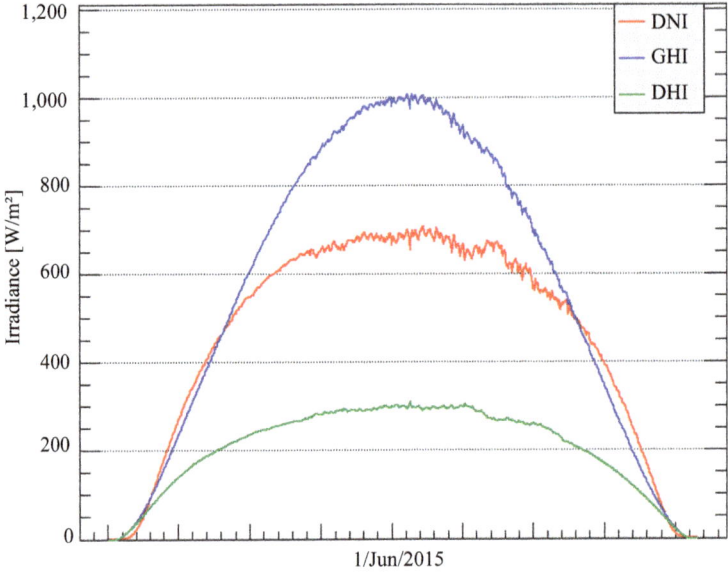

Figure 8.6   A clear day in summer with typical max radiation

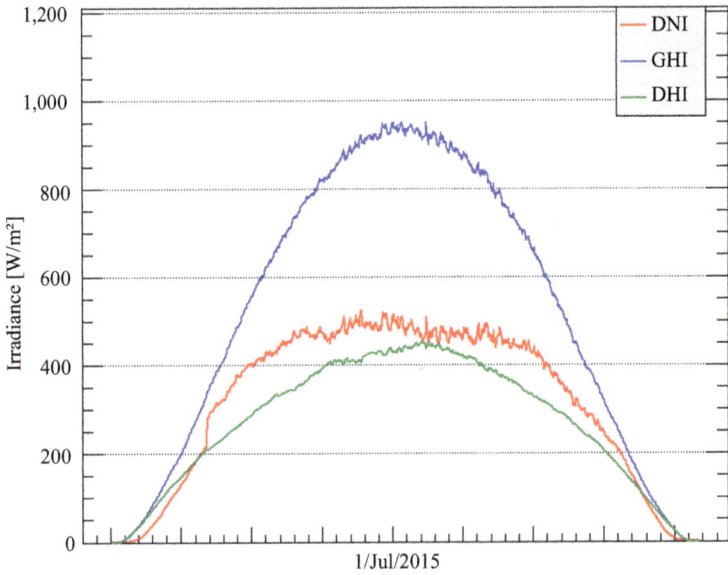

Figure 8.7   A day in summer with dust

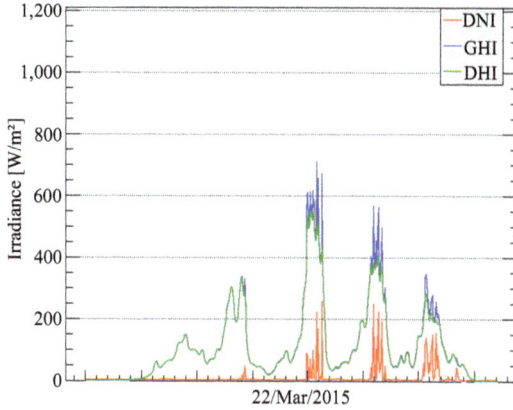

Figure 8.8   A day with clouds and rain

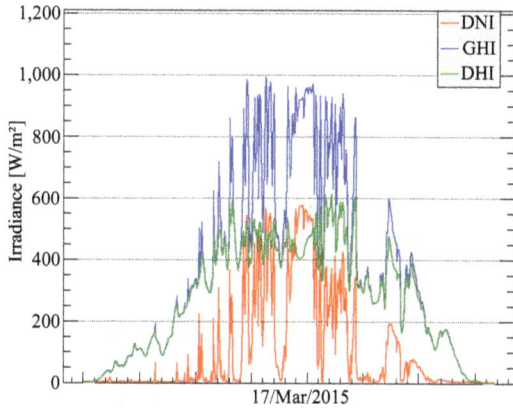

Figure 8.9   A day with cloud, haze, and dust

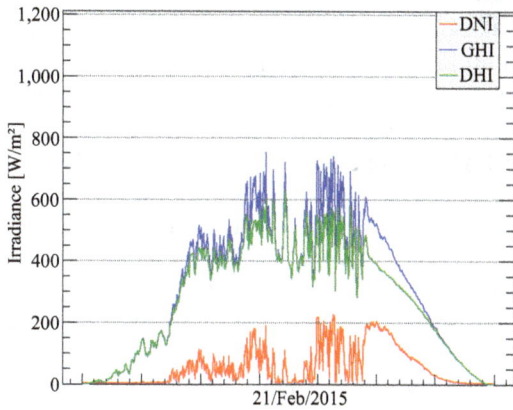

Figure 8.10   Day in winter with a dust storm

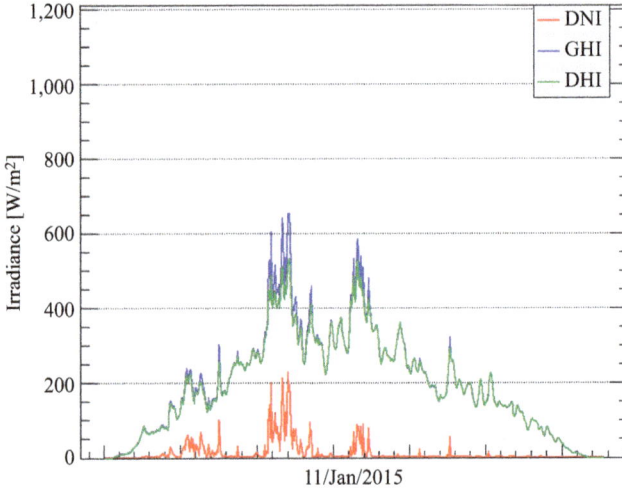

Figure 8.11   A day with dust, haze, and clouds

8.3   PV strings and arrays

PV cells are connected together to make a PV module. In typical PV installation, the used modules have voltage in the range of 12–40 V and the rated current is in the range of 5–12 A. Despite the fact that power electronic systems can be used to interface each PV module to the power system, for economic and technical reasons, PV modules are interconnected together at the input of power electronic converters. PV modules are connected in series forming PV strings to establish an appropriate voltage. The number of these strings can be connected in parallel to make PV arrays [6,7].

8.3.1   Partial shading impact on PV system performance

Within a PV module, the cells are connected in series which indicates that the same current flows over all of them. Due to manufacturing variability, ageing, shading, or accumulation of dirt/dust, the electrical characteristics of serially connected cells can be mismatched. Figure 8.12 shows the *I–V* curve of a cell (cell$_A$) that suffers partial shading and another one (cell$_B$) that does not. Clearly, cell$_B$ can supply current higher than what can be produced by cell$_A$ and in that case, the operating point of cell$_A$ will be in the negative voltage side [8]. Accordingly, significant amount of power is dissipated in the shaded cell which can lead to a permanent damage in the cell and subsequently to module failure [9,10]. For that purpose bypassing diodes are usually connected across each set of 16–20 cells (sub-module) such that whenever there is partial shading impact, the set that has underperforming cells will have its bypassing diode turned ON before operating in the negative voltage zone.

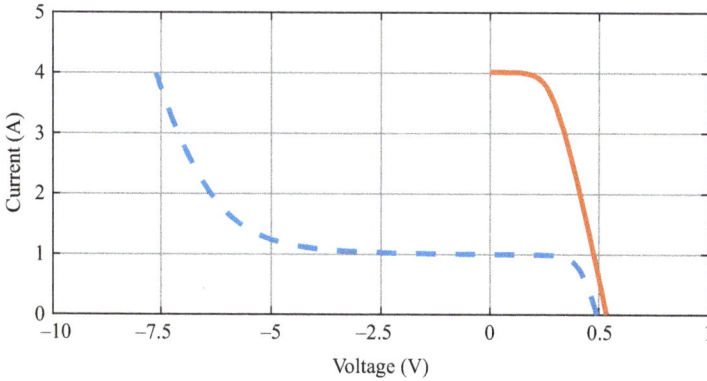

Figure 8.12 Partial shading impact on I–V *curve of serially connected PV cells*

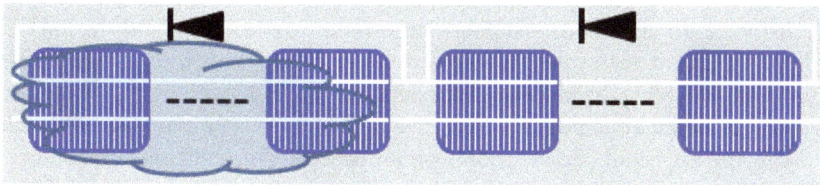

(a)

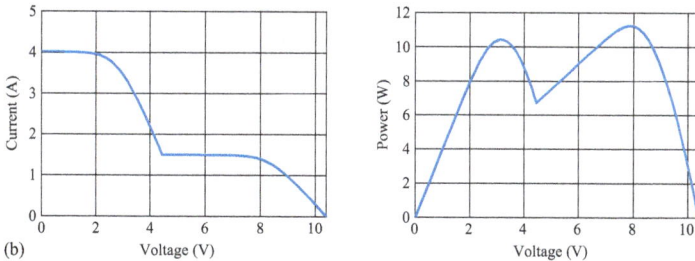

(b)

Figure 8.13 Impact of partial shading on PV modules with bypassing diodes. (a) PV module with partial shading. (b) I–V *and* P–V *curves for PV module with partial shading.*

The addition of bypassing diodes protects the PV modules from partial shading caused failure. However, the power–voltage characteristics of the module might have multiple peaks in this case. Figure 8.13(a) shows a simple case of a two sub-modules PV connected in series where each one has it's own bypassing diode. The current supplied by the module is determined by the connected load across its terminals. If a high current is supplied by the unshaded sub-module; the shaded sub-module will be bypassed by its diode making the power–voltage characteristics of the entire module follows that of one sub-module. However, if the current drawn by the load is reduced the shaded-submodule will operate at the positive voltage at a certain current, which makes the power–voltage characteristics look like the one in Figure 8.13(b) which has two peaks rather than one.

When several PV modules are connected in series in a string it is indicating a large number of sub-modules connected in series. Accordingly, multiple peaks in the power–voltage characteristics of a PV string become a real problem that needs to be addressed.

As PV modules/strings have nonlinear characteristics, power electronics converters (PECs) are usually connected across their terminals to maintain the operation at the maximum power point. To achieve that maximum power point tracking (MPPT) algorithm is implemented in the PECs control logic. Several algorithms have been proposed for that purpose such as perturb and observe [11], incremental conductance [12], and hill climbing [13] where these algorithms operate by searching for the peak point. Accordingly, when the PV system becomes with multiple peaks the MPPT might become trapped in a local minimum, which reduces the system yield and effectiveness.

In the desert environment soiling can be classified as soft or hard one [14]. Soft soiling is uniform across the module and it affects only the module current. Hard soiling on the other hand is an accumulation of dust over part of some cells resulting in the multiple peak effect discussed before. This indicates the importance of taking the impact of partial shading on the desert PV installation to maximize its effectiveness.

The selection of the installed PV inverter types, and their configurations are very critical to deal with partial shading issues. In the following section, these considerations are analyzed and discussed in detail.

8.3.2   *Performance of PV inverters in desert environments*

As stated before, PV modules, strings, and arrays produce DC current and they have a nonlinear voltage–current relationship. To achieve the highest yield from a PV system, its operation should be maintained at the maximum power point (MPP). The MPP voltage and current vary with the atmospheric condition and time of the day and it is also affected by the impact of partial shading. PECs are then used to interface the PV modules with the rest of the power system which could be a utility grid, microgrid, or isolated loads. The inverter plays two important roles in this context. First, it tracks the MPP and maintains the operation at that point to maximize the PV system power yield. Second, it converts the DC voltage/current at the PV modules output into voltage/current form that matches the output side characteristics such as standard voltage and frequency of the utility grid or the voltage level needed by the isolated loads. Therefore, PECs are an essential part of PV system operation.

The PECs constitute 8–12% of the PV system's total cost over its lifetime. As the number of PV modules interfaced with the same PEC increases, its percentage cost in the system decreases. However, other considerations besides the cost determine the type and configuration of PECs to be used in the system. It is always important to differentiate between three types of installations which are: large-scale, commercial, and residential PV systems.

Large-scale PV installations have capacities at the Megawatts level, and they are installed in flat areas. The name large-scale is used in this text to include both grids connected utility-scale as well as off-grid installations. As large-scale PV installations take place over large areas of flat surfaces, the impacts of shading and mismatches in modules production are very limited if not eliminated. This provides opportunities for grouping more modules into the same PECs which helps to reduce the system cost. For this reason, large-scale PV installations use a type of inverter called a *central inverter* where large PV arrays are attached to the same inverter to reduce the overall system cost.

In commercial installations, large buildings such as offices and buildings for business PV installations of a few kilowatts to a few Megawatts can be found. In such systems, more flexibility in sizing PVs system is needed to match the available space. Moreover, not all PV modules have the same orientation as they are installed on the buildings' rooftops. As the partial shading is expected to occur in these types of installation, the "*string inverters*" are thus used with 10–30 PV modules, which are connected per unit inverter and thus, per watt. The cost associated to this type of installation is usually higher than that of the large-scale PV systems.

In the residential installation, the system size is limited to a few kilowatts. As the rooftops of residential buildings can have several surfaces with different orientations, mismatches in the module's power production are highly expected. Furthermore, the impact of partial shading is present in most of these types of installations. The safety concerns in residential buildings also play role in deciding the best PEC technology to be used. e.g., regulations have been set to limit the distance of the PV DC side cable such that fast DC voltage discharge is secured in response to system shutdown due to fire hazards [15]. As off-grid PV installation increases, more requirements of this kind could be imposed to protect the ordinary people around the system. In this type of installation, each PV module is connected to a dedicated inverter called a *microinverter*. Using microinverters enables harnessing the highest amount of power, but on the other hand, it has a higher per cost than other types of installations.

8.4   Loads and energy consumption in Qatar

Agricultural farms form a major power consumer in a rural desert environment. Being away from utility grids and usually scattered geographically, desert-based farms are mostly powered as off-grid systems [16]. Many countries in the Middle East and North Africa have a significant portion of their land as desert areas within which agricultural farms can usually be found, e.g., there are more than 1,300 farms in Qatar and most of them are off-grid based farms [1]. Moreover, the farms in desert areas are expected to increase in the near future to meet the expanding need for food supply as the population worldwide keeps on increasing. Besides that, the energy demand in these farms increases over time

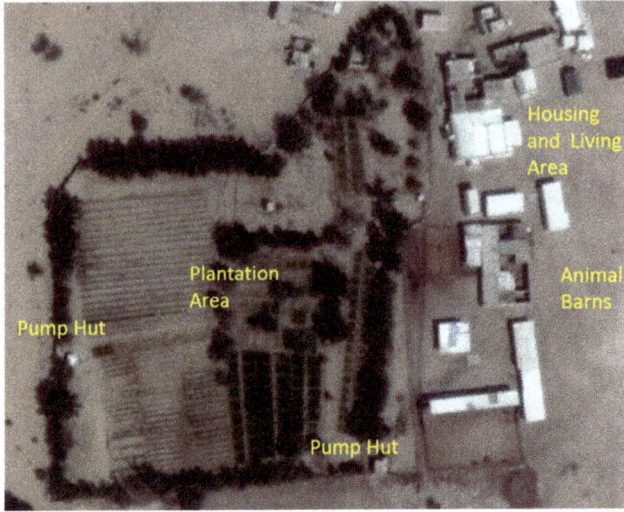

Figure 8.14   An aerial view of a typical desert farm

as farmers are moving more toward greenhouses which extends the farming season, allows the introduction of new types and plants, and increases power consumption accordingly. For this isolated farm, the electricity is supplied by a diesel generator. Some of these farms are already connected to grid electricity through an electrical network connection.

The energy consumption in the remote farm is mainly the water pump for pumping underground water to be used for irrigation and domestic use. The energy is also needed for lighting and other house appliances for the house or building that co-exist with the farm. The typical aerial view of the farm in the desert is as in Figure 8.14 shown. Just like a typical farm, electrical energy is needed to power the farm. Just like any other farm, there is a need for water pumps, air conditioners, fans, and lighting. A good system's safety and reliability aspects are as important as they can result in the farm to be less tolerance to power interruptions.

8.4.1   Animal barns

Similar to agricultural farms, large numbers of animal barns can be found in desert areas with grid power supply. Up to 5,000 off-grid barns exist in a small country in Qatar. The animals that are kept in desert-based farms are very diverse. The electrical power needed in these barns differs depending on the type of animals that are hosted within. For example camels need the power to get a supply of water; on the other hand, horses need a properly cooled spaced using air conditioners. Having a continuous supply of regulated power is very critical in this type of load as the power interruption in such an environment might jeopardize some of the animal lives [16].

8.4.2   Water pumps

Desert areas usually suffer a lack of water supply and in most cases, underground water represents the main source of water supply. For those reasons, water pumps are among the most common load in these areas. Depending on what depth the water exists, different kinds of pumps and pumping technologies can be used. For shallow wells not deeper than 30 m, surface pumps can be used. These pumps are placed outside the well which indicates that their installation and maintenance are easy. On the other hand, deeper wells required submersible pumps that need to be installed submerged in the water increases the installation and maintenance cost. The energy requirement of the pump depends on the amount of daily water extraction, the depth of the well and pump efficiency.

Pumps can be driven by combustion engines which are based on burning fuel or by motors that require electrical energy. Different kinds of motors have been used to drive pumps DC motor or AC motor which includes induction, permeant magnet, and brushless DC motors. Solar PV based water pumps are very common in off-grid systems and the system complexity determines their effectiveness. In the simplest case, PV modules operate the DC motor directly without any power conversion. These systems have the lowest cost, but their performance could be very far from being optimal as the DC motor consumption could be far from the PV maximum power point. Moreover, DC requires frequent maintenance and replacement for the brushes. Another type of PV-based pump uses induction motors that are driven by PV connected to 50/60 Hz inverters without maximum power point tracking. A more efficient system uses AC motors with an inverter that uses variable frequencies to track the PV maximum power point all the time. This system has the highest efficiency and it can also be integrated with batteries to store any surplus energy generated by the PV to be used later when needed.

8.4.3   Water desalination

Besides the need for energy for water pumping, in some cases, energy is needed to improve the water quality and water desalination represents the major energy consumer in this context. Despite being reserved inland, underground water could suffer from salinity levels higher than what exists in seawater [17]. The salinity could be due to seawater trapped on the landscape or due to the dissolution of some minerals into underground fresh water. In either case, there is a need for water desalination especially if the water salinity level exceeds 500 mg/L [18]. In other applications, small islands and coastal areas do not have access to fresh and desalinating sea water becomes a convenient way to get sufficient water supply.

Water desalination systems consume a large amount of power and in most cases, they are established as large desalination plans which are grid-connected [19]. For the off-grid location in a desert environment, it is highly desirable to have small-scale desalination units that use renewable energy sources. Despite the fact these kinds of systems are very common currently, they are expected to be widespread in the near future [20]. To be run from renewable sources, energy storage for nighttime operations remains to be the major obstacle [21]. However, the dropping

Table 8.6   Typical energy consumption of different desalination techniques [19]

Desalination technique	Multistage flashing	Multi-effects	Vapor-compression	Reverse osmosis	Electro-dialysis
Total equivalent energy consumption (kWh/m^3)	13.5–25.5	6.5–11	7–12	3–7	2.6–5.5

cost of batteries is expected to accelerate the spread of small-scale off-grid water desalination systems.

Number techniques are used for water desalination and they differ in their need for energy, the most common among them are multistage flashing, multi-effects, vapor-compression, reverse osmosis, and electro-dialysis [22]. In multistage flashing, multi-effects and vapor-compression the use of thermal energy is needed to desalinate the water. In reverse osmosis, membrane is used to remove particles and ions from water and the energy is used to apply pressure that makes the water flow through the membrane against various natural pressures that cause the water to go in the opposite direction. Electro-dialysis uses electric potential difference within electrochemical cells to separate ions from the water, but despite its high efficiency, it can only remove low modular weight particles from water. Table 8.6 shows the energy need of the various water desalination techniques, however, capital and maintenance costs need also to be considered when a specific technique is to be selected for any application.

8.4.4   HVAC systems

The harsh weather condition in desert environments is a major challenge people usually face. Therefore, the need to acquire and operate systems to provide heat, ventilation, and air conditioning (HVAC) is essential in such environments.

Evaporative cooling systems have low costs and require relatively low electrical energy to operate. They operate by passing dry air through wet surfaces where the water absorbs heat from the passing air to evaporate causing its temperature to drop. These kinds of systems work perfectly in a dry climate in areas with sufficient water availability. However, in some desert environments, these conditions might not be met during the hot season and in these cases compressing cooling systems are usually deployed.

In compression cooling, compressors are used to compress certain gases called refrigerant till it condenses. The condensed refrigerant is then cooled by air or water at ambient temperature before releasing the pressure. When the pressure is released, the refrigerant temperature drops causing the required cooling effect. Compression cooling systems work effectively in all climates, but they consume more electrical energy than evaporative cooling systems. However, in the context of off-grid power systems, compression cooling systems have a distinct advantage. Unlike evaporative cooling, the generated cooling effect can be used to produce chilled water or ice that can be stored to be used later when needed. The ability to store chilled water or ice can have a great value to reduce the need for battery storage in off-grid systems that are supplied by renewable energy sources.

8.4.5 Additional loads

Besides the ones mentioned above, a number of loads are usually found in a desert environment such as lamps, fans, electronic devices, and home appliances as listed in Table 8.7. Unlike the ones described above, most of these loads do not have an energy storage capability and their operation depends on the user's need and some of which is needed during the night times. Therefore, battery storage is essential to operate these systems in off-grid systems that are supplied solely by renewable resources.

8.4.6 Urban energy consumption of the Qatar Desert

The climate in Qatar influences greatly the energy consumption which makes it unique from the rest of the world. As the weather is intensely hot in the summer, air conditioning is intensively used during this season. In the current days, most of the electricity consumption is for the air conditioner. The air-conditioner is run continuously during the summer season and still most of the time during the winter. In addition, in the winter, the water heater is mostly turned on due to colder weather. The consumption that was monitored during one of the days during July and November is shown in Figures 8.15 and 8.16. Figure 8.15 is depicting the consumption of a two-bedroom apartment. Figure 8.16 is depicting the consumption of four bedrooms villa.

Something need to be noticed is that during summer, there is not much variation on energy consumption throughout the day, especially for a four-bedroom villa. The level of energy consumed is almost flat. For a two-bedroom apartment, the consumption is slightly swelled during midday. This is the same for both winter and summer.

Table 8.7 Typical load in the farm and its energy consumption

Load type	Power consumption	Quantity	Hour per day	Total consumption	Comments
Pump	4 kW	2	2–6	16–48 kWh	Well pumping and irrigation
Lighting	60 W	30	6	10.8 kWh	For the farm
Air conditioners	2.4 kW (2 tons)	3	6	43.2 kWh	For animals
Fans	75 W	6	6	2.7 kWh	For animals
Other appliances	2 kW	–	0–24 kW	0–48 kWh	For human

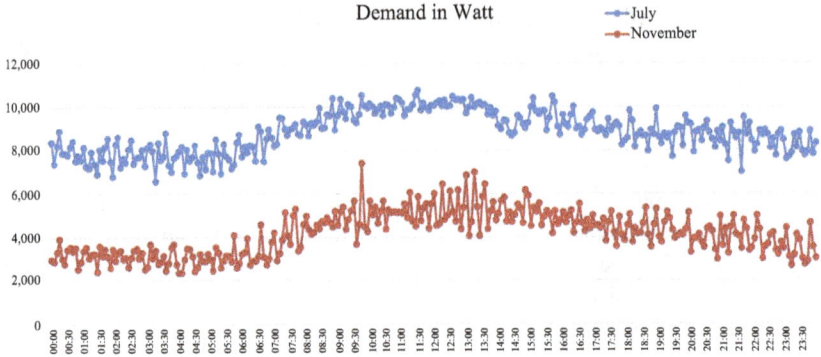

Figure 8.15   Load profile of typical day for two-bedroom apartment

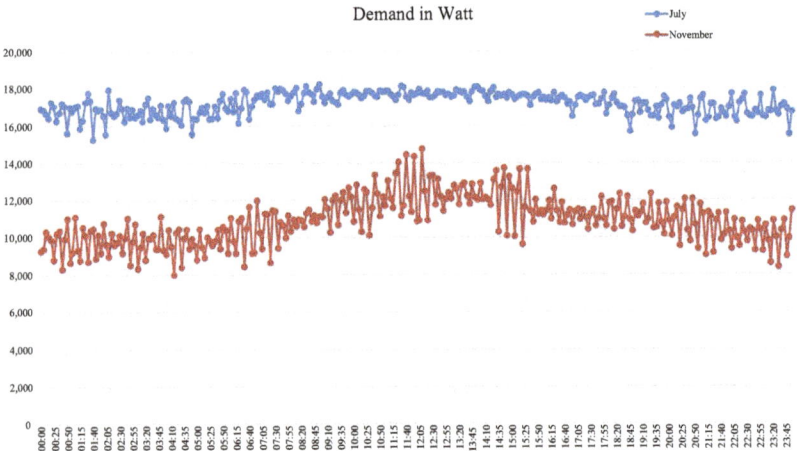

Figure 8.16   Load profile of a typical day for four-bedrooms villa

8.5   PV energy system design, control and operation

A photovoltaic (PV) system can exist in many different forms. It can be broadly classified into a few categories. For application in hot and arid areas such as in the desert, the physics of the PV system remains the same. However, special considerations need to be taken into account, including the device operating temperature, the high concentration of dust, the wind storm, and other meteorological factors. For desert application, a few types of PV system option deployment are available depending on the application requirements.

A stand-alone PV system is a system that supplies power independently of other power generation sources [15]. These include PV-powered water pumping systems, PV-powered lighting systems, and remote residential PV systems. Each of these can provide power to DC loads. By utilizing a power inverter in the system, each can also supply AC loads.

8.5.1 PV-powered water pumping system

Water pumping is one of the potential applications for a PV system in the desert. Typically, this system as depicted in Figure 8.17 comprise of a ground-mounted array, a pump controller, an inverter for AC pump motors, and the pump/motor assembly operating off either DC or AC. Water is normally pumped only during daylight hours and is usually stored in a water tank to cover periods of bad weather. Energy storage such as battery banks can be incorporated into this system as well to increase reliability.

8.5.2 PV-powered lighting systems

A PV-powered lighting system as depicted in Figure 8.18 is an option for providing area lighting and sign lighting in lieu of extending utility services. These systems are sold as packages including array, batteries, battery enclosure, change controller, lighting controller, light fixture, ballast, and lamp. The system is typically made with a small number of PV modules with a total power output not surpassing 250 W. The arrays are usually pole-mounted or mounted to the sign structure and should be equipped with vandal resistance hardware. High-pressure sodium, low-pressure sodium, and fluorescent fixtures are popular choices for these lights. Protection of the batteries from significant temperature variations is an important installation issue with these systems. Enclosure provisions for the batteries should moderate any temperature excursions to extend the lifetime and capacity of the batteries.

8.5.3 Remote residential PV system

PV systems can be used for powering complete remote residences and other small facilities where utility power is not available or desired. These systems typically utilize a roof or ground-mounted array, a battery charge controller, battery storage, and an inverter to supply electrical services. These systems may also come with an auxiliary source of power such as small wind generators and/or engine generators to meet electrical needs during a period of unavailability of the sun. This system can also be configured as a portable power generator, either skid or trailer mounted

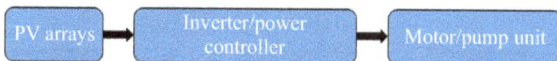

Figure 8.17   The PV-powered water pumping system

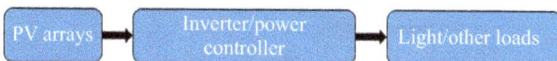

Figure 8.18   PV-powered lighting system

and is a complete package with integrated components. When the system is having more than one generation of sources, it is called a hybrid system. This kind of system is described further in the next section.

8.5.4   PV-hybrid system

A hybrid PV power generation system as shown in Figure 8.19 combines a renewable energy source with other forms of generation, usually a conventional generator powered by diesel or even another renewable form of energy like wind. Such a hybrid system serves to reduce the consumption of non-renewable fuel [23].

8.5.5   A system with energy storage

This kind of system as shown in Figure 8.20 is popular where backup power is required for standby loads such as a refrigerator, water pumps, lighting, and other necessities. Under the availability of the sun, the system supplies power to the loads and stores the excess power in the battery. When there is no sun or the radiation level is low, energy storage will be used to ensure enough power is supplied to the load.

8.5.6   Off-grid vs. grid connected system

Grid-connected system as shown in Figure 8.21 is designed to operate in parallel with the electric utility grid. The component in a grid-connected PV inverter or power-conditioning unit converts the DC power produced by AC power consistent with the voltage and power quality requirement of the utility grid and automatically

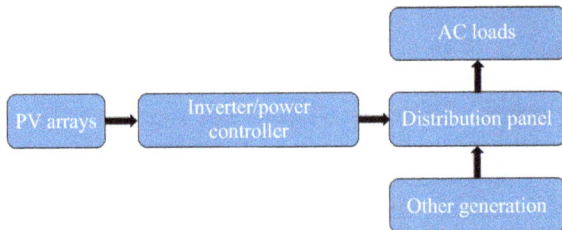

Figure 8.19   Off grid PV-hybrid system

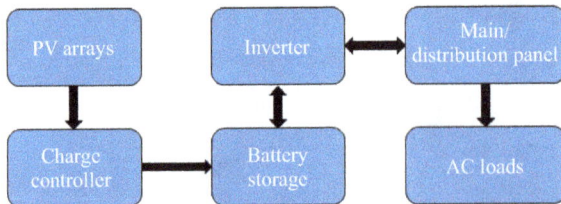

Figure 8.20   Off-grid PV-battery system

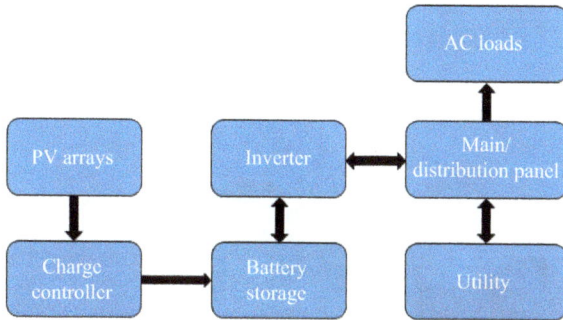

Figure 8.21   Grid connected PV-battery system

stops supplying power to the grid when the utility grid is not energized. A bidirectional interface is made between the PV system AC output circuit and the electric utility network, typically at the onsite distribution panel or service entrance. This allows the AC power produced by the PV system to either supply on-site electrical loads or to supply power to the grid when the PV system output is greater than the on-site demand. At night and during other periods when the electrical loads are greater than the PV system output, the balance of power required by the loads is received from the electric utility. When the utility grid is down, this system automatically shut down or disconnects from the grid. This safety feature is required in all grid-connected PV systems and ensures that the PV system will not continue to operate and feedback onto the utility grid when the grid is down for service or repair.

8.5.7   Energy management and control

PV power generation is directly related to the irradiance level of the sun. The load, however, depends mainly on human characteristics and behavior. Being a small and flat country, the climate condition is uniform throughout the region.

For the correlation between load and PV generation comparison, the case study for the farm is used. There is a 20 kW PV system installation at the farm. There is no measurement at the farm so the load profile from four bedrooms villa is used with additional consideration of the pump load. For simplicity of explanation, the PV irradiation profile from a day with a very high irradiation yield is used. The profile of the load and PV generation profile is plotted together as shown in Figure 8.22.

For this case scenario, it is assumed that the pump is in use for 7 h only starting from 7:00 am to 13:00 pm. In the traditional farm, all the load demand is supplied by a diesel generator. There is some farm that already has access to the electric utility supply and the diesel generator is no longer used but still exists in the farm in case there is a utility power interruption. For this case scenario, it is assumed the farm is not yet having a utility service connection and only using a diesel generator for its power need.

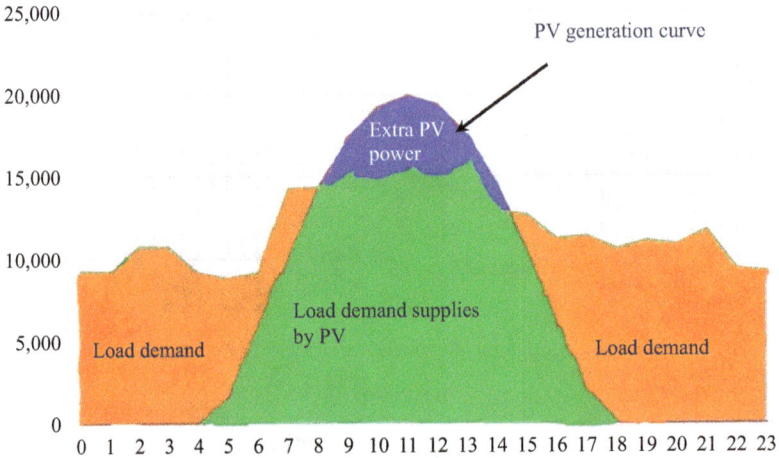

Figure 8.22   Load demand and potential PV generation

Before there is power generated by PV on days with an irradiance profile like this, the load demand needs to be supplied by another source of generation such as a diesel generator. Starting from 4:00 am, when the PV system starts generating power, the diesel generator can reduce its generation to give way for PV. Based on this profile, there is an excess generation from the PV for a few hours from 8:00 am to 15:00 pm. This excess power can be stored in energy storage if it is available.

8.6   Techno-economic benefits and case study

The techno-economic analysis is performed to find the optimized system to satisfy load demand and to find the most economical system [23]. If the system is grid-connected, then additional evaluation is on the optimized system for enhancing the grid reliability and amount of power fed to the grid. For performing a techno-economic analysis, a few important information are needed such as load demand, solar radiation, and component of the energy system. For example, if we are going to PV installation, one of the thinking is either to have it as a stand-alone, having PV-battery or having a hybrid PV-diesel generator-battery. In order to reach a decision on the final energy system, typically techno-economic analysis is performed.

When installing the PV system a few costs have to be considered. These include component costs, cost tradeoffs, labor costs, and other relevant costs [23,24].

- Component costs – Solar modules are the most expensive component of the system. In a system without batteries, inverters are the second most costly component. Batteries increase the complexity of the system and thus impact both material and labor costs.

- Cost tradeoffs – Consider cost tradeoffs when planning a PV system. For example, while less efficient modules may have a lower cost per installed watt than modules with higher efficiency, they also will take up more space and require a larger racking system. Mounting costs will be considerably greater and should be considered against the cost savings of the modules.
- Labor costs – The labor required to install a PV system varies depending on the installation type. For example, an experienced crew can install a 2 kW non-battery PV system in two-to-four-person days. Systems with large solar arrays require relatively less effort per watt of power than smaller systems. If batteries are included in the PV system, both material and installation costs will be greater. Batteries can add 50–100% to the time required for the installation.
- Other costs – Incentive programs can substantially reduce installed costs.

There are a number of software tools for designing a PV system available online. These software tools can be used to evaluate the technical and economic benefits of the system in planning. A few software that is available for free include:

- PVWatts^{TM} (http://www.nrel.gov/rredc/pvwatts/ or http://www.pvwatts.org/) is a free, Internet-based model that calculates electrical energy produced by a grid-connected PV system.
- RETScreen$^{®}$ (www.retscreen.net/ang/g_photo.php) is a free spreadsheet-based model for grid-connected and off-grid systems. Developed with the support of Natural Resources Canada.

Some other software that is available with a license fee includes:

- HOMER Pro (https://www.homerenergy.com/products/grid/index.html)
- HOMER Grid (https://users.homerenergy.com/homer-grid/download)
- PVsyst (http://www.pvsyst.com/en/)

To demonstrate the example of techno-economic analysis, the energy system of 30 kW PV and a diesel generator for the farm is compared. The farm is a real farm located in Rawdat Rashid about 30 km South West of Doha. The option is open whether to include battery energy storage or not in the system. The energy system architecture is as depicted in Figure 8.23. The diesel generator and PV system are connected to the AC Line. Here the PV system is considered as PV panels and its inverter system. The storage is a bank of 1 kWh lead–acid battery. As the battery produces and absorbs DC current, the converter is needed for interfacing it with the AC system.

The typical daily load profile that is assumed for the farm is depicted in Figure 8.24. July represents the typical daily load profile for the summer season. During this season, normally there is no agricultural activity. Most of the load is for energizing houses and building loads. Nov represents the load profile during the winter season. During the winter season, the farm is active in agriculture activities. It is assumed that the water pump is on for 7 h each day from 6:00 am to 1:00 pm.

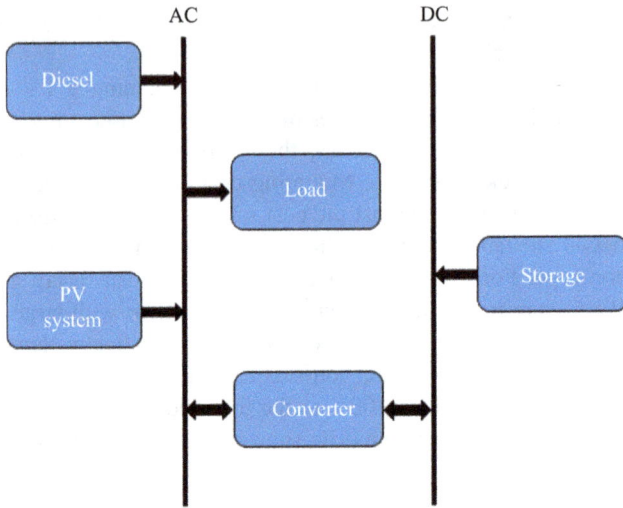

Figure 8.23   Commercial farm energy system architecture

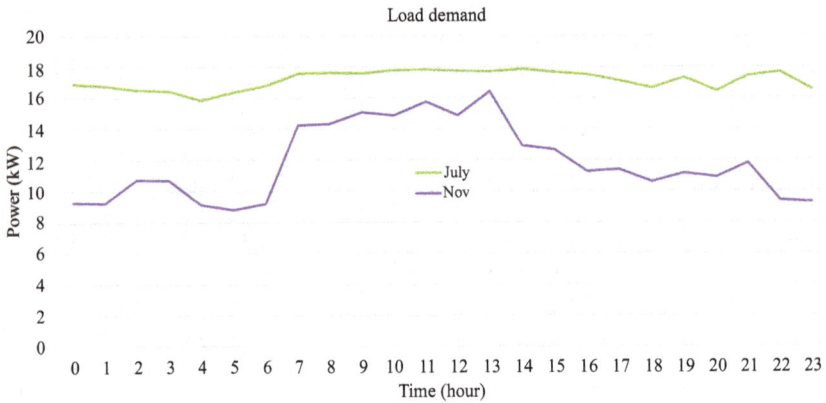

Figure 8.24   Daily load profile for the farm

It is assumed that the project life is 25 years, the discount rate is 6%, and the annual capacity shortage is 5%.

For the PV design, a load of a year is normally needed. Based on the daily load profile, the estimation of load demand for the entire year is calculated. Figure 8.25 shows the calculated average load demand for the entire year from January to December. The load demand is higher between March and September due to higher consumption.

The other information required is climate data. For this study, the NASA surface meteorology and Solar Energy data are used as the climate data. The monthly

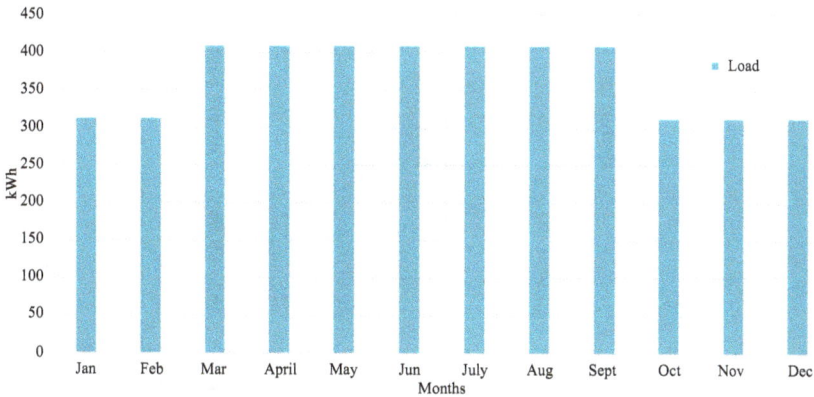

Figure 8.25   Average monthly load demand for the farm

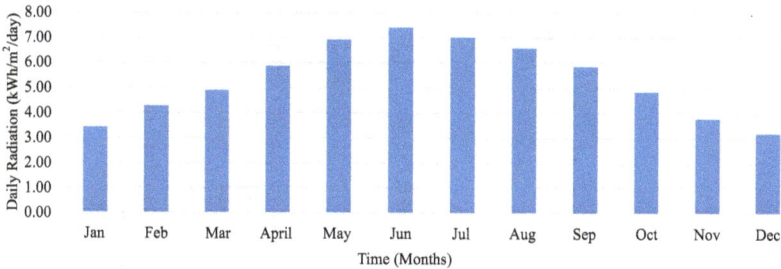

Figure 8.26   The monthly average solar global horizontal irradiance

average solar GHI is as shown in Figure 8.26. In this analysis, the solar panels are assumed fixed mounted with a slope degree of 25.15° with zero Azimuth. The effect of temperature is not considered in this case.

The PV capital cost is assumed USD 3,000 per kW with operation and maintenance cost is USD 10 per year. The diesel generator's initial capital cost is assumed USD 500 per kW with an O&M cost is USD 0.03 per operation hour. The lifetime hours are assumed 15,000. The capital cost of a 1 kWh battery is assumed USD 300 per unit with an O&M of USD 10 per year. The lifetime of the battery is assumed 10 years. The capital cost of the converter unit for the battery is assumed USD 300 per kW with the same amount of replacement cost after 15 years. The efficiency of the converter is set at 95%.

The techno-economic analysis is performed using HOMER Pro Software. The energy system cost is as tabulated in Table 8.8. The analysis result shows that the capital cost for the system with PV, diesel generator and the battery will cost more than the system with only PV and diesel generator. However, if compared to the Net Present Cost, the system with a battery is more attractive from an economic point of view. There are many assumptions made in performing this analysis as

Table 8.8 Techno-economic analysis of PV-diesel-battery

PV (kW)	Gen (kW)	1 kWh LA (Units)	Converter (kW)	Dispatch (cycle charging/ load following)	NPC ($)	COE ($)	Operating cost ($/yr)	Initial capital ($)	Remarks
0	**34**	–	–	CC	**580,152.00**	**0.370**	**44,054.00**	**17,000.00**	Base case
0	34	1	5	CC	582,771.00	0.372	44,118.00	18,800.00	
0	34	3	0.25	CC	582,077.00	0.371	44,128.00	17,973.00	
0	34	5	5	LF	585,150.00	0.373	44,210.00	20,000.00	
5	34	–	–	CC	582,928.00	0.372	43,097.00	32,000.00	
30	34	–	20	CC	634,124.80	0.400	41,235.25	107,000.00	
30	34	70	20	CC	551,172.00	0.352	32,634.00	134,000.00	
30	34	81	20	CC	544,046.30	0.350	31,815.70	137,334.90	
35	**34**	**82**	**20**	**CC**	**542,745.00**	**0.346**	**30,520.00**	**152,600.00**	**Lowest cost** system
112	0	628	55	CC	832,033.00	0.549	22,696.00	541,900.00	
125	0	566	39	CC	842,251.00	0.556	22,423.00	555,609.00	

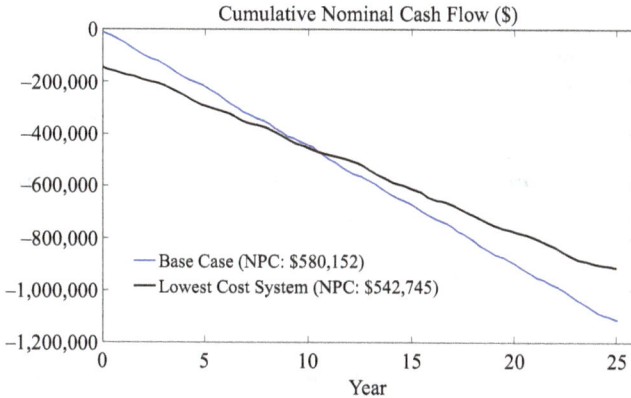

Figure 8.27 NPV between the base case and lowest cost system

much real data is not available. With more accurate data, the analysis results will be more accurate.

Based on the calculation, IRR is calculated at 8.9%, ROI is calculated at 5.9%, and the simple payback year is calculated at 10.4 years. The chart that shows how the economical best system is projected to save money is depicted in Figure 8.27. The base system here is the energy system that only has a 34 kW diesel generator. The lowest cost system is the hybrid system that has a 34 kW diesel generator, 25 kW PV, 82 kWh battery storage. It is also interesting to found out that the energy system is not economically attractive if the conventional diesel generator is removed.

8.7 Conclusions

This chapter has presented different aspect on desert PV applications in the state of Qatar. The climate of the desert was described first including the availability of solar resources and the performance of PV in the desert. The electricity consumption in the desert farm describing the type of typical load available was also discussed. The chapter also discussed the various options of system design, control and operation for the PV system. Finally, with the different PV energy system options applicable, the techno-economic analysis for the optimal energy system for the farm is presented and discussed. The techno-economic analysis prevails the hybrid energy system that coupled PV-battery-diesel generator is the best option economically.

8.8 Term definitions

Grid-connected PV systems: An electrical power generating system that uses a PV array as the primary source of electricity generation and is intended to operate synchronously and in parallel with the electric utility network. Such systems may also include battery storage, other generating sources, and may operate on-site loads independent of the utility network during outages.

Stand-alone PV system: A solar PV system that supplies power independently of other generation sources.

PV hybrid systems: Hybrid power systems that combine solar power from a PV system with another power generating energy source.

Net present cost (NPC): A life cycle cost. It is the present value of all the costs of installing and operating the component over the project lifetime, minus the present value of all the revenues that it earns over the project lifetime.

Levelized cost of energy (COE): The average cost per kWh of useful electrical energy produced by the system.

Operating cost: The operating cost is the annualized value of all costs and revenues other than initial capital costs.

Initial capital cost: The initial capital cost of a component is the total installed cost of that component at the beginning of the project.

Annualized cost of a component: The annual cost that occur evenly in each year of the project lifespan, would provide the same net present cost as the real cash flow sequence associated with that component.

Return on investment (ROI): The annual cost savings in relative to the initial capital cost.

Discount rate: The interest rate use to determine the present value of future cash flows.

Annual capacity shortage: A shortfall that occurs between the required operating capacity and the actual amount of operating capacity the system can provide.

Project lifetime (years): The length of time over which the costs of the system occur.

Internal rate of return (IRR): A degree of an investment's rate of return. The term internal denotes the computation ignores external factors, such as the risk-free rate, inflation, the cost of capital, or various financial risks.

Discount factor: A fraction used to compute the present value of a cash flow that occurs in any year of the project lifespan.

Salvage value: The value residual in a component of the energy system at the expiration of the project lifespan.

Dispatch-cycle charging: A dispatch approach whereby whenever a generator is required to serve the main load, it runs at full capacity. Excess electrical production goes toward the lower-priority demand, for example, serving the deferrable load or charging the energy storage.

Dispatch-load following: A dispatch approach whereby whenever a generator running, it generates only sufficient power to feed the primary load. Lower-priority aims such as charging the energy storage or feeding the deferrable load are left to other energy sources.

References

[1]   A. Kannan, *Global Environmental Governance and Desertification: A Study of Gulf Cooperation Countries*, Concept Publishing Company, New Delhi, 2012 (ISBN-13:978-81-8069-848-4)

[2]   World Travel Guide. *Qatar Weather, Climate and Geography*. https://www.worldtravelguide.net/guides/middle-east/qatar/weather-cl imate-geography/

[3]   N. Thomas and S. Nigam, Twentieth-century climate change over Africa: seasonal hydroclimate trends and Sahara desert expansion. *Journal of Climate* 31, 3349–3370 (2018).

[4]   NASA Surface meteorology and Solar Energy. http://eosweb.larc.nasa.gov/sse/.

[5]   M.Z.C. Wanik, M.M. Bukshaisha, and S.R. Chaudhry. PV generation in distribution network and its impact on power transformer on-load tap changer operation. In *2017 IEEE PES Manchester Powertech* (2017).

[6]   B. Paridaa, S. Iniyanb, and R. Goicc. A review of solar photovoltaic technologies. *Renewable and Sustainable Energy Reviews* 15, 1625–1636 (2011).

[7]   FSEC STANDARD, Procedures for Photovoltaic System Design Review and Approval, FSEC Standard 203-10, Florida Solar Energy Center, January 2010.

[8]   Washington State University Extension Energy Program, Solar Electric System Design, Operation and Installation, October 2009.

[9]   S. Dezso and Y. Baghzouz. On the impact of partial shading on PV output power. In *WSEAS/IASME International Conference on Renewable Energy Sources*, 2008.

[10]   V. Quaschning and R. Hanitsch. Numerical simulation of photovoltaic generators with shaded cells. *Simulation* 2(4), 6 (1995).

[11]   F. Nicola, P. Giovanni, and S. Giovanni. Optimization of perturb and observe maximum power point tracking method. *IEEE Transactions on Power Electronics* 20(4), 963–73 (2005).

[12]   F. Liu Shanxu, D. Fei, L. Bangyin, and L.Y. Kang. Variable step size INC MPPT method for PV systems. *IEEE Transactions on Industrial Electronics* 55(7), 2622–2628 (2008).

[13]   X. Weidong and W.G. Dunford. A modified adaptive hill climbing MPPT method for photovoltaic power systems. In *IEEE Annual Power Electronics Specialists Conference*, 35, 1957–1963 (2004).

[14]   M.M. Reza, H. Hashim, G. Chandima, *et al.* Power loss due to soiling on solar panel: a review. *Renewable and Sustainable Energy Reviews* 59, 1307–1316 (2016).

[15]   B. Bill and S.K. White. Section 690.12 rapid shutdown. *Photovoltaic Systems and the National Electric Code*. 58(73), 58–73 (2018). ROUTLEDGE in association with GSEResearch, 2018.

[16]   T. Ryan. Electricity Supply Issues for Farmers. (2013).

[17]   V. Weert, F. Jac Van der Gun, and J. Reckman. Global overview of saline groundwater occurrence and genesis (Report number: GP 2009-1).

International Groundwater Resources Assessment Centre, 105 (2009). Available at: 2009_Global_Overview_Saline_Groundwater.pdf

[18]   T. Subramani, L. Elango, and S.R. Damodarasamy. Groundwater quality and its suitability for drinking and agricultural use in Chithar River Basin, Tamil Nadu, India. *Environmental Geology* 47(8), 1099–1110 (2005).

[19]   A. Al-Karaghouli and L.L. Kazmerski. Energy consumption and water production cost of conventional and renewable-energy-powered desalination processes. *Renewable and Sustainable Energy Reviews* 24, 343–356 (2013).

[20]   N.G. Prakash, M.H. Sharqawy, E.K. Summers, *et al.* The potential of solar-driven humidification–dehumidification desalination for small-scale decentralized water production. *Renewable and Sustainable Energy Reviews* 14(4), 1187–120 (2010).

[21]   A. Al-Karaghouli, D. Renne, and L.L. Kazmerski. Solar and wind opportunities for water desalination in the Arab regions. *Renewable and Sustainable Energy Reviews* 13(9) 2397–2407 (2009).

[22]   A. Alkaisi, R. Mossad, and A. Sharifian-Barforoush. A review of the water desalination systems integrated with renewable energy. *Energy Procedia* 110, 268–274 (2017).

[23]   T. Nacer, A. Hamidat, and O. Nadjemi, Techno-economic impacts analysis of a hybrid grid connected energy system applied for a cattle farm, *Energy Procedia* 75, 963–968 (2015).

[24]   K. Sami and C. Dahl, The economics of hybrid power systems for sustainable desert agriculture in Egypt. *Energy* 30(8), 1271–1281 (2015).

Chapter 9

PV systems in Australia: market evolution and performance in desert applications

Jose Bilbao[1], Sharon Young[1] and Bram Hoex[1]

9.1 Introduction

The evolution of the PV market in Australia has been the result of a blend of international efforts to reduce PV module prices and local policy that was designed to build and maintain the PV industry in Australia, until it was able to become self-sustaining. Section 9.1 will focus on the details of the evolving PV market in Australia, while Section 9.2 will delve more deeply into some of the policies utilized in Australia in the last 25 years and the lessons learned. Due to Australia's long-standing participation in the IEA Photovoltaic Power Systems Program (PVPS) [1], further details of the uptake of PV in Australia are readily available as curated by the Australian Photovoltaic Institute [2].

9.2 Evolution of the PV market in Australia

The evolution of the PV market in Australia can be broken down into a series of stages that are in part dictated by the cost of PV systems at the time stated. PV in Australia was initially dominated by off-grid applications, mainly in remote locations, where it was uneconomical to supply electricity to the required loads by expanding the existing grid. Close to the start of the millennium, there was a push towards the development of a grid-connected PV industry, but the high price of PV at the time necessitated the use of subsidies to maintain growth. Initially, these subsidies took the form of direct rebates but progressed over time to feed-in tariffs and certificates. In the late 2000s, the decreasing price of PV accelerated the approach of the stage where grid-connected PV systems reached parity with residential electricity prices. The ongoing cost decline continued to widen the market for PV from residential to commercial and finally utility-scale installations. In recent years, the cost of PV has continued to decline, such that the cost of utility-scale PV is now below the average wholesale market prices in Australia, creating a new wave of large-scale PV investments.

---

[1]School of Photovoltaics and Renewable Energy Engineering, University of New South Wales, Australia

9.2.1   International context

While this work focuses on the Australian context, it is worth considering the broader international context at several stages within the lifetime of the Australian PV industry. In the late 1990s, there were a few PV manufacturers globally, and the total production capacity was comparatively small – a total of just over 300 MW by the end of 1999 [3]. In the early 2000s, a large scale manufacturing capacity began to come online in overseas markets, particularly in Europe and Japan, leading to almost 3,000 MW of manufacturing capacity by the end of 2006 [4]. With the later opening of manufacturers in the US (First Solar) and particularly China (Suntech, Jinko Solar, Trina Solar, etc.), the global PV production capacity increased by more than 500% between 2006 and 2010 [5]. This rapid expansion in manufacturing capacity provided substantial economies of scale, driving costs down. Initially, the price of polysilicon peaked resulting in an increasing PV price when the PV market became the main user of silicon overtaking from the semiconductor industry. Between 2007 and 2010, PV dedicated polysilicon production came online which significantly reduced the price of polysilicon, placing further downward pressure on module prices [6]. These externalities were contributing factors to the declining cost of PV modules globally, as shown in Figure 9.1, which had knock-on effects in Australia. This also influenced the policy mechanisms that were likely to be most effective in the Australian context at a given time.

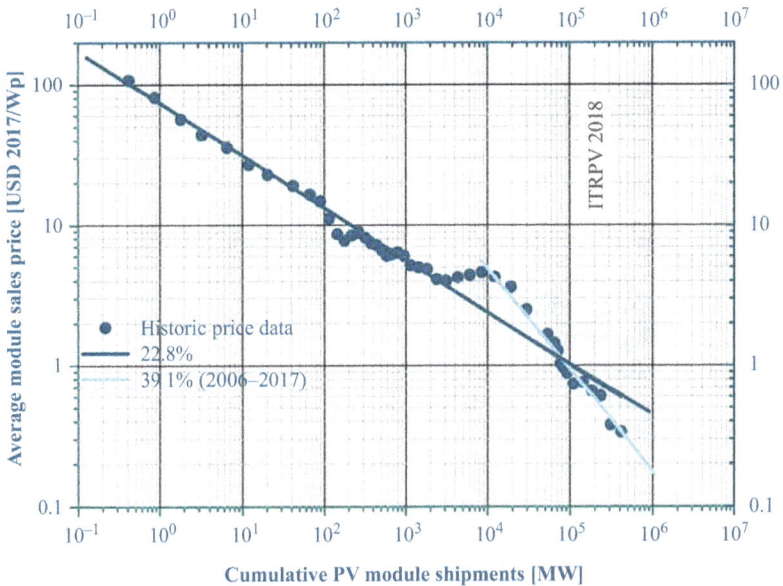

Figure 9.1   *Learning curve for module price as a function of cumulative shipments. Source: International Technology Roadmap for Photovoltaic, Ninth Edition, September 2018.*

9.2.2   *Off-grid uses (pre-2000)*

Prior to the year 2000, the PV market in Australia was dominated by off-grid installations. The extremely high cost of modules (Figure 9.3) meant that they were rarely deployed in grid-connected locations. However, the unique sprawling geography of Australia, coupled with its low population density outside of urban areas [7] made it a place where the self-sufficiency provided by off-grid PV was highly prized. Early market uses of PV were dominated by the telecommunications technology, DC water pumping, and lighting applications, in remote or inaccessible areas. This is evident in the high proportion of off-grid, non-domestic uses, which comprised over 80% of all installations at this time. In these early times, the high cost of PV modules (over $8/Wp) was such that there was a solid market for second-hand modules at the end of their initial use. The market was comparatively small, and Australia was exporting most of the PV modules it manufactured [8].

9.2.3   *Development of a grid-connected industry*

The early 2000s saw the fruition of several programs designed to encourage more grid-connected PV installations and develop the emerging Australian PV industry. Demonstration projects such as the Olympic Village and the Olympic park assisted in raising the profile of solar power in an urban context (Figure 9.2), which contributed to the success of other actions. There were a large scale of consumer engagement models, such as GreenPower [9] where consumers could pay a small premium on their electricity bill to be certain a percentage of their electricity came from renewable sources; major demonstration projects like solar cities, which encouraged solar power in conjunction with energy efficiency and smart metering; and a substantial rebate programme to homes that installed solar power.

In 2000, less than 5% of all PV power installed was grid-connected and the volumes installed were quite small, considering the current context. In the full year, 3.89 MW of PV was installed, an increase of 15% on the previous year, driven by several of the projects mentioned above. Of this, 80% was installed in off-grid locations, but it was the beginning of a turning point towards grid-based installations. Growth in installations continued at approximately 15% p.a. for more than 5 years, largely as a result of supporting policies, as typical module prices at this time remained relatively constant at $8/Wp (see Figure 9.3), while total system costs were approximately $12/Wp, down from $14/Wp in 2001.

9.2.4   *The dominance of grid-connected PV*

In 2007, the amount of grid-connected PV installed equaled the off-grid installations for the first time in Australia. This was a turning point in PV installations as it also marked a rapid acceleration in the rate of PV installations from this point on. The year 2007 saw an increase of approximately 20% on the previous year's installations, but 2008 installations were 80% higher than those in 2007; the majority of it (69%) were grid connected [10]. The Solar Homes and Communities Plan (SHCP) was one of the major drivers of this. In 2006, the SHCP (initially called the PV Rebate Program) offered rebates at a rate of $4/Wp, close to half of the module cost of a system of up

Figure 9.2   Sydney Olympic Village PV. Source: IEA-PVPS-Task 10 Community-Scale PV: real examples of PV-based housing and public developments.

to 1 kW, not including the balance of system (BOS) costs [11]. In 2007, this rebate was doubled to $8/Wp, capped at $8,000, substantially reducing the total cost to consumers. Coupled with the beginning of the decline in PV prices, it raised the attractiveness of grid-connected PV to consumers substantially, which was reflected in the total number of installations, as shown in Figure 9.4.

While the average price of PV in 2007 remained close to the $8/Wp price it had sustained for the previous decade, the increased global manufacturing capacity started to create a spread in module prices. In 2007, the difference between the average module price and the best available price was $1/Wp. In both 2008 and 2009, the difference between the best and average price was approximately $3/Wp,

Average PV Module Price in Australia $/Wp

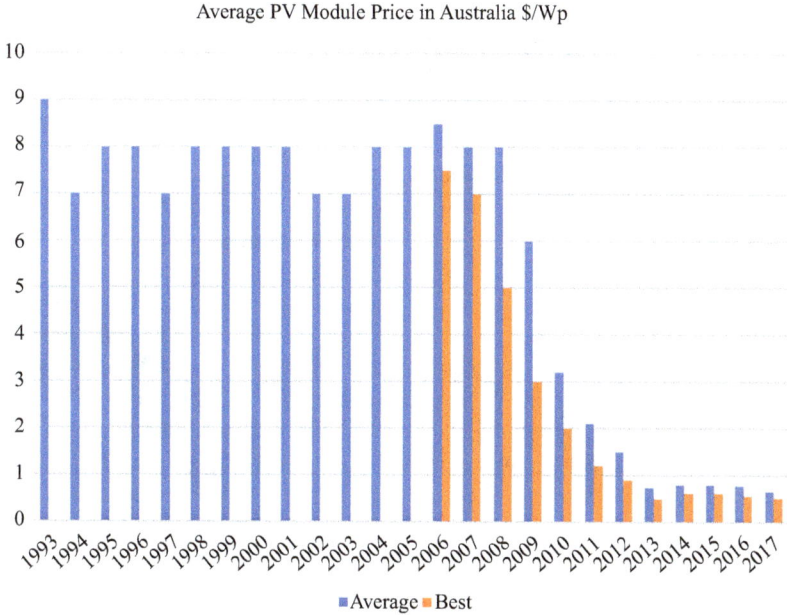

Figure 9.3 Average module prices in Australia over the last two and a half decades

Annual Australian PV installations MW

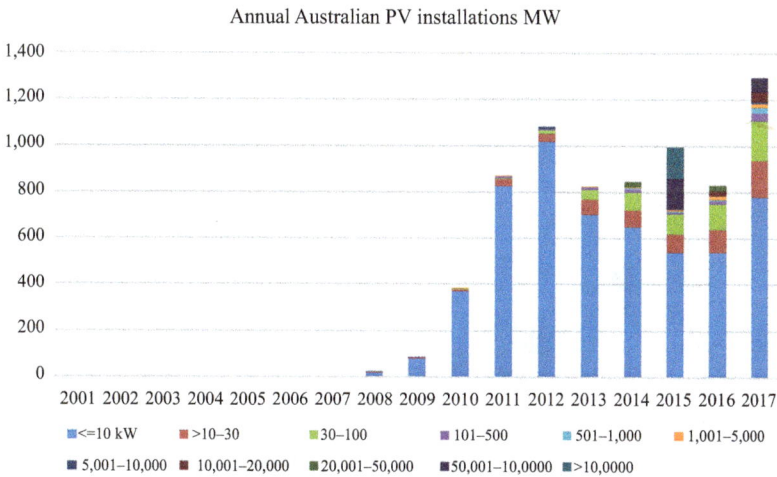

Figure 9.4 PV installations in Australia since 2001 by system size

reflecting an increasingly competitive market. As module price continued to decline in price, the difference between the average price and the best also reduced but is still in the order of approximately 20%.

9.2.5   Grid parity

In Australia, grid parity has progressed in a series of stages which have been reflected in the installation patterns. Grid parity is defined here as when the levelized cost of electricity from a PV system is equal to the retail price of electricity. In the first instance, grid parity was achieved with residential electricity prices, followed by commercial electricity rates, and finally wholesale market prices. In some regions, grid parity can also refer to the wholesale electricity price. To distinguish between the retail electricity price and the wholesale market price, the phrase "socket parity" is used to refer to retail electricity prices. Between 2007 and 2011, average PV module prices dropped from $8/Wp to $2.10/Wp, with the best available prices as low as $1.2/Wp. This rate of decline was significantly faster than predicted. Hence, socket parity, which in 2010 was expected to be reached in Australia between 2015 and 2020, was achieved in some Australian jurisdictions in 2011, with the remainder of jurisdictions following in 2012, several years ahead of expectation.

By 2013, rising electricity prices had caused commercial electricity rates to approach grid parity with PV prices. Coupled with the upfront discount provided by the small-scale renewable energy scheme (SRES) to systems under 100 kW, systems for commercial premises became economically attractive for businesses. This became evident in the increased uptake of PV systems in the medium size ranges (10–30 kW and 30–100 kW), a market group that is continuing to expand even as the residential market remains stable. This boom in the so-called "rooftop solar" area has driven the Australian solar energy market until 2017, after which "utility-scale" solar makes a main entrance in the Australian Market (Figure 9.5).

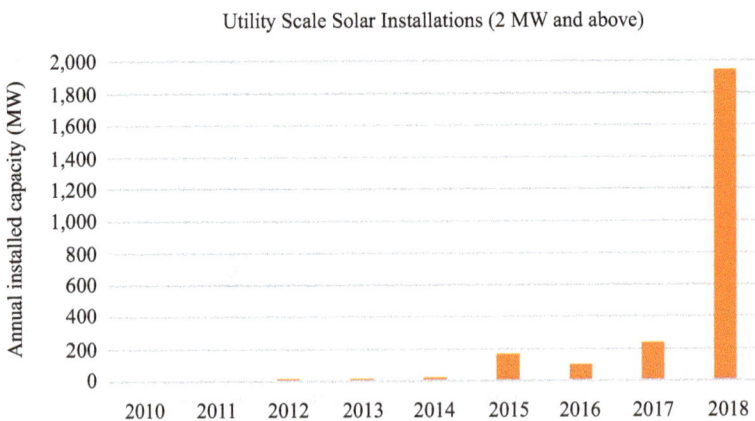

Figure 9.5   Utility scale solar installations

9.2.6   The new investors

Grid parity is not just a function of the module or system price, it is also a function of the electricity price. From 2011 onwards, retail electricity prices in Eastern Australia rose sharply as a consequence of factors including rising network costs, fuel costs, wholesale market prices, and retail costs. A lack of definitive federal energy policy since 2013 has also contributed to an investment deficit in generation capacity, reducing generation reserves and raising wholesale market prices. This became an economic risk for large energy users in Australia who were exposed to these wholesale market prices. As a result of the above, PV systems achieved grid parity with wholesale market prices sooner than expected and became a way to firm wholesale prices to large customers through Power Purchase Agreements (PPAs), driving an unprecedented growth in the market after 2017 (Figure 9.5).

The modular nature of PV generation makes it readily scalable to any given capacity. As such, PV has far lower barriers to entry or use than almost all other forms of electricity generation. With sufficiently low financing costs, the LCOE of PV is now below the average wholesale market price. Thus, while two of the first three MW scale solar farms in Australia were owned and operated by an electricity supplier, this is now the exception, rather than the rule. The majority of new solar farm installations in Australia have been partnerships with developers and large energy users. As an example, a large energy user such as Sun Metals aluminum refinery reduces their exposure to the wholesale market by directly contracting for the output of a solar farm [12]. There are many alternative ownership and operation strategies, but the main current investors in large scale solar are the energy users, not electricity retailers of old.

9.3   Policies and their effects on the market: lessons learned

The policies employed within Australia reflect the evolution of the solar industry over 20 years, from an expensive technology only economically viable in niche opportunities, to one of the cheapest available forms of electricity generation. PV in Australia has transitioned from a fringe technology requiring extensive support, to a mainstream form of generation requiring little or no subsidy to be competitive with other technologies.

9.3.1   Small-scale programs

Prior to 2000s, the Australian solar industry was largely dominated by research and development, with a small amount of manufacturing. In the late 1990s, funding began to shift away from R&D and toward further industry development and commercialization. There was at the time a level of strategic focus on developing the PV industry, as evidenced in the Australian Renewable Energy Action Agenda, and the Australian Renewable Energy Commercialization Program, both of which were released in 2000.

Out of these came subsidy programs such as the Photovoltaic Rebate Program, which was a long-running rebate program for residential consumers. Initially open to both off-grid and grid-connected installations, the program decreased its support for off-grid projects over time to better develop the grid-connected PV market. This sustained funding helped to develop the solar industry in Australia, opening the market up, creating local capacity and reducing installation costs. High profile installations like that at the Olympic Park increased awareness of the technology in urban areas.

This created an excellent platform for further subsidy projects like the Solar Schools program but also created an opportunity for demonstration projects like the Solar Cities program.* It is difficult at times to conceive of how small the PV industry at the time was, despite the rapid growth it was seeing. Ambitious projects like Solar Cities tested PV on a wide scale and helped to make it even more appealing to ordinary households.

9.3.2   The Renewable Energy Target: a national initiative

The Renewable Energy Target (RET) has been Australia's longest standing ongoing renewable energy policy, designed to encourage investment in renewable energy technology at both large and small scales. It is a policy that has undergone many revisions over time, but the cornerstone has always been to achieve a nominated amount of electricity generation delivered by renewable energy resources. The proportion of generation to be sourced from renewable energy sources has fluctuated over time, but the concept of a nationwide target has remained.

Originally known as the Mandatory Renewable Energy Target (MRET), the RET began in 2001, with an initial target set at 2% of total electricity demand in Australia [13]. Electricity retailers were required to source an equivalent of 2% of their electricity sales from renewable energy sources. The renewable energy was supplied through renewable energy certificates (RECs), where one REC is equivalent to one MWh of generation by a renewable energy source. Wholesale purchasers of electricity such as electricity retailers were obligated to purchase and surrender these certificates to meet their renewable energy requirements [14–16]. In 2009, the target generation to be supplied by the RET was lifted from 2% of national demand, approximately 4,000 GWh–41,000 GWh. The aim of this was to encourage more large-scale renewable energy generation. At the time this occurred, PV was still non-competitive with other forms of renewable energy at a utility scale.

Eventually, in 2011, the RET was split into the SRES for systems below 100 kW [14–16] and the large-scale renewable energy target (LRET), for systems above that threshold [14–16]. This allowed household PV and solar hot water installations to be included in the overall target and for the RECs generated by these systems to count towards the greater total.

Certificates from large systems (over 100 kW in size) were deemed for every metered MWh of electricity generated from a renewable energy source [14–16].

---

*http://www.townsvillesolarcity.com.au/portals/0/pdf/factsheet1-final.pdf

These certificates were then on-sold into the RECs clearing house, providing a second stream of income to large-scale renewable energy generators, beyond the revenue earned on the wholesale market. This second income stream provided a form of revenue certainty that assisted in obtaining the finance necessary for project development. The certificates for small systems, also known as small technology certificates (STCs), are provided up front to the owner of the system for its expected generation over 15 years, or from the year of installation until 2030 [14–16]. Frequently, the system owner would assign the right of their certificates to the system installer in exchange for an upfront discount (reducing the capital investment on the PV system). The installer would then sell the STCs in the clearing market to recover its costs.

9.3.3 Certificate multipliers and adaptability

At the time that small-scale renewable energy was included in the RET through the SRES scheme, an additional incentive was offered to households – the certificate multiplier. The STCs generated for a system up to 1.5 kW in size were multiplied by the solar multiplier to reduce the payback time on the investment for households [6]. The multiplier was initially set at five and was designed to be stepped down incrementally over time. This is an example of a subsidy program that was designed with an adaptation mechanism built in. The value of this was demonstrated when the rapid uptake of solar in 2010, due to multiple stimulus mechanisms and a plummeting module price, created a glut of RECs on the market, pulling down the market price of the certificates [17]. By reducing the multiplier, it was possible to rebalance the STC market and restore the economic value to the RECs. The amount of STCs offered to households has been designed to taper off as the RET draws to a close in 2030. The multiplier had clearly defined progression steps that were able to be accelerated when installations of solar panels rapidly exceeded expectations. Unfortunately, these two elements that enable the management of a graceful exit from a subsidy programme were lacking in the Australian experience of feed-in tariffs.

9.3.4 Feed in tariffs: boom and bust

In the late 2000s, the individual states within Australia began implementing feed-in tariffs (FITs) to improve the payback times of PV installations, with the aim of making them more economically attractive to households. Households would be paid a nominated amount per kWh of electricity generated (gross) or exported to the grid (net), and this payment would be guaranteed for a number of years. There are several lessons to be learned from the planning and execution of FITs as a mechanism for building the PV industry in Australia. The first lesson: an appropriate FIT must be well balanced between the retail price of electricity and the desired system payback time. In 2009, a gross feed-in tariff was implemented in New South Wales (NSW) that paid 60 c/kWh for all generation, a value which was substantially higher than the existing residential electricity tariff at the time, thus leading to the scheme being substantially oversubscribed [6]. A similar scheme in Queensland had the same outcome: due to the exceptionally high FIT price offered to consumers, concurrent with the sudden decline in module costs as a result of international factors, there was a boom in residential installations.

This then highlights the second lesson to be learned with respect to FITs, or indeed any form of subsidy – a means of gradually decreasing subsidies is essential to prevent excessive ongoing costs. This is something that has been done particularly well in the German context, where there are known, incremental changes in the state FIT that occur on a well-publicized and predictable basis [18]. In contrast, the over-subscription of the NSW FIT led to an abrupt 60% cut in the FIT offered, down to 20 c/kWh, before the scheme was closed in June 2011. The installation boom, that reached the heights of 16,000 installations in a single calendar month, crashed to a bust of barely 2,000 installations the following month as a result of the closure of the program [19]. This had a large impact on the PV industry, forcing many of the small PV installers to bankruptcy and a consolidation face with mergers and acquisitions by the larger players. Indeed, while the installed capacity per month has recovered to the levels seen in June 2011, the sheer volume of installations at that time has never come close to being matched. However, one benefit still remains: as a consequence of developing the capacity to install so many systems in such a short time, Australia had, and still has, the cheapest PV installation costs of any developed nation [1].

A third lesson in policy, particularly subsidies, that can be drawn from the matter of Australian FITs and other programs is that of the value in continually checking the appetite for a given policy – if it is too high, a scheme may be over-subscribed and exhaust its allocated budget well ahead of time. This would suggest that either the scheme needs to be redesigned to reduce the number of participants or that a different mechanism may be suitable. However, Australian subsidy schemes regularly underestimated the enthusiasm of participants towards solar. Examples include but are not limited to: the National Solar Schools Program received such a high level of applications in 2008–09 that the program had to be suspended in 2009 [20] the Solar Cities project in 2009, a scheme to support 100 rooftops going solar in 100 days was completed in 50 days; in the Australian Capital Territory (ACT), a generous feed-in tariff was opened to households on 12 July 2011 and was capped at 15 MW of capacity, but the allocation was exhausted by 13 July 2011 [6]. It is not easy to perfectly design any form of a subsidy scheme, but when considering future uptake, it is well worth looking at the past.

Further to this is perhaps necessary to be willing to adjust, close, or abandon a given policy if the circumstances surrounding it change. As previously discussed, Australia saw a sharp decline in PV prices between 2007 and 2011. Feed-in tariffs are a policy designed to reduce the payback time of a household PV system, but this would also be delivered by the decline in module prices. FITs started in 2008, such as those in South Australian and Queensland, were addressing high module prices, which made up approximately 60% of system costs when modules were $8/Wp and had been for over a decade. However, by the time the NSW FIT was introduced in January 2010, module prices had dropped by at least 25%, and the lowest cost modules were more than 60% cheaper than had been the case in the previous two years. It is worth questioning then if the NSW and West Australian FITs would have gone ahead had there been sufficient awareness of PV price trends at the time.

9.3.5   The cost of finance

Two Australian finance initiatives, the Solar Flagships program and the Clean Energy Finance Corporation (CEFC), had the same objective, support large-scale renewable energy projects, but had quite different levels of success. The Solar Flagships Program launched in late 2009 was specifically targeted at large scale solar projects [20]. Accepted projects would have to find part of their funding from conventional banking sources and would then be awarded federal government funding to go ahead. While there were projects that were accepted, they failed to secure additional funding outside of government and thus failed to proceed. While the low value of RECs as a result of the certificate glut at the time will have been a contributing factor, another matter was the lack of financial experience in investing in large-scale renewable energy projects. Such projects had never been built in Australia before, thus making them appear to be high-risk projects. In 2010, large-scale solar in Australia was simply not economically viable.

However, the CEFC did learn from this experience and the policy framework addressed the same issues differently. The critical element was that the CEFC effectively operated as a government-backed investment "bank," with a priority of investing in clean energy projects. As a government investor, it had different risk/return requirements, while investing with a public policy purpose. This therefore made it possible for large-scale renewable energy projects to receive low-cost financing. With the government taking the majority of the economic risk, further private sector finance could be obtained for large-scale projects, enabling them to be built. The CEFC is also highly unusual as a government policy – in its first year, it delivered a profit of 7%, providing a measurable benefit to taxpayers. The first solar farm funded by the CEFC was commissioned in 2015 [21]. As since that time PV prices have declined further and the industry has grown sufficiently comfortable with solar power, new solar farms no longer need this level of support, as shown by the large increase in installed solar farms in 2018 in Figure 9.5.

9.3.6   The importance of the political environment

A discussion on the lessons learned concerning Australian solar policy would be incomplete without addressing the political climate in which those policies are implemented and sustained. Australia is itself an excellent example of the importance of the political environment and its influence on the success or failure of policies. Within the space of a single decade, the rhetoric surrounding climate change, and hence the attitudes to associated matters like renewable energy technologies shifted from one extreme to another. In 2007, climate change was declared to be "the greatest moral challenge of our generation" [22] by the Australian Prime Minister. In contrast, the Australian Prime Minister elected in 2013 is on record as stating that "global warming may be a good thing" [23]; which is an unfortunate misinformed remark for a national leader.

The effects of this are evident in the policy actions that took place and their implications on the wider renewable energy industry. In a context where there was substantial federal support for renewable energy, this was reflected in both the large number of supporting programs and the sustained growth of the industry in the midst of the global financial crisis. The RET was increased and then further opened up to small-scale installations. Household subsidy schemes and feed-in tariffs were generous. A pricing mechanism on carbon emissions was developed and put into place. Then, as the industry continued to mature, further long-term support was added in the form of the CEFC and ARENA.

The transition to a government that was strongly anti-renewable energy had stark impacts on the renewable energy industry. This was most evident in the collapse of investment into large scale renewable energy by 88% [24] as a direct result of the actions of the federal government of the time. Further to this, the RET was substantially reduced, the carbon pricing mechanism removed, and efforts were made to eliminate or defund both the CEFC and ARENA. Such actions had detrimental effects on the renewable energy industry and the broader electricity industry, from which it has taken years to begin to recover, highlighting the importance of not just the policies, but also the social environment in which they operate.

However, the success of solar PV as a low-emission and now low-cost electricity generation technology has ensured its path as one of the main energy sources in the future. In Australia, PV is being embraced by utilities and energy companies like never before, with many of them stopping or closing coal-fired power plants, in spite of the current lack of long-term energy policy, particularly around renewable energies.

9.4 Australian climate: challenges and opportunities

Australia, the large island-continent which makes most of Oceania, is located in the Southern hemisphere and surrounded by the Pacific Ocean to the East, the Indian Ocean to the west, the Timor Sea to the north, and the Tasman Sea and Southern Ocean to the South. As expected for a landmass of 7.69 million kilometers (the sixth largest country on earth), it has a wide range of climates, from desert to tropical and from grasslands to temperate (see Figure 9.6).

Although the desert and grasslands cover most of the territory in the center of Australia, the majority of the population can be found in the coastal regions, where the major cities are located, particularly in the east (Sydney and Brisbane), south (Melbourne, Adelaide, and Hobart), and west (Perth) borders. However, if we analyze the average solar exposure presented in Figure 9.7, it is easy to conclude that most of the Australian population are located away from the best solar resources, situated in the center and north of the country. Nevertheless, the solar resource across Australia is one of the best in the world. To put things in context, Hobart, the southernmost capital city in Australia and hence with the lowest solar resource, has a solar irradiance about 23% higher than the sunniest cities in Germany (like Munich) and similar to Lyon, France. This is mainly due to the comparatively low latitudes (closeness to the equator) of most of Australia when compared to Europe, North America, and most of Asia.

On the other hand, although the center or Australia provides the best solar resource, accordingly it is expected to find the highest temperatures, as shown in

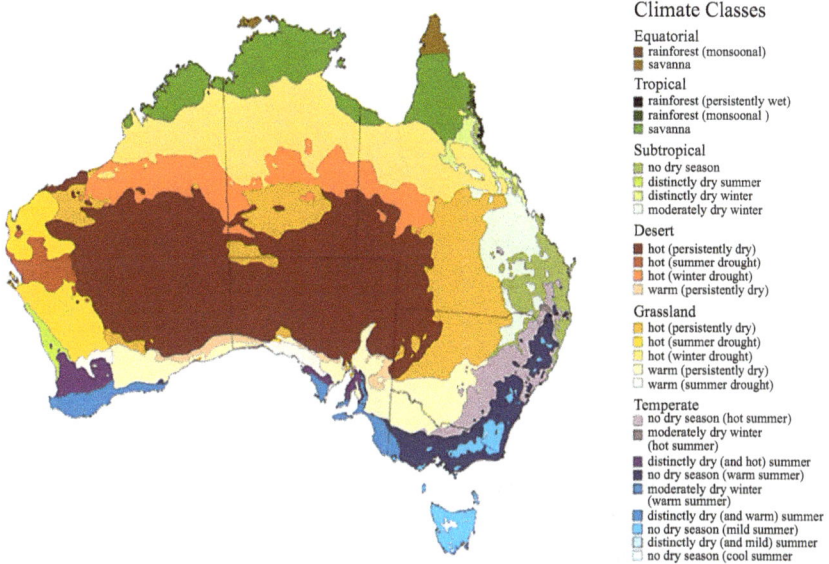

Climate Classes

Equatorial
- rainforest (monsoonal)
- savanna

Tropical
- rainforest (persistently wet)
- rainforest (monsoonal)
- savanna

Subtropical
- no dry season
- distinctly dry summer
- distinctly dry winter
- moderately dry winter

Desert
- hot (persistently dry)
- hot (summer drought)
- hot (winter drought)
- warm (persistently dry)

Grassland
- hot (persistently dry)
- hot (summer drought)
- hot (winter drought)
- warm (persistently dry)
- warm (summer drought)

Temperate
- no dry season (hot summer)
- moderately dry winter (hot summer)
- distinctly dry (and hot) summer
- no dry season (warm summer)
- moderately dry winter (warm summer)
- distinctly dry (and warm) summer
- no dry season (mild summer)
- distinctly dry (and mild) summer
- no dry season (cool summer)

Figure 9.6   Australia climate zones based on a modified Köppen classification system using a standard 30-year climatology dataset (1961–1990). Source: Australian Bureau of Meteorology (BOM).

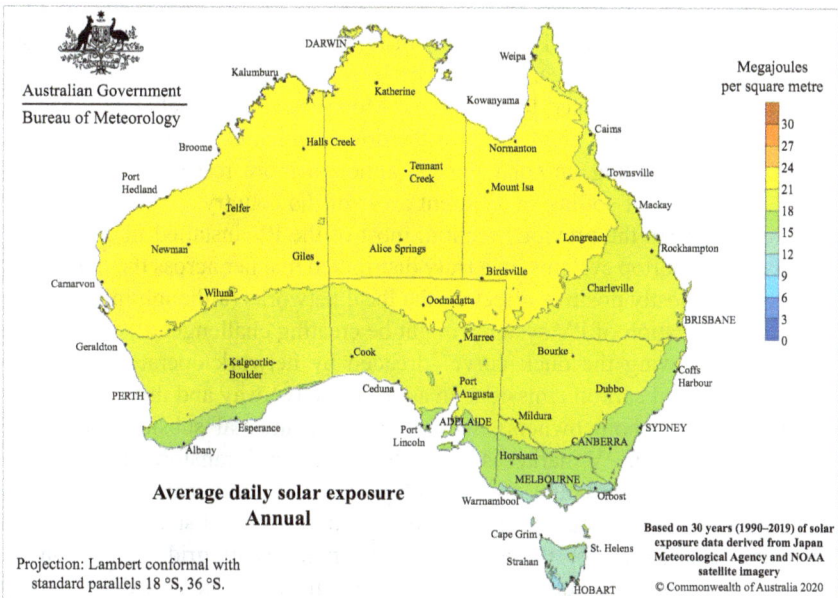

Figure 9.7   Average daily solar exposure in MJ/m². Source: Australian Bureau of Meteorology (BOM).

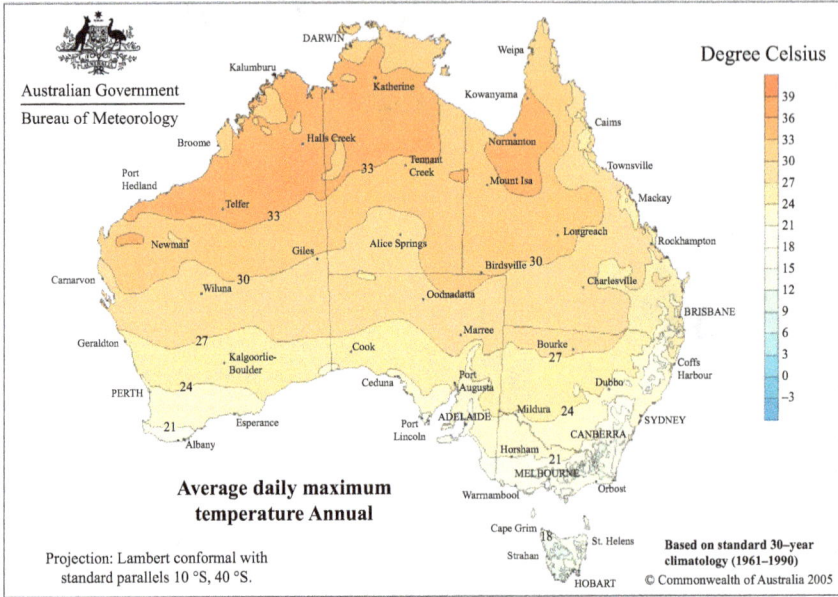

Figure 9.8   Average daily maximum temperature in degrees Celsius. Source: Australian Bureau of Meteorology (BOM).

Figure 9.8. This means that PV systems installed in these areas, although would produce more energy per area, will also be less efficient electricity generators, as they will experience elevated temperatures. Mounting to this challenge is the fact that the higher irradiance areas are also the driest, which can result and increased levels of dust depositing on solar panels, although more research is required to quantify the effects of soiling in different areas of the country.

As discussed in the previous section, most of the PV installed in Australia so far has been on rooftop systems, in a truly distributed manner across the main cities. This has had a mostly positive effect on the local networks (although in some areas the high concentration of PV systems might be creating challenges at some periods of the day, following the duck curve[†] dreaded by network operators), reducing energy demand and carbon emissions in a cost-effective way and driven by private investment (mostly homeowners), with some governmental subsidies. However, the current Australian PV market is going through a rapid transformation, and it is ready to install a large amount of PV in utility-scale systems that are far from the main population centers, but close to the areas with the highest solar resource. This means new challenges for these solar farms, in terms of grid connection and transmission to the load centers and the performance degradation due to higher

[†]https://www.nrel.gov/news/program/2018/10-years-duck-curve.html

temperatures, soiling, wildlife, and remote access, which will be discussed in the following sections.

9.5 PV systems across the Australian network

As discussed previously, Australia is a large country with low population density in its centre and high population density in its coastal regions. This creates a challenge for the transmission and distribution of electricity, due to the large distances between population and load centers, as can be observed in Figure 9.9. The National Energy Market (NEM) is the largest interconnected grid in Australia, with more than 40,000 km of transmission lines and 54 GW of generation capacity[‡] across the east and south coast of Australia (including Tasmania), but it does not

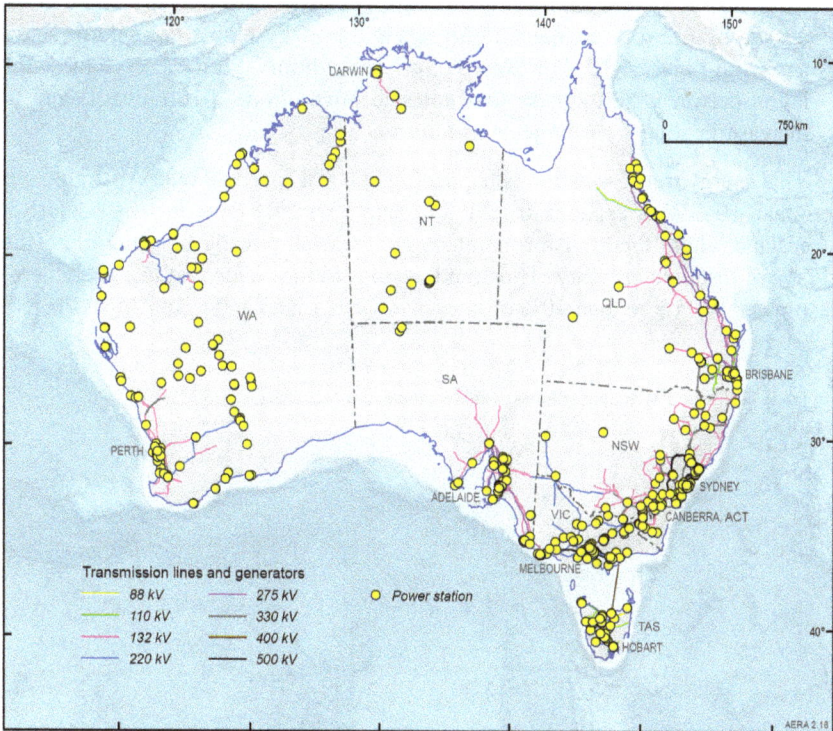

Figure 9.9   Australia's electricity infrastructure. Source: Australian Energy Resource Assessment, Geoscience Australia and BREE, 2014, 2nd Ed.

[‡]https://www.aemo.com.au/Electricity/National-Electricity-Market-NEM

include the state of Western Australia nor the Northern Territory, which have their own independent networks.

Furthermore, the NEM is now facing three additional challenges:

1. how to deal with high levels of penetration of distributed PV generation in urban areas (a challenge for the distribution network);
2. how to provide enough connection capacity with low losses to the new utility-scale PV systems (or solar farms) built in the middle of the country where the solar resource is better, away from the urban centres (an energy transmission challenge); and
3. how to control the grid as a whole, given the new generation and load profiles and the inherent intermittency of renewable energy sources (demand response and storage challenge). These challenges are occurring because of the rapid installation of PV systems in Australia, which can be explained by the early subsidies provided by the government, the drastic reduction in PV system costs, but also lately by the large increase in electricity prices, driven in the last decade by network augmentation costs and now by high wholesale prices, as shown in Figure 9.10. For example, it is straightforward to observe the correlation between the increase in wholesale prices from 2016 onwards and the increase in solar farm projects in Figure 9.5.

This energy transformation is happening at a fast speed all over Australia, with more than 10 GW of capacity installed to date (Oct 2018), but it is particularly in Queensland where this transformation is the greatest, already boasting more than 3.5 GW of installed capacity. This trend extends to the residential market, where it is estimated that more than 30% of houses in Queensland and South Australia (SA)

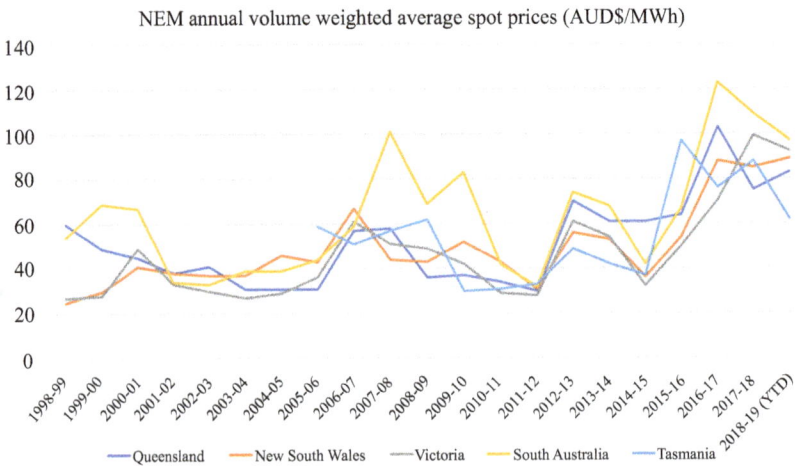

Figure 9.10   Historical weighted average spot prices. Source: Australia Energy Regulator (AER).

Installed PV capacity (MW) per state

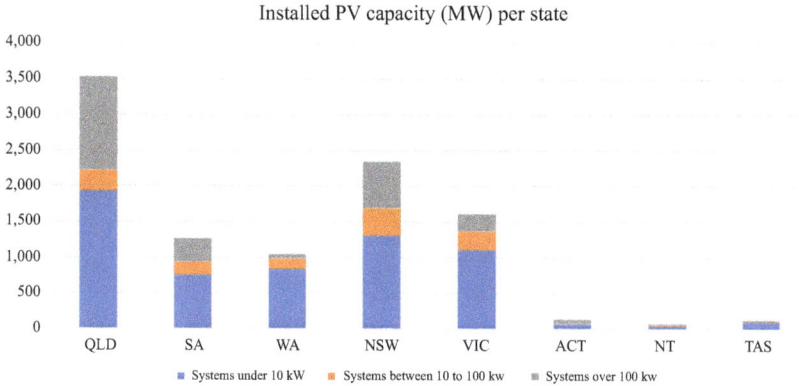

Figure 9.11   Installed PV capacity per state by system size. Source: APVI.

have a PV system installed, followed by Western Australia (WA) with 26% and New South Wales (NSW) with 18%. Figure 9.11 shows the Installed PV capacity per state by system size.

Furthermore, Figure 9.12 shows the (a) density (percentage of households with PV) and (b) capacity of PV systems per postcodes across Australia. As expected, it can be seen that the number of PV installations is greatest closer to the urban areas (coastal regions), with some exceptions in north-west New South Wales and in the centre of Western Australia, while the capacity is now greatest in the areas with higher solar irradiance (towards the centre of the country) due to the recent increase in the number of utility-scale projects, although, as shown in Figure 9.9, these projects are limited by the presence of transmission lines.

Finally, an intent to compare the performance of systems across Australia was made, using the available data, by calculating the specific yield (SY) of small-scale PV systems (<100 kW) for all the states, as shown in Table 9.1. Although the data is limited and aggregated to a high level (hence, potentially resulting in increased error margins), it is possible to see that the SY aligns with the latitude of the states and the solar map (Figure 9.7), as expected. The only exception is NSW, with an average SY lower than expected, which can be due to errors or omissions in the data, geographical location of the PV systems within the state, or to technical problems or failures of the PV systems installed in this state. This is only speculation, as there is no available data to support or discard these options. In any case, one could argue that even though higher temperatures reduce the efficiency of PV systems in warmer places, this reduction is more than overcome by the increased solar resource, given the high levels of clearness index (low cloud coverage of two to three oktas) in most of Australia[§] during the year.

[§]This refers to the possibility in some places of the world where there is a high average ambient temperature but reduced levels of solar irradiance due to overcasts conditions across the year.

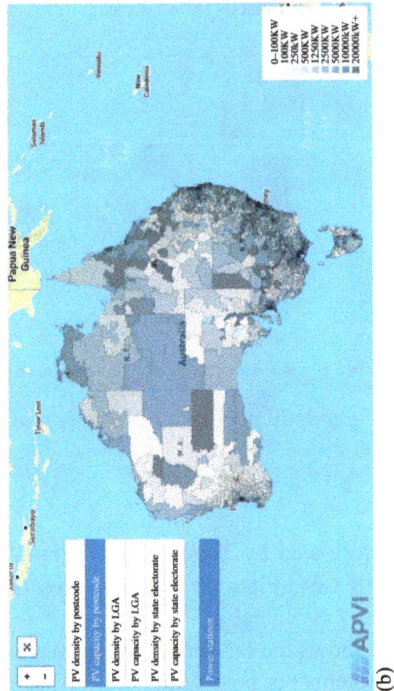

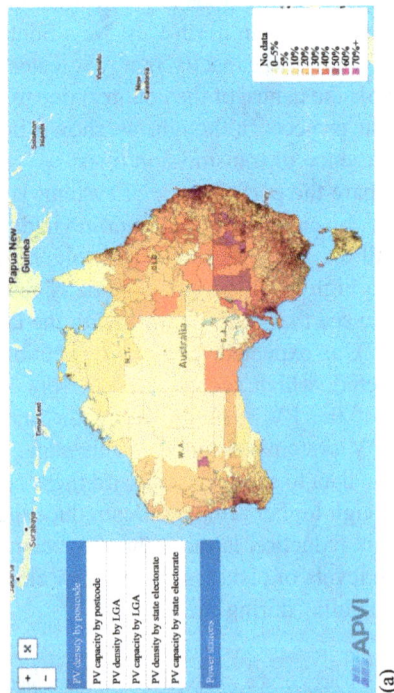

Figure 9.12 Australia PV installations per postcode based on (a) PV density and (b) PV capacity. Source: APVI.

Table 9.1   Average specific yield for small-scale PV systems (<100 kW) in 2018

State	Average specific yield (kWh/kWp)
QLD	1,282
SA	1,132
WA	1,231
NSW	1,095
VIC	1,110
NT	1,623
TAS	1,151

The next section will focus on systems installed near Alice Springs (Northern Territory) where high levels of solar recourses are available, as confirmed by the high specific yields obtained by the average systems in this territory in Table 9.1.

9.6   Case study: Desert Knowledge Australia Solar Centre

The Desert Knowledge Australia Solar Centre (DKASC) is a solar technology demonstration facility for commercialized PV modules located in Alice Springs, NT, latitude 23.762287 south and longitude 133.875205 east.

The DKASC was stablished in 2007 and commissioned in 2008, with the aim of showcasing PV systems installed in desert climate and sharing the data produced by these systems. Currently, the center has 38 PV systems (Figure 9.13) with a range of installation type (fixed, tracking, and roof mounted) and technology type (mono-Si, a-Si, Copper indium gallium selenide (CIGS), etc.).[||] The data provided in the site is freely available in the website dkasolarcentre.com.au and a great resource for anyone wanting to study the performance of PV systems in desert conditions.

The climate in Alice Springs can be categorised as a hot desert climate (BWh) in the standard Köppen climate classification and as a hot persistently dry grassland in the modified BOM classification. Nevertheless, it is clear that Alice Springs has an arid climate with low precipitations (an average of 280 mm per year) warm summers and mild winters, as shown in Figure 9.14. Alice Springs also has a very good solar resource, averaging a GHI of 6.1 kWh/m^2 per day, which is equivalent to 2,227 kWh/m^2 per year.

This short case study will be focused on the performance of fixed systems due to the wide range of systems installed in this way. All systems in fixed position were installed due north and with a tilt angle of 20°. A total of 25 systems were included in this study and a sample of the specifications for some of those systems

[||]http://dkasolarce<tab />ntre.com.au/locations/alice-springs/graphs

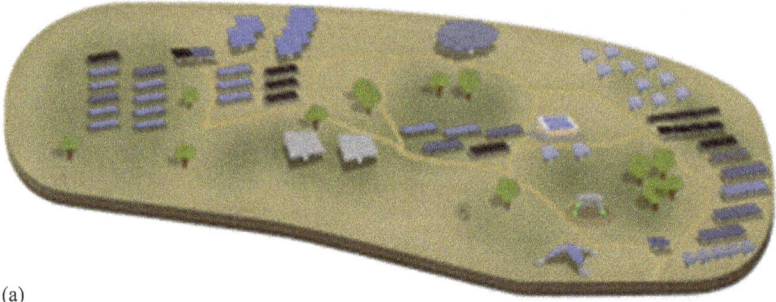

(a)

(b)

Figure 9.13   (a) Schematic of the DKASC demonstration facility. Source: http://
dkasolarcentre.com.au/locations/alice-springs and (b) aerial image
of the site. Source*: Google Earth.*

are included in Table 9.2. It can be seen that there is a good range of efficiencies and temperature coefficients among the systems, although most of the PV modules were installed several years ago, making them obsolete by current standards in terms of efficiency.

In order to study the performance of each system, two metrics were selected, Performance Ratio (PR) and the Specific Yield (SY). The PR metric is defined as the ratio between the yield of a PV system with the yield of an ideal system, with no PV or systems losses, but including the amount of irradiance at the plane of the

Alice Springs Temperature and Solar Profiles

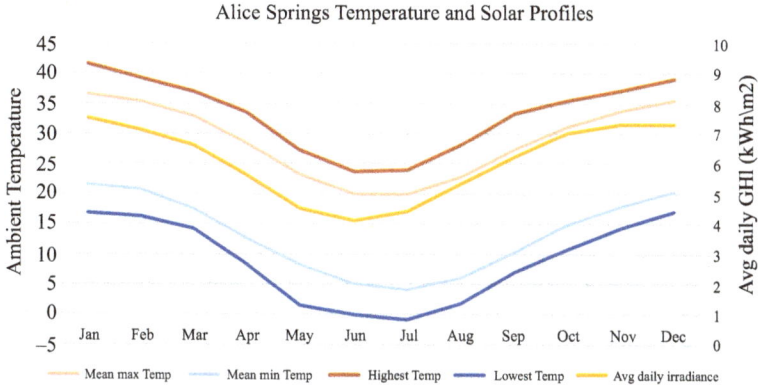

Figure 9.14 *Average monthly temperatures and daily irradiance for Alice Springs. Source: BOM.*

Table 9.2 *Sample of systems installed in DKASC*

Manufacturer	System size (kW)	Cell type	Module model	Module Power (W)	Eff	P_{coeff} (%/C)	Date
BP Solar	4.95	poly-Si	BP3165	165	13.11%	−0.50	2008
First Solar	6.96	CdTe	FS-272	72.5	10.07%	−0.25	2008
Kaneka	6.00	a-Si	G-EA060	60	6.31%	−0.23	2008
BP Solar	5.10	mono-Si	BP4170	170	13.51%	−0.50	2008
Trina	5.25	mono-Si	TSM-175DC01	175	13.70%	−0.45	2009
Sanyo	6.30	HIT	HIP-210NKHE5	210	16.70%	−0.30	2010
SunPower	5.24	mono-Si	SPR-238-WHT-D	238	19.10%	−0.38	2011
Evergreen Solar	4.92	poly-Si	ES-A-205	205	13.10%	−0.43	2010
Calyxo	5.40	CdTe	CX-50	50	6.94%	−0.25	2010
Hanergy	5.61	CIGS	SL1-85	85	11.17%	−0.38	2010

CdTe: Cadmium Telluride; HIT: Heterojunction with Intrinsic Thin Layer.

array (POAI). On the other hand, the SY is defined as the amount of energy produced over a year per one kW of PV cells installed (kWp or sometimes kWdc) and it is expressed in terms of kWh/kWp (this is one of the few metrics that combine AC energy to DC power). The SY and PR are related as the SY can be defined as the PR times the total POAI for the year.

Table 9.3   Average PR per cell type

Year	Average PR						Number of systems				
mono-Si	poly-Si	CdTe	a-Si	HIT	CIGS	mono-Si	poly-Si	CdTe	a-Si	HIT	CIGS
2009	0.770	0.765	0.803	0.823			3	5	1	1	0	0
2010	0.785	0.779	0.820	0.792	0.905	0.848	4	6	2	1	1	1
2011	0.773	0.763	0.791	0.766	0.869	0.852	4	6	2	1	1	1
2012	0.776	0.780	0.787	0.764	0.868	0.890	4	8	2	1	1	1
2013	0.768	0.782	0.792	0.754	0.829	0.862	5	11	3	1	1	1
2014	0.774	0.784	0.785	0.734	0.813	0.864	6	12	3	1	1	1
2015	0.761	0.776	0.771	0.719	0.809	0.837	6	12	3	1	1	1
2016	0.761	0.775	0.768	0.729	0.812	0.823	6	12	3	1	1	1
Avg.	**0.779**	**0.785**	**0.801**	**0.760**	**0.841**	**0.854**	**6**	**12**	**3**	**1**	**1**	**1**

The average PR per PV cell type is presented in Table 9.3 with the number of systems included in the calculation. It is important to notice that the number of systems increases over the year, which bumps up the average performance ratio as new systems tend to have better PR because of no degradation and better technology. A clear example of this effect can be seen for the poly-Si systems. As discussed above, the PR is an indicator of the losses in the system, which in this case are dominated by the temperature of the module and the annual degradation of the system. For example, although a-Si systems have a better temperature coefficient than its crystalline counterparts, its annual degradation reduces the PR considerably to levels below the other technologies. In general terms, however, the data shows that technologies with better temperature coefficient (CIGS, HIT, and CdTe) have better PR on average, which is expected. Nevertheless, most of the systems obtain reasonable performance ratios for their age and they do not seem to "suffer" in the extremes conditions of the desert climate beyond the expected temperature degradation. Although on average the numbers look within the expected ranges, there are certain systems that have deteriorated more than expected in all technologies, reaching two or three times the warranted degradation rate. Only one system (based on early CIGS technology) reached extreme levels of degradation to the point that the system was only producing 22% of its original output and had to be replaced in late 2016.

A similar picture can be observed in Table 9.4 summarizing the calculated averages specific yields. The technologies to stand out in terms of efficiency are HIT and CIGS.¶ All other technologies (with the exception perhaps of a-Si) seem on average to provide similar levels of degradation and performance over the years. This would seem to indicate that on average, the technologies with a lower temperature coefficient (besides HIT and CIGS) have degraded at a higher rate than

---

¶Notice that the CIGS system that presented extreme degradation was not included in the values presented in the tables.

non-HIT crystalline silicon PV systems but have managed to mitigate that degradation with the higher efficiency obtained for the systems with the lower temperature coefficient technologies, balancing the long-term performance between different cell types.

More detailed information regarding the degradation of different systems is presented in Table 9.5, listed in descending order within each technology (notice that each system has been anonymised). The values were calculated using linear regression for systems with more than 7 years of data. It can be seen that around 50% of the systems have higher degradation rates than expected, based on current industry standard warranty terms. On the other hand, it is positive to see that several systems have been able to keep annual degradation rates below 0.7% and in the

Table 9.4   Average SY per cell type

Year	Average SY (kWh/kWp)						Number of systems					
	mono-Si	poly-Si	CdTe	a-Si	HIT	CIGS	mono-Si	poly-Si	CdTe	a-Si	HIT	CIGS
2009	1924	1912	2005	2056			3	5	1	1	0	0
2010	1806	1793	1886	1822	2083	1950	4	6	2	1	1	1
2011	1759	1738	1802	1744	1978	1940	4	6	2	1	1	1
2012	1901	1911	1927	1871	2125	2179	4	8	2	1	1	1
2013	1867	1900	1926	1833	2015	2095	5	11	3	1	1	1
2014	1905	1928	1932	1806	2000	2126	6	12	3	1	1	1
2015	1912	1949	1935	1805	2033	2102	6	12	3	1	1	1
2016	1727	1760	1745	1656	1843	1869	6	12	3	1	1	1
Avg.	**1850**	**1861**	**1895**	**1824**	**2011**	**2037**	6	12	3	1	1	1

Table 9.5   Average annual degradation (%) calculated using linear regression for systems with at least 7 years of data

Cell technology	Avg annual degradation (%)	Cell technology	Avg annual degradation (%)
a-Si	2.02	CdTe	2.70
HIT	2.13	CdTe	1.89
mono-Si	1.48	poly-Si	1.66
mono-Si	1.33	poly-Si	1.01
mono-Si	0.51	poly-Si	0.98
mono-Si	0.20	poly-Si	0.52
CIGS[**]	11.72	poly-Si	0.47
CIGS	0.42		

[**]This is the CIGS system that was replaced in 2016 due to extreme degradation.

case of four systems circa or below 0.5%. This shows that the relatively high average degradation is not inherent of the cell technology but of the manufacturing and quality control processes of each system. The data has also shown that new systems (i.e. systems installed after 2010) tend to have lower annual degradation rates, which is a good sign and a result of the rapid and continuous improvement that manufacturing processes have undergone in the PV industry in the last 10 years or so.

From the data in Table 9.5, however, it would seem however that CdTe systems do not adhere to this conclusion, as its degradation is well above the standards. This is a case of not having enough data points, as demonstrated in Figure 9.15, which shows three CdTe systems, including one installation that has experienced very low degradation, but was not included in Table 9.5 as it was installed 4 years ago and hence it has too few data points. However, it can be seen that if the degradation trend continues, this system will be below the 0.5% annual degradation rate. It is also possible that the system with the medium degradation (orange color) stabilized its degradation rate in the last 2 or 3 years, after a high degradation rate following its installation. This behavior is expected in some thin film technologies, so a long-term dataset would be needed to draw firm conclusions.

To conclude this case study and the discussion on degradation, Figure 9.16 presents the average performance of systems based on standard crystalline silicon cells (poly-Si and mono-Si) – no average degradation rates for other technologies were calculated as there are not enough data samples. The figure shows that on average, the c-Si systems installed in the DKASC have an annual degradation rate of around 0.82% for mono-Si cells and 0.86% for poly-Si cells. Although this is higher than the standard 0.7%/year degradation rate, it is important to notice that

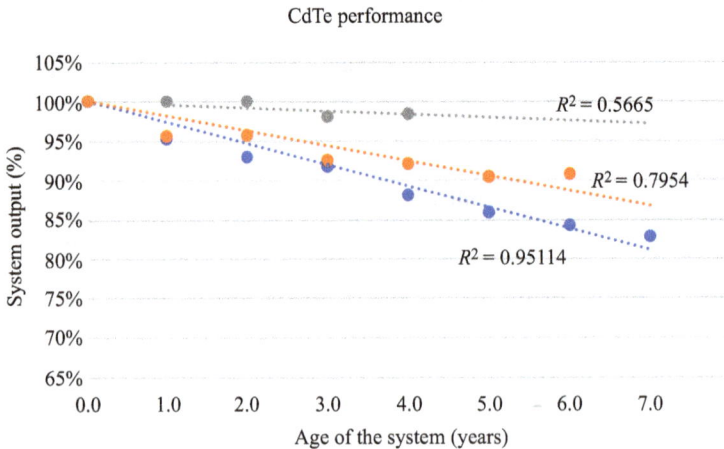

Figure 9.15   Performance of CdTe systems from the DKASC

Average crystalline Si Performance

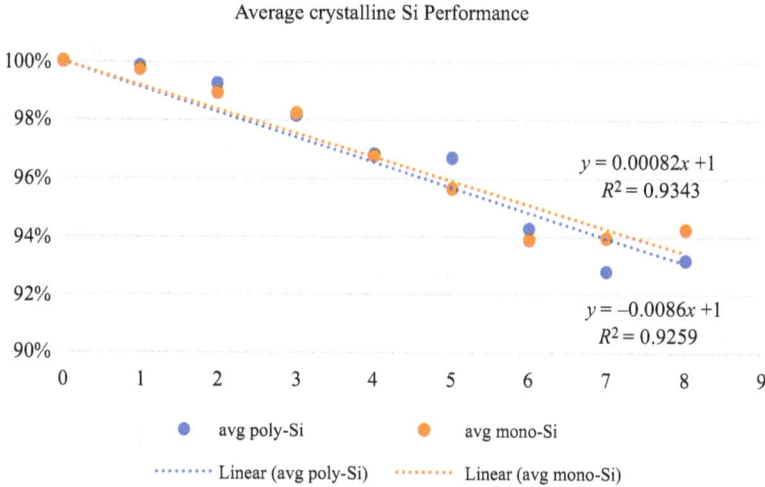

$y = 0.00082x + 1$
$R^2 = 0.9343$

$y = -0.0086x + 1$
$R^2 = 0.9259$

avg poly-Si   avg mono-Si

Linear (avg poly-Si)   Linear (avg mono-Si)

Figure 9.16   Average performance of crystalline silicon systems

the data used here is AC data, so it also includes the degradation of any other part of the system or malfunctions not detected during the data cleaning process. It is not clear if the higher degradation is due to the extreme weather conditions of the location or to other errors in the system or the data. For example, Copper *et al.* [25] showed that using irradiance data from a BOM station close to the site instead of the DKASC pyranometer, the degradation rates reduced importantly, with an average close to the 0.7%.

The conclusions that can certainly be drawn are two: (1) that good manufacturing and quality assurance processes, together with good design and diligent installation of PV systems can result in low degradation rates (<0.5%/year) even in extreme desert conditions like Alice Springs, for most of the PV cell technologies; and (b) it is clear that PV modules and systems of less quality might be affected severely by the thermal cycles of desert locations and the general rougher conditions, accelerating the degradation process of systems.

9.7   Conclusions

In this chapter, we have first explained the development of the Australian PV market with a focus on local policies and the global market. At the moment of writing, Australia has among the highest distributed PV penetration on the world, resulting in unique challenges for its electricity grid. In 2019, PV is the cheapest way to produce electricity in most parts of Australia resulting in a boom in PV deployment including large utility-scale installations. The best solar resource is available in Australia's interior with a desert climate. However, the unique design of the Australian electricity grid in combination with the fact that most people are living in highly concentrated areas at its coast, means that most PV will be deployed closer to the coastal regions. Field-testing in the interior has revealed that recent PV installations show a degradation rate which is in line with their warranty despite the harsh climate conditions including a lack of rain to keep the PV systems clean. Although more research is needed in this area, this is comforting news as Australia is intensively investigating options to use the immense solar resource in its interior for the production of solar fuels, such as hydrogen and ammonia, to fuel the next "renewable" phase of resource export economy.

References

[1]   International Energy Agency. (2018, January 20). Trends in Photovoltaic Applications. Retrieved from International Energy Agency: http://www.iea-pvps.org/index.php?id=trends

[2]   Australian Photovoltaic Institute. (2019, February). PV in Australia Reports 1997 to 2012. Retrieved from Australian Photovoltaic Institute: http://apvi.org.au/pv-in-australia-reports-1997-to-2012/

[3]   International Energy Agency. (2000). *Trends in Photovoltaic Applications 1992 to 1999*. Paris: IEA International Energy Agency.

[4]   International Energy Agency. (2007). *Trends in Photovoltaic Applications*. Paris: IEA International Energy Agency.

[5]   International Energy Agency. (2011). *Trends in Photovoltaic Applications: 1992 to 2010*. Paris: IEA International Energy Agency.

[6]   Australian Photovoltaic Association. (2012). *PV in Australia 2011*. Sydney: Australian Photovoltaic Association.

[7]   Australian Photovoltaic Institute. (2013). *PV Integration on Australian Distribution Networks*. Sydney: Australian Photovoltaic Institute.

[8]   Watt, M. (2002). *National Survey Report of PV Power Applications in Australia 2001*. Sydney: Interanational Energy Agency.

[9]   GreenPower. (2018, December). GreenPower . Retrieved from GreenPower: https://www.greenpower.gov.au/

[10]   Australian Photovoltaic Association. (2009). *National Survey Report of PV Power Applications in Australia 2008*. Sydney: APVA.

[11]   Australian Photovoltaic Association. (2007). *National Survey Report of Photovoltaic Power Applications in Australia 2006*. Sydney: Australian Photovoltaic Association.

[12]   Gifford, J. (2017, January 6). Australian zinc refinery to build 100MW solar plant. Retrieved from RenewEconomy: https://reneweconomy.com.au/australian-zinc-refinery-build-100mw-solar-plant-50245/

[13]   Australian Government. (2016, November 30). History of the Scheme. Retrieved from Renewable Energy Target: http://www.cleanenerg

[14]   Australian Government. (2018, May 31). About the Renewable Energy Target. Retrieved from Renewable Energy Target: http://www.cleane-nergyregulator.gov.au/RET/About-the-Renewable-Energy-Target

[15]   Australian Government. (2018, October 08). Large Scale Renewable Energy Target. Retrieved from Renewable Energy Target: http://www.cleane-nergyregulator.gov.au/RET/About-the-Renewable-Energy-Target/How-the-scheme-works/Large-scale-Renewable-Energy-Target

[16]   Australian Government. (2018, May 30). Small-scale Renewable Energy Scheme. Retrieved from Renewable Energy Target: http://www.cleane-nergyregulator.gov.au/RET/About-the-Renewable-Energy-Target/How-the-scheme-works/Small-scale-Renewable-Energy-Scheme

[17]   Johnston, W. (2012, June 26). Australia's Renewable Energy Target, Enhanced or Anhillated. Retrieved from SunWiz: http://www.sunwiz.com.au/index.php/2012-06-26-00-48-25/115-australias-renewable-energy-target-enhanced-or-annihilated.html

[18]   NREL. (2010). A Policymaker's Guide to Feed-in Tariff Policy Design. National Renewable Energy Laboratory.

[19]   APVI. (2019, February). PV Postcode Data. Retrieved from Australian Photovoltaic Institute: http://pv-map.apvi.org.au/postcodeyregulator.gov.au/RET/About-the-Renewable-Energy-Target/History-of-the-scheme

[20]   Australian Photovolatic Association. (2010). *National Survey Report of PV Power Applications* 2009. Sydney: Australian Photovoltaic Association.

[21]   Clean Energy Finance Corporation. (2017, April). Financing Large Scale Solar. Retrieved from Clean Energy Finanace Corporation: https://www.cefc.com.au/media/267820/large-scale-solar-presentation.pdf

[22]   Rudd, K. (2007, March). Climate Change: Forging a New Consensus. Retrieved from Parliament Information: https://parlinfo.aph.gov.au/parlInfo/download/media/pressrel/H4OM6/upload_binary/h4om62.pdf

[23]   Bickers, C. (2017, October 10). Tony Abbott Tells Climate Sceptics Forum Global Warming May Be Good. Retrieved from news.com.au: https://www.news.com.au/national/politics/tony-abbott-tells-climate-sceptics-forum-global-warming-may-be-good-and-climate-science-is-crap/news-story/dc42c5598f4c63e0e9689d6eacaf3b07

[24]   Climate Council. (2015, June 06). Investment in Renewable Energy in Australia Dropped 88% in 2014. Retrieved from Climate Council: https://www.climatecouncil.org.au/new-report-finds-investment-in-renewable-energy-in-australia-dropped-88-in-2014-1/

[25]   Copper JK, Jongjenkit K, and Bruce A. (2016). Calculation of PV system degradation rates in a hot dry climate. In *Asia Pacific Solar Research Conference*, Canberra, Australia.

Chapter 10

Conclusions, learned lessons and outlook into the future

Brahim Aïssa[1]

The solar power systems cover more than 3,000 km^2 worldwide, with PV modules accounting for a big majority. In 2016, the Paris Agreement acknowledged the need to limit the maximum global average temperature rise to 1.5°C. However, the current policies to limit global greenhouse gas emissions due to the burning of fossil fuels are still insufficient to maintain the temperature increase below 2°C. Therefore, the decarbonization of energy systems is currently the only option to achieve this target. Towards such a goal, we are now welcoming the new terawatt era of photovoltaic (PV) solar energy. In fact, the world's cumulative installed solar PV capacity grew by 22% to reach 940.0 GW by the end of 2021 representing 56% share of all renewable energies [1]. By May 2022, the installed capacity surpassed the milestone of 1TW, lead mainly by the deployment of large utility-scale solar plants, being several of them located in desert environments.

The use of photovoltaic technology in desert environments presents both opportunities and challenges. One of the main advantages of using PV in deserts is the high levels of solar radiation that are available, which can lead to higher PV energy generation compared to other locations. Additionally, the vast expanses of land available in desert regions can provide a space for large-scale PV installations. However, desert environments also pose several challenges for PV systems. The high temperatures, soiling, irradiance and high humidity in these areas can lead to increased degradation of PV panels, which can shorten their lifespan and reduce their efficiency. To mitigate these issues, PV systems in desert environments often require specialized design and maintenance practices, such as the use of anti-reflective coatings on PV panels and regular cleaning to remove dust and sand.

Another challenge for PV in desert environments is the potential for limited access to water for cleaning and maintenance. Water is an important resource for maintaining PV systems, as it is used to clean panels and cool inverters and other equipment. In desert regions where water is scarce, PV systems may need to rely on

[1]Qatar Environment and Energy Research Institute (QEERI), Hamad Bin Khalifa University (HBKU), Qatar Foundation, Qatar

alternative cleaning methods, such as the use of robotic dry cleaning or air blasters (also known as forced airflow).

Despite these challenges, the use of PV in desert environments has the potential to make a significant contribution to the global energy mix. With continued research and development, it is likely that PV technology will become even more efficient and cost-effective in desert environments, making it an increasingly attractive option for power generation.

Indeed, to cite a representative example, the state of Qatar, like other Middle East and North Africa (MENA) countries, has a large potential for the deployment of solar energy production due to the very high solar irradiations with the lowest LCOE, becoming an emerging PV market [2–4]. The average annual global horizontal solar irradiation goes up to 2,286 kWh m^{-2} with an associated annual PV electricity production potential of 2,100 kWh/kWp for c-Si modules [5]. Solar tender results provide testimony for the growing competitiveness of solar technology around the world, with new record-low solar tariffs registered in 2021 with the top four lowest bids (1.04–1.48 USD cents/kWh) in desert areas [1]. In 2008, the state of Qatar launched the Qatar National Vision 2030, setting a national target of 20% renewable energy by 2030 and a 10 GW target for solar energy, and aimed at increasing PV adoption as part of the global transition towards a net-zero carbon emission. Qatar has a clear commitment with the implementation of PV solar plants in the country as proven by the inauguration in October 2022 of one of the largest PV plants in the world, Al-Kharsaah PV power plant with 800 MWp, and accounting for 10% of the country demand. Two more projects starting in 2023 in Ras-Laffan and Mesaieed will account for 875 MWp, and will be finished by end of 2024.

10.1 Summary of the book

This book "Photovoltaics in Desert Environments" addresses the multiples challenges and opportunities of deploying solar energy systems in arid, desert regions, including the harsh and extreme weather conditions that can affect the performance and reliability of PV panels, as well as the difficulty of transporting and installing PV systems in remote locations. Despite these challenges, there are many benefits to be gained from implementing PV systems in desert environments. These benefits include the potential for large-scale energy production, as well as the economic and social benefits that can come from creating jobs and bringing electricity to remote areas.

Introducing the solar energy resources and harvesting technologies are detailed in Chapter 1. The sun releases on Earth a power of about 3.8×10^{26} W where only 1.7×10^{17} W are received, corresponding to an irradiance of 1,367 W/m^2. The desert regions known as the *Sun Belt* has an average daily insolation exceeding 6 kWh/m^2. Photovoltaics (PV) deployment in the desert regions face the issues of a dusty atmosphere, high ambient temperatures, high UV component of the light spectrum and high humidity in the regions close to the sea. These characteristics induce a reduction of the energy yield of the solar installations and a faster degradation over time which leads to an increased levelized cost of energy (LCOE). The impact of these environmental factors were discussed in detail throughout the book and Chapter 1 reviewed the basic characteristics of the solar light and the techniques to assess the insolation in a given place and harvest the solar energy.

In Chapters 2 and 3, the solar cell fundamentals and the thermodynamics of solar energy conversion were presented. The first silicon solar cell was produced at Bell Lab in 1954 consisting of a silicon p–n junction. The physical processes of light conversion into electricity and the working principle of a solar cell were described along with the key parameters that are commonly used to characterize solar cells, namely the short-circuit current I_{sc}, the open circuit voltage V_{oc} and the fill factor FF. The impact of the materials properties and environmental factors such as the temperature and insolation effect on the solar cell performance are discussed. The theoretical limit of cell efficiency was detailed and our recently developed computational approach to expand the materials space for the fabrication of solar cells of high performance was also presented. Indeed, the more detailed the thermodynamic model is, the more realistic upper bounds are obtained. However, the increase in the model's complexity is accompanied by more involved calculations. Readers are hence supplied with the relevant information to enable them to calculate the explicit values for a broad class of processes. Throughout, general principles are illustrated using idealized models, and end-of-chapter materials for solar cells are described and practical examples are presented so as to compare reality with theory.

The various solar cell technologies are reviewed in Chapter 4. Discussion about the main parameters affecting the silicon solar cell device performance under desert conditions (V_{oc}, temperature coefficient TCη) is presented followed by a description of the main silicon-based photovoltaic technologies (Al-BSF, PERC, SHJ), their configuration and fabrication, and the device operation under various temperatures, including our own experimental findings. Thin films technologies and organo-metal halide perovskites-based solar cells' architectures and designs are presented along with the stability and toxicity issues.

In Chapter 5, we compared the performance and reliability of different silicon PV module technologies after 5-years of operation in a desert climate (in Qatar). Out of various silicon manufacturers, multi-crystalline silicon has shown the lowest annual degradation rate (below 1 %/year) while silicon heterojunction technology showed the lowest temperature coefficient. Having a large sample size for indoor measurements was found to be a prerequisite to perform a comprehensive statistical analysis to enable accurate determination of module degradation rate and possible roots factors of module failure. Back sheet cracking and encapsulant yellowing were pointed out as the most common failure mode observed in the field during these first 5-years of operation. The high UV irradiance and high operating temperature have been suggested to be responsible for the occurrence of these failures. Electroluminescence imaging of silicon PV modules returned from the field have revealed the presence of solar cell cracking for some of the modules, and the associated power loss was correlated to a combination of low shunt resistance, high series resistance, solar cell cracking, and cell mismatch.

Chapter 6 revolved around the bifacial solar technology and module installation. This included the general concepts, and an in-depth analysis of our own field data. As a matter of fact, in-array mounted bifacial modules, we witnessed an annual average energy yield gain of 8.6% for a panel with bifaciality of 65% and 16.3% gain for bifaciality of 90% when compared with monofacial PV. In addition, four synthetic and natural albedo materials were used for 22° tilt and vertical bifacial. Local desert sand showed an increase of 21% in power output relative to monofacial panels under STC for standalone systems. Grass albedo gained a minimum of 8.4% whereas white cement performed best with a bifacial gain of 28.5%. One of the most important installation parameters for bifacial PV was pointed out to be the height of the panel as it determines the amount of irradiation on the rear side. In our experiments, tilted panels were found to be more sensitive to mounting height than vertical bifacial panels. This effect is due to self-shadowing of the panels and non-homogeneity of irradiation at rear side which was found to saturate after $\sim$100 cm for tilted panels.

On the other hand, soiling of solar collectors has been recognized as the main issue for solar energy systems operating in Middle East and North Africa (MENA) region, which results in significant losses of solar power generation and an increase of associated plant operational and maintenance (O&M) costs. Chapter 7 has dealt with this specific issue. A comprehensive review summarizing our experience with field-measurements key findings and challenges in addressing soiling research obtained from our last 10 years of testing at the outdoor test facility was made available for the book' readers. Although the soiling problem is far from being solved and further research efforts are still needed to understand and to tackle the issue, many practical solutions are made already available. For example, in the short term, commercially available PV cleaning machines and real-time scheduling of the cleaning based on atmospheric forecasts are found to be promising approaches. In the long term, anti-soiling coatings may further reduce the cleaning frequency and cost and are highly encouraged to be implemented.

In Chapters 8 and 9, two case studies representing two different desert climates, namely Qatar and Australia, were, respectively, presented. The climate of the Qatar' desert was first described. The electricity consumption in the desert farm describing the type of a typical load available was also discussed. Chapter 8 also discussed the various options of system design, control and operation for the PV system. A techno-economic analysis for the optimal energy system for the farm was presented and discussed and recommended the use of a hybrid energy system that coupled PV–battery–diesel generator as the best option economically. The study performed in Australian's desert has demonstrated that on average, the c-Si systems installed in the DKASC have an annual degradation rate of around 0.82% for mono-Si cells and 0.86% for poly-Si cells, which has bene found to be higher than the standard 0.7%/year degradation rate.

10.2   Looking into the future: PV in harsh environments at large

Looking into the future, it is important to consider the potential environmental impacts of PV systems in desert environments. While PV technology is a renewable and relatively low-impact form of electricity generation, the construction and operation of large-scale PV installations can have an impact on desert ecosystems and the wildlife that inhabit them. It is crucial to carefully consider these impacts and take steps to minimize them, such as using habitat restoration and conservation measures.

There has been a trend towards increased deployment of PV systems in harsh environments in recent years. This is due in part to the growing demand for renewable energy integration in the power system and the increasing awareness of the environmental impacts of traditional fossil fuels and its impact on the accelerated climate change patterns we are witnessing. PV systems can be an attractive option in harsh environments where other forms of energy generation may be difficult or impossible to build and/or operate.

One of the main advantages of PV in harsh environments is that they do not require a constant supply of fuel, as is the case with fossil fuel-based power plants. This makes them well suited for remote locations where it may be difficult to transport fuel or where the infrastructure needed to support traditional forms of energy generation is lacking. PV electricity can be consumed at the generation point.

There are several types of harsh environments in which PV systems have been deployed, including desert, polar and marine environments. Each of these environments presents its own set of challenges and opportunities for PV deployment. In desert environments, high temperatures and high humidity can lead to increased degradation of PV panels, reducing their lifespan and efficiency. To mitigate these issues, as-discussed, PV systems in desert environments often require specialized design and maintenance practices. In polar environments, extreme cold temperatures and long periods of darkness can present challenges for PV systems. To address these issues, PV systems in Polar Regions may need to be designed with specialized materials and insulation to withstand the cold temperatures and may need to be paired with storage systems such as batteries to provide power during periods of darkness. Marine environments present a unique set of challenges for PV deployment, as PV systems must be able to withstand the corrosive effects of saltwater and the constant motion of waves. To address these issues, PV systems in marine environments may need to be designed with corrosion-resistant materials and specialized mounting systems to ensure stability in rough seas. Take steps to minimize them, such as using habitat restoration and conservation measures.

Finally, by learning lessons from the exposure of PV panels to the harsh environment of this region, Qatar has a unique opportunity to develop innovative mitigation strategies that will benefit other arid regions of the GCC and the world. Indeed, innovative building and operating solutions, cleaning concepts and strategies, and novel coatings are continually developed, addressing new functionalities. Researchers and engineers are continually working on developing new materials and technologies that can improve the durability and reliability of PV systems in desert environments. These efforts include the development of more durable and efficient PV cells and panels, as well as the use of advanced tracking and shading systems to optimize the performance of PV systems. A constant and continual effort on advanced soiling research is thus necessary. As a matter of fact, the establishment of a QEERI National Dust and PV Soiling Network of soiling monitoring stations will include the capability to directly measure soiling and all the other physical factors that influence it. This information will then be used to develop AI-based solutions for forecasting soiling on PV solar panels at different spatial and temporal resolutions for the whole Qatar. From an operational point of view, it is cheaper and more efficient to perform maintenance cleaning operations than cleaning when is needed. In addition, it builds the capability of providing this information in synergy with the Solar Resource Assessment and Forecasting of PV production. Moreover, the deep understanding of the impact of desert climate factors on PV generation in the state of Qatar will serve us to interact with other institutions internationally and assist the stakeholders to accurately conduct techno-economic feasibilities, and to identify the optimal parameters for devices and/or PV installation construction and operations.

Overall, the outlook for PV in desert environments will be key for the future developments of the industry. As the demand for renewable energy continues to grow and the technology improves, it is likely that we will see an increasing number of PV projects being implemented in desert regions around the world. These projects have the potential to provide clean and reliable electricity, as well as economic and social benefits, to the communities that they serve.

References

[1]   SolarPower Europe: Global Market Outlook for Solar Power 2022–2026, 2022.
[2]   K. Ilse, L. Micheli, B. W. Figgis, *et al.*, Techno-economic assessment of soiling losses and mitigation strategies for solar power generation. *Joule*, 3, 2019, 2303–2321, https://doi.org/10.1016/j.joule.2019.08.019.
[3]   A. Sayyah, M. Horenstein, and M. Mazumder, Yield loss of photovoltaic panels caused by depositions, *Solar Energy* 107, 576–604, 2014, http://dx.doi.org/10.1016/j.solener.2014.05.030
[4]   M. Maghami, H. Hizam, C. Gomes, *et al.*, Power loss due to soiling on solar panel, *Renewable and Sustainable Energy Reviews* 59, 1307–1316, 2016.
[5]   Betak, J., Súri, M., Cebecauer, T., and Skoczek, A., Solar resource and photovoltaic electricity potential in EU-MENA region. In: *Proceedings of the 27th European Photovoltaic Solar Energy Conference and Exhibition*, 2012, pp. 2–5.

Index

abrasion effects 232–3
absorbing layer 23
accelerated lifetime tests (ALTs) 83
Adachi model 47
adhesion forces 210, 218
Aerosol Optical Depth (AOD) 204
Alice Springs 256, 303, 309
Al Kharsaah solar power plant 241
aluminum back surface field (Al-BSF) cells 63–4, 112
aluminum doped zinc oxide (AZO) 74
aluminum oxide (Al$_2$O$_3$) 33
amorphous silicon cells 59
amorphous silicon nitrides (a-SiNx) 33
angle-of-incidence effects 227–9
animal barns 255, 268
anti-reflection coating (ARC) 28, 63
anti-soiling coating 113, 233–5, 316
Arabian desert 256–7
Arabian Peninsula 256–7
arid climate 10, 256, 303
a-Si:H film deposition 66
ASTM G-173 (ASTM stands for American Society for Testing and Materials) 4
atomic layer epitaxy 77
Auger recombination 29–33, 64
Australia Energy Regulator (AER) 300
Australian Capital Territory (ACT) 294
Australian Renewable Energy Action Agenda 291
Australian Renewable Energy Commercialization Program 291
azimuth angle 2–3, 222–4

back-surface field (BSF) 15, 153
balance of system (BOS) 17
base 23
beam extraterrestrial (ET) radiation 2
beam horizontal radiation 6
Beer–Lambert law 3
bifaciality 125, 155, 157
bifacial PV 153
bifacial solar cell technology 113
bifacial solar technology
    bifacial performance parameters 158–60
    bifacial perovskite silicon tandem 165–7
    cell to module to field performance 169–72
    material characterization model for device assessment 167–9
    module installation 173
        bifacial versus monofacial energy yield gain 175–6
        effect of azimuth orientation 181–2
        effect of device parameters 182–5
        effect of module temperature 179
        effect of mounting height 176–8
        effect of natural and synthetic ground albedo 178–9
        optimization of height, tilt, and albedo 179–81
        outdoor module performance monitoring 174–5
        standalone versus in-array mounted losses 175–6
    potential applications 160–1
    solar cell device optimization 161–2
    technological progress 156–8

theoretical and practical efficiency limits
   of solar cells 163–5
Bill of Material (BOM) 145
Boltzmann's constant 25, 30, 41
broadband radiation 5
buffer layer 73
building integrated PV (BIPV) 115, 161

cadmium sulfide (CdS) 162
cadmium telluride (CdTe) 13, 161, 165
cadmium tin oxide (CTO) 74
capillary aging 212–15
capillary forces 210–11
carbon-based perovskite solar cells
   (CePSCs) 88–9
carriers mobilities (μ) 47
cell cracking 145
cell to module (CTM) 114
   electrical 118
   electrical losses 120–4
   geometrical 117
   geometry-related losses 118–20
   optical 118
   optical losses and gains 124–5
cementation 212–15
central inverter 267
cesium (Cs) 88
chemical bath deposition (CBD) 77
Clean Energy Finance Corporation (CEFC)
   295–6
cleaning economics 232
close-space sublimation (CSS)
   technique 77
coefficient of thermal expansion
   (CTE) 114
cold deserts 256
collector geometry 221–4
computational fluid dynamics model 222
concentrated solar power (CSP) 1, 17–19
concentrators 35
Copernicus Atmosphere Monitoring
   Service (CAMS) 204

copper indium gallium selenide (CIGS)
   72–93, 156, 165
cost of finance 295
crystalline silicon technology 111
Cunningham slip factor 219
Czochralski (Cz) growth process 15
Czochralski process 59

dangling bonds 32
data sets 11–13
deep levels 31
deposition techniques 77
desalination techniques 270
descriptor 54
Desert Knowledge Australia Solar Centre
   (DKASC) 303–9
desert PV applications
   desert climate and solar resource 256
      Arabian and Qatar desert climate
         256–7
      daily PV profile in desert environment
         258–64
      general weather profile of desert 256
      solar resource in Qatar desert 257–8
   loads and energy consumption in Qatar
      267
      additional loads 271
      animal barns 268
      HVAC systems 270–1
      urban energy consumption of Qatar
         desert 271–2
      water desalination 269–70
      water pumps 269
   PV energy system design, control and
      operation 272
      energy management and control 275–6
      off-grid *vs.* grid connected system
         274–5
      PV-hybrid system 274
      PV-powered lighting systems 273
      PV-powered water pumping system
         273
      remote residential PV system 273–4

system with energy storage 274
PV strings and arrays 264
    partial shading impact on PV system
        performance 264–6
    performance of PV inverters in desert
        environments 266–7
    techno-economic benefits and case study
        276–81
dew 212–15
diborane (B_2H_6) 66
diffuse horizontal irradiance (DHI) 4–6,
    9–10
diffusion constant 28
diffusion current 25
direct or beam normal irradiance (DNI)
    4–6, 8–9, 19
    clearness index 204
DLVO theory 211
doped a-Si:H films 66
dust belt 205
dust storm (DS) 23

effective masses (m^*) 47
Einstein equation 28
electrical grid 255
electrical losses 120–4
electrically conductive adhesive (ECA) 114
electrodeposition 77
electro-dialysis 270
electrodynamic screen 236
electrodynamic shields (EDSs) 236–9
electroluminescence (EL) imaging 135,
    145–6, 315
electron affinity 47–8
electron mobility 29
electron transport layer (ETL) 88–9
electrostatic forces 211–12
emitter 23
emitter wrap through (EWT) 120
energy band gap (eV) 35
energy consumption 255, 268
energy gap (*Eg*) 29–30, 46

energy management and control 275–6
Energy Sector Management Assistance
    Program (ESMAP) 11
ethyl vinyl acetate (EVA) 16, 113, 143, 154
exponential function 29
external quantum efficiency (EQE) 80, 93

feed-in tariffs (FITs) 293–4
field of view (FOV) 4
fill factor (FF) 27, 29, 121
flat plate module 35
fluorine doped tin oxide (FTO) 74, 169
forced airflow 314
formamidinium (FA) 88
forward bias 25

gallium arsenide (GaAs) 30
global clearness index (GHI) 6
global horizontal irradiance (GHI) 4–6, 9,
    233
graded bandgap absorber 75
grain boundaries (GBs) 84
GreenPower 287
grid-connected PV systems 281–2
grid parity 290
ground coverage ratio (GCR) 129–30

Hamad Bin Khalifa University (HBKU)
    255
Hamaker constant 211
harvesting technologies 13
    CSP 17–19
    photovoltaics 14
        concentrated photovoltaic 17
        PV technology from cell to panel
        14–17
    solar energy conversion 13–14
heat, ventilation, and air conditioning
    (HVAC) 270–1
heterojunction intrinsic thin-layer (HIT)
    solar cells 157
hetero-junction solar cells (HJSC) 15

heterojunction thin film solar cells
    (HTFSC) 72
high concentrated photovoltaic (HCPV) 17
highly resistive transparent (HRT) layer 74
HOMER Grid 277
HOMER Pro 277
hybrid inorganic–organic perovskite solar
    cells (HPSCs) 85
hybrid perovskites solar cell (PSC) 157
hybrid system 274
hydrofluoric acid (HF) 63

IEC 61646 standard test 83
indium tin oxide (ITO) 74
infrared (IR) imaging 112
insolation maps 11–13
initial capital cost 282
interdigitated back contact (IBC) 112, 120
internal rate of return (IRR) 282
International Civil Aviation Organization
    (ICAO) 4

Klassen's model 47
Köppen climate classification 303

Lambert law 169
large-scale renewable energy target (LRET)
    292
levelized cost of energy (LCOE) 1, 111,
    120, 165, 282, 315
light soaking effect 81
light transmission 225–6

Mandatory Renewable Energy Target
    (MRET) 292
maximum power point (MPP) 26, 266
maximum power point tracking (MPPT)
    algorithm 266
metallurgical silicon (MG-Si) 15
Metal Oxide Field Effect transistors
    (MOSFET) 15
metal wrap through (MWT) 114
MeteoSat 10

methylammonium (MA) 88
methyl-ammonium lead triiodide
    (MAPbI$_3$) 52
microinverter 267
Middle East and North Africa (MENA)
    region 23, 314, 316
Mie scattering 3
module encapsulation 78
molecular beam epitaxy (MBE) 92
molecular organic chemical vapor
    deposition (MOCVD) 92
monocrystalline (c-Si) 14, 59
mono-crystalline solar cells 59
monolithic interconnections 78
multi-crystalline silicon (mc-Si) 14
multi-effects 270
multi-filter rotating shadow band
    radiometer (MFRSR) 204
multistage flashing 270

National Oceanographic and Atmospheric
    Administration (NOAA) 3
National Solar Schools Program 294
net present cost (NPC) 282
net soiling rate 220–1
New South Wales (NSW) 293, 301
nominal operating module temperature
    (NOMT) 129, 173
nonabsorbing material 162
number of Suns 34

ohmic losses 121
open circuit voltages (*V*oc) 26, 94–6, 315
operating cost 282
operational and maintenance (O&M) 233,
    316
optical losses and gains 124–5
optimum load resistance 27

particle caking 212–15
particle deposition 215–17
particle rebound and resuspension 217–20

passivated emitter rear contact (PERC) 153

passivating contacts 65

Performance Ratio (PR) 304

perovskite solar cells (PSCs) 52

phosphine (PH3) 66

phosphorus oxychloride (POCl3) 63

photocarriers 24–5

photocurrent 24, 29

photo electrochemical (PEC) cell 86

photons 24

Photovoltaic Geographical Information
System (PVGIS) 13

photovoltaic (PV) modules 111
  CTM 114
    electrical 118
    electrical losses 120–4
    geometrical 117
    geometry-related losses 118–20
    optical 118
    optical losses and gains 124–5
  design 113–15
  materials 113
  module energy yield and reliability under
      desert environment 125
    effect of azimuth orientation 130–2
    effect of ground albedo 128
    effect of ground coverage ratio 129–30
    effect of module temperature 129
    effect of mounting height 127
    effect of tilt angle 128–9
    module installation parameters for
        mono-facial and bifacial modules
        125–7
  performance and reliability of
      crystalline-silicon photovoltaics in
      desert climate 139
    electroluminescence imaging 145–6
    PV performance monitoring and
        degradation rate 139–41
    temperature coefficients (TC)
        measurement 141–2
    visual inspection 142–5

reliability issue for hot arid desert 132
  infrared thermography of PV modules
      to identify defects 132–9

Photovoltaic Power Systems Program
    (PVPS) 285

Photovoltaic Rebate Program 292

photovoltaics (PV) 1
  hybrid systems 274, 282
  module output 226–7

photovoltaics (PV) market in Australia 285
  Australian climate 296–9
  development of grid-connected industry
      287
  DKASC 303–9
  dominance of grid-connected PV 287–90
  grid parity 290
  international context 286
  new investors 291
  off-grid uses (pre-2000) 287
  policies and their effects on the market 291
    certificate multipliers and adaptability
        293
    cost of finance 295
    feed in tariffs 293–4
    political environment 295–6
    RET 292–3
    small-scale programs 291–2
  PV systems across the Australian
      network 299–303

photovoltaic soiling (PV soiling) 236
  analysis of PV field data 227
    angle-of-incidence effects 227–9
    surface imaging 229
  on attenuation of solar radiation in a
      desert environment 203–5
  cleaning and soiling mitigation 229
    abrasion effects 232–3
    anti-soiling coating 233–5
    cleaning economics 232
    electrodynamic shield/screen 236–9
    TiO_2-based self-cleaning coating
        235–6

description of OTF 237–9

dew, cementation, particle caking, capillary aging 212–15

dust characteristics in different countries 207

  particle size distribution 207–9

  particle-surface adhesion forces 209–12

field measurement of 224

  light transmission 225–6

  PV module output 226–7

  soil mass 224–5

impact on global solar power production and energy costs 239–41

mechanics of dust accumulation 215

  collector geometry 221–4

  net soiling rate 220–1

  particle deposition 215–17

  particle rebound and resuspension 217–20

renewable energy 241–3

soiling rates in different countries 205–6

photovoltaic (PV) solar energy 313

  in harsh environments at large 316–18

photovoltaics (PV) system

  component costs 276–7

  cost tradeoffs 277

  labor costs 277

Planck' constants 48

plane of array (POA) 155

plane of the array (POAI) 304–5

plasma-enhanced chemical vapor deposition (PECVD) 63

polycrystalline (Poly c-Si) 59

polycrystalline silicon 59

potential-induced degradation (PID) 115

power conversion efficiency (PCE) 16, 27, 52, 89, 126, 157

power electronics converters (PECs) 266

Power Purchase Agreements (PPAs) 291

PV Rebate Program 287

PVsyst 277

PVWatts™ 277

pyranometers 7, 204

pyrheliometers 7

Qatar 255, 314, 317

  loads and energy consumption in 267

    additional loads 271

    animal barns 268

    HVAC systems 270–1

    urban energy consumption of Qatar desert 271–2

    water desalination 269–70

    water pumps 269

Qatar desert

  solar resource in 257–8

  urban energy consumption of 271–2

Qatar Environment and Energy Research Institute (QEERI) 255

Qatar Science & Technology Park (QSTP) 237

quantum wells (QWs) 88

radiant flux 2

radiometers 7

Rayleigh scattering 3–4

reaction etching (RIE) 28

recombination-center 31

refrigerant 270

relative air mass 4

relative humidity (RH) 234

remote residential PV system 273–4

renewable energy 241–3

renewable energy certificates (RECs) 292

Renewable Energy Target (RET) 292–3

resistive losses 121

RETScreen 277

return on investment (ROI) 282

reverse osmosis 270

rooftop solar 290

rotating shadow band radiometers (RSRs) 9

Ruddlesden–Popper with the formula (RNH3) 88

Sahara Desert 256
satellite-based data 10–11
screen-printing 77
shallow levels 31
Sharma's model 47, 48
Shockley diode equation 45
Shockley–Queisser limit (SQ limit) 30, 35–6
Shockley–Read–Hall (SRH) 30, 44, 68, 163–4
short-circuit current 315
silicon (Si) 165
silicon heterojunction (SHJ) 33, 166
silicon nitride (SiNx) 28
silicon oxide (SiO$_2$) 63
silicon solar cells 59, 112
single junction solar cells 43
Siraj-1 solar power plant 241
small-scale renewable energy scheme (SRES) 290
small technology certificates (STCs) 293
smart wire connection technology (SWCT) 172
soft soiling 266
soil mass 224–5
solar cell cracking 315
solar cell device optimization 161–2
solar cell mismatch 315
solar cells 14, 23, 113, 153
    basic operational concept of 25
    cell structure and light conversion 23–5
    CIGS and CdTe thin film solar cells 72
        device configuration 72–5
        device fabrication 75–8
        device performances 78–83
        device reliability 83–5
    current–voltage (I–V) characteristics 25–7
    device operation under various tempera-tures 68
        comparison of different technologies 68–71

factors limiting cell performance 27
    Auger recombination 31–2
    current losses through defects 30–1
    effect of temperature 33–5
    energy gap 29–30
    surface recombination 32–3
general device structure of 162
materials for 50
    computational materials design for 53–5
    conventional materials 50–1
    emerging materials 51–3
multijunction solar cells 89
    basic principles of 89–92
    fabrication of 92–3
    under high temperature operation 93–6
organo-metal halide perovskites based solar cells 85
    architectures and designs 86–7
    device fabrication 87–9
    stability and toxicity issues 89
Shockley–Queisser limit 35–6
silicon-based photovoltaic technologies 61
    aluminum back surface field cells 63–4
    configuration and fabrication 66–8
    device fabrication 87–9
    PERC 64
    silicon heterojunction solar cells 64–6
    solar cell architectures and designs 86–7
silicon solar cell device performance under desert conditions 60–1
solar constant 2
solar cycles 2
solar energy 2
    working principle of 315
Solar Flagships Program 295
solar light components 4–6
solar modules 276
Solar Position Algorithm (SPA) 3

solar resources 1
  effects of earth's atmosphere on sunlight 3–4
  estimating insolation 6
  ground measurements, instruments 6–10
  insolation maps and data sets 11–13
  satellite-based data 10–11
  solar light components 4–6
  standard solar spectrum 4
  sun–earth system 1–3
solar spectrum 14
Solar Test Facility (STF) 23
solar thermal 4
solid bridge bonds 212
Specific Yield (SY) 304
sputtering 77
stand-alone PV system 282
standard solar spectrum 4
standard testing conditions (STC) 60, 116, 155
Stefan–Boltzmann constant 2
Stefan–Boltzmann law 2
stringing method 119
string inverters 267
substrate structure 73
Sun Belt 1, 314
Sun–Earth distance 2
Sun's azimuth angle 2
Sun's zenith angle 2
surface defect layer (SDL) 74
surface imaging 229
surface recombination velocity 32

techno-economic analysis 256, 276, 279
temperature coefficient (TC) 125, 126, 157, 182–5
thermalization 41
thermodynamics of solar energy conversion AM1.5g 40–3
  effect of increased intensity in hot areas 43–5

hot climate implications on common solar cell technologies 46–50
materials for solar cells 50
  computational materials design for 53–5
  conventional materials 50–1
  emerging materials 51–3
  single junction solar cells 43
tilt angle 221–2
time-resolved photoluminescence (TRPL) 167
TiO_2-based self-cleaning coating 235–6
top-of-atmosphere radiation 2
transparent conductive oxide (TCO) 67, 87
trap-state 31
trimethylboron (TMB) 66
true dessert 256
turbulence intensity 19

United Arab Emirates (UAE) 19

vacuum-based techniques 76
Van der Waals (VdW) forces 211
vapor–compression 270
vapor transport deposition (VTD) 77
Vasileff model 47
vdep 216

water contact angle (WCA) 210
water desalination 269–70
water pumps 269
water vapor transmission rate (WVTR) 113
Western Australia (WA) 301
wet chemical etching 28
wide bandgap buffer layer 74–5
window layer 72
wind speed (WS) 23

Yankee Environmental Systems (YES) 204

zenith angle 2–3
zirconium oxide 87

www.ingramcontent.com/pod-product-compliance
Lightning Source LLC
Chambersburg PA
CBHW050508190326
41458CB00005B/1474